MOLECULAR BIOLOGY OF ACUTE LUNG INJURY

MOLECULAR AND CELLULAR BIOLOGY OF CRITICAL CARE MEDICINE

Robert S. B. Clark and Joseph A. Carcillo, Series Editors

---

1. H. R. Wong and T. P. Shanley (eds.): Molecular Biology of Acute Lung Injury. 2001. ISBN: 0-7923-7434-7

MOLECULAR BIOLOGY OF ACUTE LUNG INJURY

edited by

Hector R. Wong
Division of Critical Care Medicine
Children's Hospital Medical Center and
Children's Hospital Research Foundation
Cincinnati, Ohio

and

Thomas P. Shanley
Division of Critical Care Medicine
Children's Hospital Medical Center and
Children's Hospital Research Foundation
Cincinnati, Ohio

Kluwer Academic Publishers
Boston/Dordrecht/London

Distributors for North, Central and South America:
Kluwer Academic Publishers
101 Philip Drive
Assinippi Park
Norwell, Massachusetts 02061 USA
Telephone (781) 871-6600
Fax (781) 681-9045
E-Mail <kluwer@wkap.com>

Distributors for all other countries:
Kluwer Academic Publishers Group
Distribution Centre
Post Office Box 322
3300 AH Dordrecht, THE NETHERLANDS
Telephone 31 78 6392 392
Fax 31 78 6392 254
E-Mail <services@wkap.nl>

Electronic Services <http://www.wkap.nl>

Library of Congress Cataloging-in-Publication Data

Molecular biology of acute lung injury / edited by Hector R. Wong and Thomas P. Shanley.
p. ; cm. -- (Molecular & cellular biology of critical care medicine ; MCCM 1)
Includes bibliographical references and index.
ISBN 0-7923-7434-7 (hardback : alk. paper)
1. Respiratory distress syndrome, Adult--Pathophysiology. 2. Respiratory distress syndrome, Adult--Molecular aspects. I. Wong, Hector R., 1963- II. Shanley, Thomas P., 1963- III. Series.
[DNLM: 1. Respiratory Distress Syndrome, Adult--physiopathology. 2. Molecular Biology. WF 140 M718 2001]
RC776.R38 M654 2001
616.2'407--dc21

2001038125

Printed on acid-free paper.

Printed in the United States of America

TABLE OF CONTENTS

CONTRIBUTORS

John A. Belperio
Department of Medicine, Division of Pulmonary and Critical Care Medicine, Los Angeles (UCLA) School of Medicine
Los Angeles, California

Robert D. Bongard
Medical College of Wisconsin
Milwaukee, Wisconsin

Augustine M. K. Choi
Division of Pulmonary, Allergy, and Critical Care Medicine
University of Pittsburgh Medical Center
Pittsburgh, Pennsylvania

Christopher A. Dawson
Medical College of Wisconsin, Veterans Affairs Medical Center, and Marquette University
Milwaukee, Wisconsin

Claudia C. dos Santos
Inter-Departmental Division of Critical Care
University of Toronto
Toronto, Canada

Karine Faure
Department of Anesthesia and Surgery, University of California at San Francisco
San Francisco, California

Dara Frank
Department of Microbiology and Molecular Genetics, Medical College of Wisconsin
Milwaukee, Wisconsin

Thomas K. Geiser
Division of Pulmonary Medicine
University Hospital
Bern, Switzerland

Benoit Guery
Laboratoire de Recherche en Pathologie Infectieuse
University of Lille
Lille, France

Sujata Guharoy
Department of Pathology
University of Michigan
Ann Arbor, Michigan

Cory Hogaboam
Department of Pathology
University of Michigan Medical School
Ann Arbor, Michigan

Michael P. Keane
Department of Medicine, Division of Pulmonary and Critical Care Medicine, Los Angeles (UCLA) School of Medicine
Los Angeles, California

David Kelley
Department of Medicine, Division of Pulmonary and Critical Care Medicine, Los Angeles (UCLA) School of Medicine
Los Angeles, California

Neil W. Kooy
Division of Critical Care Medicine, Children's Hospital Medical Center and Children's Hospital Research Foundation
Cincinnati, Ohio

Jeanine P. Wiener-Kronish
Department of Anesthesia and Perioperative Care, University of California at San Francisco
San Francisco, California

Steven L. Kunkel
Department of Pathology
University of Michigan Medical School
Ann Arbor, Michigan

Markus S. Huber-Lang
Department of Pathology
University of Michigan Medical School
Ann Arbor, Michigan

Ann Marie LeVine
Division of Pulmonary Biology and Critical Care Medicine
Children's Hospital Medical Center and Children's Hospital Research Foundation
Cincinnati, Ohio

Mingyao Liu
Thoracic Surgery Research Laboratory, University Health Network, Toronto General Hospital
Toronto, Canada

Nicholas W. Lukacs
Department of Pathology
University of Michigan Medical School
Ann Arbor, Michigan

Michael A. Matthay
Cardiovascular Research Institute
University of California San Francisco
San Francisco, California

G. Umberto Meduri
Division of Pulmonary and Critical Care Medicine,
University of Tennessee Health Science Center, The Memphis Lung Research Program, and Baptist Memorial Hospitals
Memphis, Tennessee

Marilyn P. Merker
Medical College of Wisconsin and Veterans Affairs Medical Center
Milwaukee, Wisconsin

Sem H. Phan
Department of Pathology
University of Michigan
Ann Arbor, Michigan

Jean Francois Pittet
Department of Anesthesia and Surgery, University of California at San Francisco
San Francisco, California

Ammar Sakkour
Department of Medicine, Division of Pulmonary and Critical Care Medicine, Los Angeles (UCLA) School of Medicine
Los Angeles, California

Teiji Sawa
Department of Anesthesia and Perioperative Care, University of California at San Francisco
San Francisco, California

Jigme M. Sethi
Division of Pulmonary, Allergy, and Critical Care Medicine
University of Pittsburgh Medical Center
Pittsburgh, Pennsylvania

Thomas P. Shanley
Division of Critical Care
Medicine, Children's Hospital
Medical Center and Children's
Hospital Research Foundation
Cincinnati, Ohio

Arthur S. Slutsky
Department of Critical Care
Medicine, St. Michael's Hospital
Toronto, Canada

Theodore Standiford
Division of Pulmonary and
Critical Care Medicine
University of Michigan Medical
School
Ann Arbor, Michigan

Robert M. Strieter
Department of Medicine, Division
of Pulmonary and Critical Care
Medicine, Los Angeles (UCLA)
School of Medicine
Los Angeles, California

Peter A. Ward
Department of Pathology
University of Michigan, School of
Medicine,
Ann Arbor, Michigan

Jeffrey Whitsett
Division of Pulmonary Biology
Children's Hospital Medical
Center and Children's Hospital
Research Foundation
Cincinnati, Ohio

Hector R. Wong
Division of Critical Care
Medicine, Children's Hospital
Medical Center and Children's
Hospital Research Foundation
Cincinnati, Ohio

Liqian Zhang
Division of Pulmonary Biology
Children's Hospital Medical
Center and Children's Hospital
Research Foundation
Cincinnati, Ohio

Jerry J. Zimmerman
Division of Critical Care
Medicine, Children's Hospital and
Regional Medical Center
Seattle, Washington

PREFACE

Visit any intensive care unit in the world, whether it be dedicated to the care of medical, surgical, pediatric, or neonatal patients, and the extent to which acute lung injury (ALI) impacts patient care will become immediately evident. The impact is measured not only in the number of patients that suffer from ALI, but also in the costs attributed to care, the number of clinical and basic researchers dedicated to ALI, and the morbidity and mortality associated with ALI and its complications.

Our collective understanding of the pathophysiology of ALI has grown immensely over the past decade. Despite this knowledge and the realization that the outcome of patients with ALI has improved modestly over the last several years, we contend that the morbidity and mortality of ALI remain unacceptably high. Thus, clinicians and researchers alike continue to face the challenge of developing a greater understanding of the biological basis of ALI. This fundamental understanding will allow us to move beyond our current supportive therapeutic strategies, toward novel strategies that specifically target the complex biologic processes involved in ALI.

This book is not a "how to" book of ALI. Rather, this book attempts to compile the most current available information regarding the plethora of molecules and mechanisms thought to be relevant to the pathophysiology and treatment of ALI. The focus of the book reflects our bias that in order to more effectively care for patients with ALI, we will need to develop a greater understanding of this process at the molecular level.

One other point should be addressed. Throughout this book the reader will note that the terms "ALI" and "ARDS" are used relatively loosely. Lest the reader be led to believe that we do not comprehend the formal definitions of "ALI" and its more severe manifestation, "ARDS," rest assured that these definitions are well understood and recognized. The use of these two terms in this manner reflects the concept that ALI and ARDS are not a single disease entity. Rather, they are manifestations of multiple insults to the lung (directly via the alveolar compartment, or relatively more indirect via the vascular compartment) leading to a remarkably similar endpoint.

It is our sincere hope that the information provided in this book serves as a valuable reference to both clinicians and researchers, and more importantly, that it stimulates the scientific and academic curiosity of future clinician/investigators.

Hector R. Wong
Thomas P. Shanley

ACKNOWLEDGEMENTS

We first thank our respective families (Lili, Caroline, Madelyn, and Eleanor Wong; and Maureen, Lauren, Molly, Ashleigh, and Matthew Shanley) who admirably support and tolerate our chaotic schedules, and our clinical and scientific passions.

We thank the contributors whose expertise and dedication not only allowed us to bring this book to fruition, but who also serve as important role models in our clinical and research efforts. Finally, we thank Brenda Robb for administrative and clerical assistance in organizing this book.

Chapter 1

SIGNAL TRANSDUCTION PATHWAYS IN ACUTE LUNG INJURY: NF-κB AND AP-1

Thomas P. Shanley and Hector R. Wong
Division of Critical Care Medicine, Children's Hospital Medical Center and Children's Hospital Research Foundation, Cincinnati, OH

INTRODUCTION

Our understanding of the intermediary events that occur between the reception of a biological signal at the cell membrane, and the eventual conversion of that signal to a change in gene expression at the nuclear level (i.e. signal transduction), has grown immensely over the last decade. Elucidation of signal transduction pathways in lung parenchymal cells and alveolar macrophages holds the promise of not only understanding the molecular mechanisms that contribute to acute lung injury (ALI), but also of providing novel therapeutic targets. Protein phosphorylation is the key regulatory mechanism in these pathways and the human genome has been estimated to encode more than one thousand protein kinases. An exhaustive review of the many signal transduction pathways in the lung is beyond the scope of this single chapter. Accordingly, we have focused this chapter on two pathways believed to be particularly relevant to the pathophysiology of ALI: the nuclear factor-κB (NF-κB) pathway and the activator protein-1 (AP-1) pathway.

THE NF-κB PATHWAY

NF-κB is a transcription factor that regulates the expression of a large number of genes involved in inflammation, immunity, and cell growth (reviewed in references 1-4). A list of genes regulated, at least in part, by NF-κB, and directly related to the biology of ALI, is provided in Table 1.

NF-κB belongs to the Rel family of transcription factors, which share common structural motifs for dimerization and DNA binding. Five known subunits belong to the mammalian NF-κB/Rel family: c-Rel, NF-κB1 (p50/p105), NF-κB2 (p52/p100), RelA (p65), and Rel B. "NF-κB" consists of two such subunits arranged as either homodimers (e.g. p50/p50) or heterodimers (e.g. p65/p50). The most common form of activated NF-κB consists of a p65 (Rel A) and p50 heterodimer. Other combinations of subunits, however, are known to exist and appear to have different DNA binding characteristics that can impart yet another level of gene regulation by the NF-κB pathway. For example, the p50 homodimer is thought to be a repressor of gene transcription, while the p65/p50 heterodimer is well known as an activator of gene transcription.

Table 1: Partial list of genes regulated by NF-κB that are involved in the biology of ALI

Cytokines and Chemokines	
	Tumor necrosis factor α
	Interleukins-1, -2, -3, -6, -8, and –12
	Macrophage inhibitory protein 1 α (MIP-1α)
	Macrophage chemotactic protein 1 (MCP-1)
	RANTES
	Eotaxin
	Gro-α, -β, and -γ
Growth factors	
	Granulocyte-macrophage colony-stimulating factor (GM-CSF)
	Granulocyte colony-stimulating factor (G-CSF)
	Macrophage colony-stimulating factor (M-CSF)
Adhesion Molecules	
	Intracellular adhesion molecule 1 (ICAM-1)
	Vascular cell adhesion molecule 1 (VCAM-1)
	E-selectin
Miscellaneous	
	Inducible nitric oxide synthase
	C reactive protein
	5-lipoxygenase
	Inducible cyclo-oxygenase 2

Regulation of NF-κB by IκB

Under basal conditions NF-κB is retained in the cytoplasm, in an inactive state, by a related inhibitory protein known as IκB (reviewed in references 1-4). Several isoforms of IκB exist: IκB-α, -β, -γ, -δ, and -ε. The IκBα and IκBβ isoforms are the best studied and will be the only isoforms discussed in this chapter.

IκB is characterized by the presence of five to seven conserved domains, known as ankyrin repeats, which confer the ability to interact with NF-κB. By way of these ankyrin repeats IκB physically masks the nuclear translocation sequences of NF-κB and retains NF-κB in the cytoplasm. The most commonly accepted mechanism leading to the activation of NF-κB involves rapid phosphorylation of IκB (specifically at serine-32 and -36 of IκBα) in response to a variety of proinflammatory signals such as endotoxin, tumor necrosis factor α (TNFα), interleukin-1β (IL-1β), oxidants, bacteria, viruses, and phorbol esters. Phosphorylated IκB is targeted for rapid polyubiquitination, and polyubiquitinated IκB is then targeted for rapid degradation by the 26S proteasome. Degradation of IκB unmasks the NF-κB nuclear translocation sequences, thus allowing NF-κB to enter the nucleus to direct transcription of target genes. This entire sequence of events occurs within 30 to 60 minutes of proinflammatory stimulation in most *in vitro* systems (5-7 and Figure 1).

Alternative pathways also exist for activation of NF-κB. For example, hypoxia and pervanadate (a tyrosine phosphatase inhibitor) can each cause activation of NF-κB, but the mechanism of this effect involves tyrosine phosphorylation of IκBα rather than serine phosphorylation (8, 9). In addition, pervandatate leads to NF-κB activation without degradation of IκBα. Another alternative pathway for activation of NF-κB involves ultraviolet radiation, which induces degradation of IκBα by the 26S proteasome, but does not cause phosphorylation of IκBα at either serine or tyrosine residues (10).

Relatively well described mechanisms exist to deactivate NF-κB after it has been activated by a proinflammatory stimulus, thus serving as molecular "brakes" to ongoing NF-κB activation (reviewed in references 1-4). The best known mechanism for deactivation of NF-κB involves IκBα. The promoter region of the IκBα gene contains three NF-κB binding sites (11). Thus, when NF-κB is activated it not only induces the expression of proinflammatory genes, but also induces *de novo* expression of its own inhibitor, IκBα. Newly synthesized IκBα is then able to attenuate ongoing NF-κB activation by remasking NF-κB nuclear translocation sequences. In addition, newly synthesized IκBα enters the nucleus to bind activated NF-κB and can then shuttle NF-κB back to the cytoplasm to terminate NF-κB-dependent transcription (12). Resynthesis of IκBα is detectable in most mammalian *in vitro* systems within 30 to 60 minutes of proinflammatory stimulation (Figure 1).

The IκBβ isoform is also involved in controlling the duration of NF-κB activation, but in a markedly different manner to that of IκBα. Whereas the IκBα isoform is involved in regulating transient activation of NF-κB, the IκBβ isoform appears to be involved in prolonged activation of NF-κB (13). Some stimuli, such as IL-1β and endotoxin, can cause rapid degradation of both IκBα and IκBβ. Both IκB isoforms are subsequently rapidly re-synthesized, but in spite of this resynthesis NF-κB activation can continue. The mechanism of this effect is believed to be caused by the resynthesis of a hypophosphorylated form of IκBβ. Hypophosphorylated IκBβ appears to have an altered conformation that binds NF-κB, but leaves the NF-κB nuclear translocation sequences exposed and not accessible to binding by the IκBα isoform, thus allowing for persistent NF-κB activation.

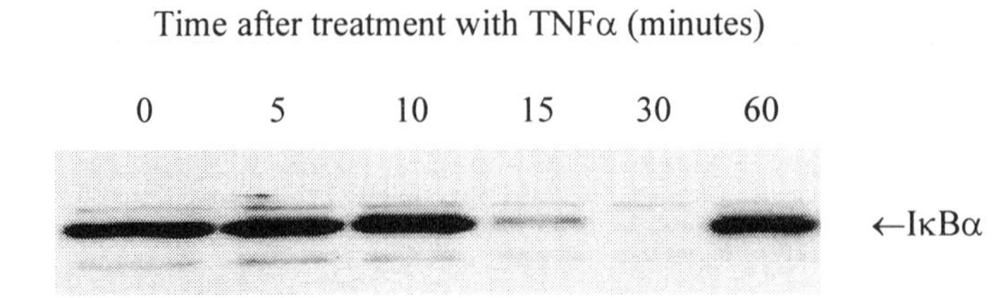

Figure 1: Western blot analysis demonstrating rapid degradation and resynthesis of IκBα after stimulation with tumor necrosis factor α (TNFα). Human respiratory epithelium were stimulated *in vitro* with 2 ng/ml of human TNFα and harvested for Western blot analysis at the indicated times. IκBα was almost fully degraded within 15 minutes of TNFα stimulation and was fully resynthesized within 60 minutes (adapted from reference 6).

IκB Kinase

Despite the central role that IκBα plays in activation and deactivation of NF-κB, the rate limiting step in activation of NF-κB appears to lie in the in the activity of the recently characterized IκB kinase (IKK). IKK consists of three subunits: the catalytic subunits IKKα and IKKβ (also known as IKK1 and IKK2), and a regulatory subunit IKKγ (14-16). IKK is now recognized as the primary kinase that specifically phosphorylates IκB at serine residues, leading to the subsequent cascade of IκB degradation and NF-κB activation.

Gene "knockout" studies have provided considerable insight regarding the function of IKKα and IKKβ with regard to proinflammatory signal transduction pathways (17-20). Stimulation with proinflammatory agents leads to normal IKK activity and IκBα degradation in embryonic fibroblasts from IKKα knockout mice. NF-κB activation, however, is reduced by 50% in these cells suggesting that IKKα plays a role in NF-κB activation, independent of its effects on IκBα phosphorylation. In contrast, IKK activity, IκBα degradation, and NF-κB activation are all drastically reduced in embryonic fibroblasts from IKKβ knockout mice exposed to proinflammatory agents. These data indicate that the IKKβ subunit is the essential control point for inflammation-associated NF-κB activation.

A current area of active investigation is focused on the factors that regulate IKK activity. It is apparent that phosphorylation of IKK subunits are key events regulating IKK activity, and that the phosphorylated level of IKK is dependent on both phosphatase activity and IKK-specific kinases. Due to space limitations the reader is referred to a recent review on this topic (14).

The Role of NF-κB in ALI

A large body of indirect and direct evidence links the NF-κB pathway to the pathophysiology of ALI. The pathophysiology of ALI involves a dysregulated inflammatory cascade, which is dependent on a complex network of humoral mediators (cytokines and chemokines) and cell adhesion molecules (21 and Chapters 2-4). With this concept is mind, it is notable that many of the genes that comprise this complex network are regulated at the transcriptional level by NF-κB (Table 1). In addition, many of the stimuli that cause ALI are well described as activators of NF-κB (Table 2).

Table 2: Activators of NF-κB that are involved in the pathophysiology of ALI

Cytokines: TNFα, Interleukins-1, -2, -6, 8
Oxidants: Hydrogen peroxide, hyperoxia, ozone
Bacteria and Viruses
Hypoxia
Ischemia-Reperfusion
Hemorrhagic shock
Mechanical distension/stretching (mechanotransduction)

Interleukin-8 (IL-8) is a primary chemoattractant and activator of neutrophils in tissues and organs subjected to an inflammatory stimulus (22). Increased levels of IL-8 have been consistently documented in the broncho-alveolar lavage fluids of patients with ALI, and have been correlated with disease severity and mortality (23-25). The promoter region of the IL-8 gene contains a functional NF-κB site and its activation has been unequivocally demonstrated to be dependent on NF-κB activation (26). TNFα, a primary cytokine involved in the pathophysiology of ALI, is known to directly activate NF-κB and consequently increase transcription of the IL-8 gene. Simultaneous exposure to hyperoxia and TNFα synergistically increases respiratory epithelial cell production of IL-8 via enhanced activation of NF-κB, when compared to the effect of either stimulus alone (27). Collectively, these data serve as relevant examples of how genes that play important roles in the pathophysiology of ALI (i.e. IL-8) are regulated by known inducers of ALI (i.e. TNFα and/or hyperoxia) via activation of NF-κB.

Animal models of ALI more directly demonstrate a role for NF-κB in the pathophysiology of ALI. Using a transgenic mouse model in which the luciferase reporter gene was placed under the control of a NF-κB-dependent promoter, Blackwell and colleagues demonstrated increased NF-κB activity in the lungs of mice injected with endotoxin (28). Lung-specific NF-κB activation has also been demonstrated in animal models of ALI caused by IgG immune complex deposition, intratracheal TNFα, hemorrhagic shock, intratracheal endotoxin, intratracheal *Streptococcus pneumoniae*, hepatic ischemia-reperfusion, or lung ischemia-reperfusion (29-36). In many of these models inhibition of NF-κB activation by a variety of strategies conferred protection against ALI. A recent report provided compelling evidence that alveolar macrophage-dependent NF-κB activation plays a central role in the pathophysiology of ALI. In these experiments, Lentsch and colleagues depleted the alveolar macrophages in rat lungs by administration of intratracheal, liposome-encapsulated dichloromethylene diphosphonate (37). Depletion of alveolar macrophages substantially reduced NF-κB activation, proinflammatory gene expression, neutrophil infiltration, and subsequent lung injury in these experiments.

Clinical studies also suggest a role for NF-κB in the pathophysiology of ALI. Schwartz and colleagues demonstrated increased activation of NF-κB in the alveolar macrophages of patients with ALI (38). Moine and colleagues also reported increased NF-κB activation in alveolar macrophages of patients with ALI compared to control patients (39). This occurred despite intact levels of cytoplasmic IκBα levels, leading this group of investigators to speculate that patients with ALI have a fundamental abnormality of NF-κB regulation. In patients with severe sepsis, which is frequently associated with the development of ALI, the degree of NF-κB activation correlated with disease severity and mortality (40). These data support the general hypothesis that increased NF-κB-dependent inflammatory activity directly contributes to the outcome of inflammation-mediated organ injury.

A more complete understanding of the role of NF-κB in ALI requires further investigation. Nevertheless, the current data strongly support a central role for this pluripotent transcription factor and hold promise for the development of therapeutic strategies targeting the NF-κB pathway.

THE AP-1 PATHWAY

One of the other principal signal transduction pathways that has been shown to play a physiologic role in inflammatory disease states such as ALI is the AP-1 pathway. The AP-1 family of transcription activating proteins consist of various homodimers and heterodimers of the Jun (e.g. c-Jun, JunD), Fos (e.g. c-fos, Fra1) or activating transcription factor (ATF2) proteins (reviewed in references 41 and 42). Several combinations of these proteins have been described, though the most well characterized AP-1 complexes are the stable heterodimers formed by members of the Jun and Fos families of proteins. In this form, AP-1 is a sequence-specific transcription factor that early on was discovered to mediate gene induction by the phorbol ester tumor promoter, TPA (43). As a result, the name TPA response element, or TRE, was given to its DNA recognition site: TGACTCA (44). Alternatively, the AP-1 complex comprised of Jun and ATF heterodimers have been shown to favor binding to the cyclic AMP-response element (CRE): TGACGTCA (45). Additional recognition sites for AP-1 have been identified including the serum response element (45).

Transcriptional Regulation by AP-1

Similar to NF-κB, AP-1 has been demonstrated to direct transcription of a number of "early" activation genes (Table 3). In examining these gene products, it is clear that AP-1 transcription and the associated upstream signaling pathways (see below), regulate a broadly diverse set of cellular functions including inflammation, cell proliferation, apoptosis and tissue morphogenesis. Therefore, it will be imperative to further our understanding of the specific mechanisms of this signaling pathway in lung inflammation in order to selectively target it for therapeutic intervention.

While beyond the scope of this chapter, it is important to note that the transcriptional activity of AP-1 is highly regulated. For example, the contribution of c-fos to the transcriptional activity of the c-Jun/AP-1 complex does not require DNA binding by c-fos. This implicates c-fos as a "co-activator" of AP-1 transcription via its binding to c-Jun. Thus, the protein directly binding to the DNA on the promoter of an AP-1 responsive gene can be regulated by a number of proteins serving as either inducers (e.g. c-fos) or suppressors (e.g. p202, reference 42) of transcriptional activity.

MAP Kinase Pathways

AP-1 transcriptional activity is the downstream result of a substantially more complex signal transduction cascade mediated by members of the mitogen-activated protein kinases (MAPK) and their upstream kinases (reviewed in references 46-50). These highly conserved protein kinases have been examined in yeast (*Saccharomyces cerevisiae*), worms (*Caenorhabditis elegans*), and mammalian species. In mammals, three major MAPK pathways have been identified (Figure 2): the c-Jun NH_2-terminal kinases (JNK), also called the stress-activated MAPK or SAPK; the extracellular-regulated protein kinase (ERK); and the p38 mitogen-activated kinase (p38 MAPK). Common to all members of these MAPK families is the phosphorylation of threonine and tyrosine residues by upstream MAPK kinases (MKK's, or MEK's) which are in turn phosphorylated and activated by an upstream MKK kinase (MKKK's or MEKK's) (47). The resulting phosphorylation and activation of these proteins identifies them as targets of regulatory phosphatases (51). As would be predicted by the complex nature of this cascade, a diverse set of stimuli (input) result in a variety of cellular functions (output) that are broadly regulated by these cascades. For the purpose of this chapter, we have focussed on those biological responses relevant to inflammation and ALI.

Table 3: Partial list of genes regulated by AP-1/MAP kinase pathways

Inflammatory Mediators	
	Tumor necrosis factor
	Interleukins-1, -2, -4, -6, -8, –12, and -18
	Inducible nitric oxide synthetase and arginine transporter
Transcriptional activators	
	c-fos/c-jun (self-activating mechanism)
	NFAT4
	MEF2C
Adhesion Molecules	
	E-selectin
	Intracellular adhesion molecule 1 (ICAM-1)
	$\alpha_m\beta_2$ integrins
	P-selectin glycoprotein ligand-1 (PSGL-1)
Miscellaneous	
	p47^{phox} (component of NADPH oxidase complex)
	Inducible cyclo-oxygenase 2 (COX 2)
	Fas ligand
	Tau (microtubule-associated protein)

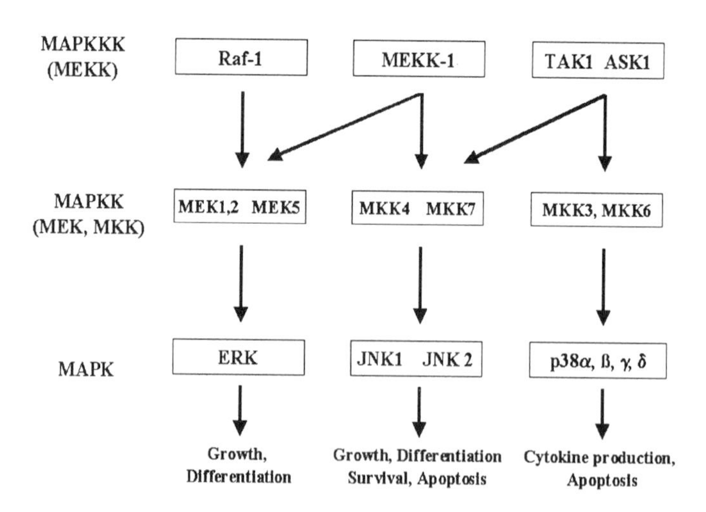

Figure 2. Schema of upstream MAP Kinase signaling pathways.

p38 MAPK

The p38 family of MAP kinases is comprised of at least six isomers: two α isoforms, β, β2, γ and δ (reviewed in reference 47). The expression of each isomer varies among tissues and cell types, but leukocytes express predominantly p38α and δ. Various inflammatory stimuli are capable of activating this pathway, notably LPS, TNF, IL-1β, IL-8, and platelet activating factor (PAF) (47). Several downstream targets of this pathway play a critical role in inflammatory disease states (Figure 3)

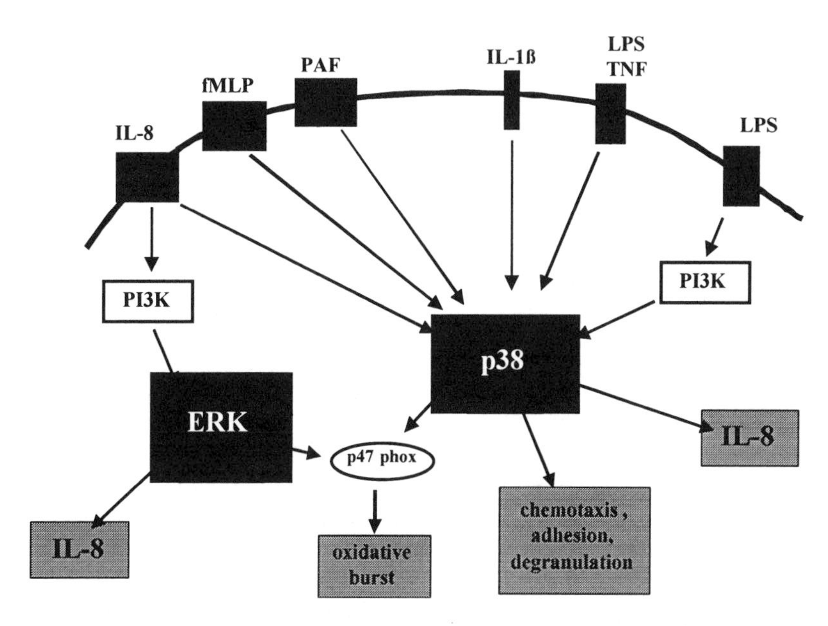

Figure 3. Stimuli activating the p38 and Erk pathways and the resultant responses.

As reviewed in Chapters 2-4, a principle characteristic of ALI is the infiltration of leukocytes to the lung mediated by the coordinated effort of cytokines, chemokines, and adhesion molecules. The p38 MAPK pathway appears to play a central role in this process. TNFα-mediated upregulation of E-selectin is regulated in part through p38 activation of ATF2 (52). Whereas selectins are necessary to initiate the "rolling" phase of the leukocyte endothelial cell adhesion cascade, the second stage of firmer adhesion is established by leukocyte integrins binding to the vascular endothelial cell adhesion molecules, ICAM-1 and VCAM-1, via a process mediated by cytokine production. Both neutrophils (53) and macrophages (54) secrete TNFα following LPS stimulation in a manner dependent upon p38 activation. The stage of emigration of the adherent leukocyte from the endothelium to the site of inflammation requires the presence of chemokines, such as IL-8 (reviewed in Chapter 2). IL-8 production from neutrophils in response to LPS and TNFα is also dependent on p38 activation (55). Finally, the tissue injury associated with neutrophil infiltration is thought to result, in part, from the release by neutrophils of injurious products such as proteases and oxygen radical species.

Production of reactive oxygen species by the oxidative burst is catalyzed by NADPH oxidase. A necessary subunit of this complex in the $p47^{phox}$ protein. Recent data suggest that phosphorylation and activation of $p47^{phox}$ by both the p38 and ERK pathways is critical to the assembly of this complex (56). Priming of neutrophils with PAF results in a p38-dependent enhancement of superoxide production (57). Furthermore, p38 activation stimulates the arginine transporter and increases iNOS gene expression, thus contributing to production of nitric oxide (47). Thus, given its ubiquitous role in mediating several events in neutrophil-mediated injury, the p38 MAPK pathway has been targeted for inhibition in the hopes of ameliorating inflammatory responses.

Several inhibitors of p38 MAPK have been developed, perhaps best exemplified by the pyridinyl imidazole compounds, such as SB203580. This compound inhibits cytokine production *in vitro*, and has modulated the *in vivo* production of TNFα in response to endotoxin (58). With regards to lung inflammation, production of TNFα and IL-6 by human alveolar macrophages was partially decreased by inhibition of p38 via regulation of gene transcription and further inhibition was observed by adding ERK inhibition (59). This study implicated the p38 MAPK pathway in mediating lung inflammation, a hypothesis that was confirmed in a mouse model of LPS-induced lung inflammation (60). In this study, inhibition of p38 resulted in significant decreases in TNFα and neutrophil accumulation in the lung following LPS challenge, though the chemokines, KC and MIP-2 were unaffected (60). It was also observed that inhibitory effects of *in vitro* cytokine production were 1000-fold greater in neutrophils as compared to macrophages and suggested that the p38 cascade may play a more critical role in neutrophil activation.

Inhibition of p38 has also demonstrated beneficial effects in models of more "chronic" inflammation, including allergic inflammatory lung disease such as asthma (61) and arthritis (62). Finally, the anti-inflammatory response that is mounted by the host in an attempt to contain inflammation is in part mediated by IL-10 (see Chapter 17). It is interesting to note that the inhibitory effect of IL-10 on LPS-mediated gene expression was associated with decreased phosphorylation of p38 MAPK (63). This suggests the MAPK pathway is targeted by the compensatory anti-inflammatory response. In light of these promising findings associated with exogenous and endogenous inhibitors of p38 MAPK, this pathway remains a valid therapeutic target.

JNK MAPK

Three JNK protein kinases have been identified. Jnk1 and Jnk2 are ubiquitously expressed, whereas Jnk3 appears restricted to the brain. Jnk protein kinases are phosphorylated on threonine and tyrosine residues by the upstream kinases MKK4 (SEK1) and MKK7 (64). While MKK4 can also activate p38 MAPK, MKK7, which is primarily activated by cytokines such as TNFα, appears to be restricted to JNK activation (46). While targeted gene disruption studies are continuing to provide further insight into the upstream MKKK's responsible for JNK pathway activation, a clear defect in JNK activation is seen in the MEKK-1 deficient cells (65). How the stimulatory signal from the cell surface to MEKK-1 occurs has been increasingly understood, and may be crucial to the pathologies of ALI and sepsis.

The TNF receptor associated family, or TRAF, adaptor proteins are key to the activation of the JNK pathway by cytokines. For example, TNF binding to TRAF2 results in receptor oligomerization and binding to MEKK-1 (66). Thus, cytokine mediated activation of the JNK pathway could serve as an "auto-amplification" mechanism to promote ongoing inflammation. Production of many inflammatory cytokines are regulated at two levels, transcription and translation. For example, TNFα production is regulated at the transcriptional level principally by NF-κB activation and binding to the TNFα promoter sequence. In addition, translation of TNFα requires the derepression of a conserved element in the 3' untranslated region via a mechanism dependent upon JNK activation (67). Relevant to our current therapeutic interventions in ALI, corticosteroids have been shown to inhibit JNK activation resulting in repression of TNF translation in monocytes (68).

It has also been demonstrated that LPS stimulation of monocytes results in JNK activation, associated with AP-1 activation and production of IL-1β via the CD14 surface receptor (69, 70). LPS-induced JNK activation appears to be regulated in part by the serine-threonine phosphatase, PP2A (69). A physical association between JNK and the regulatory subunit PP2A-A/α suggests a possible signal transduction complex comprised of MAP kinases and regulatory phosphatases in association with scaffolding proteins, as has become increasingly described (71). Future studies will continue to elucidate the mechanisms by which signals are transduced dynamically via these complexes and the crucial subunits regulating the relevant phosphoproteins. Thus, while the precise role of the JNK pathway is incompletely understood, it remains a key target in the investigations of the cellular mediation of inflammatory diseases such as sepsis and ALI.

ERK Pathway

Less is known regarding the role of the ERK pathway in acute inflammatory diseases, including ALI. However, studies suggest that this pathway may be important in the pathophysiology of inflammation. In an *in vitro* model, only ERK2 activation was observed in respiratory syncytial virus-infected respiratory epithelial cells (72). Furthermore, production of IL-8 from these cells was significantly reduced by prior inhibition of this MAPK pathway, suggesting a role for ERK in viral-induced cytokine production. Also, infection of macrophages with *Salmonella* caused activation of the ERK pathway resulting in cytokine gene expression (73). Thus, while data implicating the ERK pathway in ALI is limited, this pathway may play a role in pathogen-mediated inflammatory cell activation

Summary

While we have reviewed the NF-κB and AP-1 signaling pathways independently, it is important to note that an increasing body of data supports the hypothesis that cross-talk between the MAPK and NF-κB pathways exists. The upstream member of the MAPK cascade, MEKK-1 has clearly been shown to activate downstream MEK's and the IKK complex leading to NF-κB activation (74, 75). This observation attests to the complexity of these molecular processes.

It is anticipated that on-going basic science and clinical studies will provide further insight into how the various stimuli triggering ALI initiate these pathways and the mechanism by which subsequent signals are propagated, terminated, and regulated by modifier proteins. By gaining this knowledge, we hope to identify those molecular targets that hold the promise of more effective control of this devastating disease process

ACKNOWLEGMENTS

The authors' published and unpublished work presented in this chapter was supported, in part, by grants from the National Institutes of Health (K08HL03725, K08HL04291, and RO1GM61723) and the Children's Hospital Research Foundation.

REFERENCES

1. Baldwin, A. S. (1996) The NF-κB and IκB proteins: new discoveries and insights. *Annu Rev Immunol* 14, 649-681
2. Blackwell, T. S., and Christman, J. W. (1997) The role of nuclear factor-κB in cytokine gene regulation. *Am J Respir Cell Mol Biol* 17, 3-9

3. Ghosh, S., May, M. J., and Koop, E. B. (1998) NF-κB and Rel proteins: evolutionarily conserved mediators of immune responses. *Annu Rev Immunol* 16, 225-260
4. Abraham, E. (2000) NF-kappaB activation. *Crit Care Med* 28, (Suppl) N100-104
5. Wong, H. R., Ryan, M., and Wispé, J. R. (1997) The heat shock response inhibits inducible nitric oxide synthase gene expression by blocking Iκ-B degradation and NF-κB nuclear translocation. *Biochem Biophys Res Comm* 231, 257-263
6. Wong, H. R., Ryan, M., and Wispé, J. R. (1997) Stress response decreases NF-κB nuclear translocation and increases I-κBα and expression in A549 cells. *J Clin Invest* 99, 2423-2428
7. Mazor, R. L., Menendez, I. Y., Ryan, M. A., Fiedler, M. A., and Wong, H. R. (2000) Sesquiterpene lactones are potent inhibitors of interleukin-8 gene expression in cultured human respiratory epithelium. *Cytokine* 12, 239-245
8. Imbert, V., Rupec, R. A., Livolsi, A., Pahl, H. L., Traenckner, E. B., Mueller-Dieckmann, C., Farahifar, D., Rossi, B., Auberger, P., Bauerle, P. A., and Peyron, J. F. (1996) Tryosine phosphorylation of IκBα activates NF-κB without proteolytic degradation of IκBα. *Cell* 86, 787-798
9. Beraud, C., Henzel, W. J., and Baeuerle, P. A. (1999) Involvement of regulatory and catalytic subunits of phosphoinositide 3-kinase in NF-κB activation. *Proc Natl Acad Sci* 96, 429-434
10. Bender, K., Gottlicher, M., Whiteside, S., Rahmsdorf, H. J., and Herrlich, P. (1998) Sequential DNA damage-independent and -dependent activation of NF-κB by UV. *EMBO J* 17, 5170-5181
11. Ito, C. Y., Kazantsev, A. G., and Baldwin, A. S. (1994) Three NF-κB sites in the IκB-α promoter are required for induction of gene expression by TNFα. *Nucleic Acids Res* 22, 3787-3792
12. Arenzana-Seisdedos, F., Thompson, J., Rodriguez, M. S., Bachelerie, F., Thomas, D., and Hay, R. T. (1995) Inducible nuclear expression of newly synthesized I kappa B alpha negatively regulates DNA-binding and transcriptional activities of NF-kappa B. *Mol Cell Biol* 15, 2689-2696
13. Thompson, J. E., Phillips, R. J., Erdjument-Bromage, H., Tempst, P., and Ghosh, S. (1995) IκBβ regulates the persistent response in a biphasic activation of NF-κB. *Cell* 80, 573-582
14. Karin, M., and Ben-Neriah, Y. (2000) Phosphorylation meets ubiquiti-nation: the control of NF-κB activity. *Annu Rev Immunol* 18, 621-663
15. DiDonato, J. A., Hayakawa, M., Rothwarf, D. M., Zandi, E., and Karin, M. (1997) A cytokine-responsive IκB kinase that activates the transcription factor NF-κB. *Nature* 388, 548-554
16. Mercurio, F., Zhu, H., Murray, B. W., Shevchenko, A., Bennett, B. L., Li, J. W., Young, D. B., Barbosa, M., and Mann, M. (1997) IKK-1 and IKK-2: cytokine-activated IκB kinases essential for NF-κB activation. *Science* 278, 860-866
17. Takeda, K., Takeuchi, O., Tsujimura, T., Itami, S., Adachi, O., Kawai, T., Sanjo, H., Yoshikawa, K., Terada, N., and Akira, S. (1999) Limb and skin abnormalities in mice lacking IKKα. *Science* 284, 313-316
18. Li, Q., Antwerp, D. V., Mercurio, F., Lee, K. F., and Verma, I. M. (1999) Severe liver degeneration in mice lacking the IκB kinase 2 gene. *Science* 284, 321-325
19. Hu, Y., Baud, V., Delhase, M., Zhang, P., Deerinck, T., Ellisman, M., Johnson, R., and Karin, M. (1999) Abnormal morphogenesis but intact IKK activation in mice lacking the IKKα subunit of IκB kinase. *Science* 284, 316-320
20. Li, Z. W., Chu, W., Hu, Y., Delhase, M., Deerinck, T., Ellisman, M., Johnson, R., and Karin, M. (1999) The IKKβ subunit of IκB kinase (IKK) is essential for nuclear factor κB activation and prevention of apoptosis. *J Exp Med* 189, 1839-1845
21. Ware, L., and Mathhay, M. (2000) The acute respiratory distress syndrome. *N Engl J Med* 342, 1334-1249

22. Shanley, T. P. (1998) Cytokines in inflammatory diseases: Role and therapeutic targets in acute respiratory distress syndrome. *Emerging Therapeutic Targets* 2, 1-16
23. Schutte, H., Lohmeyer, J., Rosseau, S., Ziegler, S., Siebert, C., Kielisch, H., Pralle, H., Grimminger, F., Morr, H., and Seeger, W. (1996) Bronchoalveolar and systemic cytokine profiles in patients with ARDS, severe pneumonia and cardiogenic pulmonary oedema. *Eur Respir J* 9, 1858-1867
24. Miller, E. J., Cohen, A. B., Nagao, S., Griffith, D., Maunder, R. J., Martin, T. R., Weiner-Kronish, J. P., Sticherling, M., Christophers, E., and Matthay, M. A. (1992) Elevated levels of NAP-1/interleukin-8 are present in the airspaces of patients with the adult respiratory distress syndrome and are associated with increased mortality. *Am Rev Respir Dis* 146, 427-432
25. Donnely, S. C., Strieter, R. M., Kunkel, S. M., Walz, A., Robertson, C. R., Carter, D. C., Grant, I. S., Pollock, A. J., and Haslett, C. (1993) Interleukin-8 and development of adult respiratory distress syndrome in at-risk patient groups. *Lancet* 341, 643-647
26. Mukaida, N., Okamoto, S., Ishikawa, Y., and Matsushima, K. (1994) Molecular mechanisms of interleukin-8 gene expression. *J Leukoc Biol* 56, 554-558
27. Allen, G. A., Menendez, I., Ryan, M., Mazor, R., Wispé, J., Fiedler, M., and Wong, H. R. (2000) Hyperoxia synergistically increases TNF-α-induced interleukin-8 gene expression in A549 cells. *Am J Physiol* 278, L253-L260
28. Blackwell, T. S., Yull, F. E., Chen, C. L., Venkatakrishnan, A., Blackwell, T. R., Hicks, D. J., Lancaster, L. H., Christman, J. W., and Kerr, L. D. (2000) Multiorgan nuclear factor kappa B activation in a transgenic mouse model of systemic inflammation. *Am J Respir Crit Care Med* 162, 1095-1101
29. Amory-Rivier, C. F., Mohler, J., Bedos, J. P., Azoulay-Dupuis, E., Henin, D., Muffat-Joly, M., Carbon, C., and Moine, P. (2000) Nuclear factor-kappaB activation in mouse lung lavage cells in response to *Streptococcus pneumoniae* pulmonary infection. *Crit Care Med* 28, 3249-3256
30. Blackwell, T. S., Lancaster, L. H., Blackwell, T. R., Venkatakrishnan, A., and Christman, J. W. (1999) Differential NF-kappaB activation after intra-tracheal endotoxin. *Am J Physiology* 277, L823-L830
31. Lentsch, A. B., Jordan, J. A., Czermak, B. J., Diehl, K. M., Younkin, E. M., Sarma, V., and Ward, P. A. (1999) Inhibition of NF-kappaB activation and augmentation of IkappaBbeta by secretory leukocyte protease inhibitor during lung inflammation. *Am J Pathol* 154, 239-247
32. Lentsch, A. B., Czermak, B. J., Jordan, J. A., and Ward, P. A. (1999) Regulation of acute lung inflammatory injury by endogenous IL-13. *J Immunol* 162, 1071-1076
33. Moine, P., Shenkar, R., Kaneko, D., Tulzo, Y. L., and Abraham, E. (1997) Systemic blood loss affects NF-kappa B regulatory mechanisms in the lungs. *Am J Physiol* 273, L185-L192
34. Shenkar, R., and Abraham, E. (1999) Mechanisms of lung neutrophil activation after hemorrhage or endotoxemia: roles of reactive oxygen intermediates, NF-kappa B, and cyclic AMP response element binding protein. *J Immunol* 163, 954-962
35. Yoshidome, H., Kato, A., Edwards, M. J., and Lentsch, A. B. (1999) Interleukin-10 inhibits pulmonary NF-kappaB activation and lung injury induced by hepatic ischemia-reperfusion. *Am J Physiol* 277, L919-923
36. Lentsch, A. B., Czermak, B. J., Bless, N. M., Rooijen, N. V., and Ward, P. A. (1999) Essential role of alveolar macrophages in intrapulmonary activation of NF-kappaB. *Am J Respir Cell Mol Biol* 20, 692-698
37. Ross, S. D., Kron, I. L., Gangemi, J. J., Shockey, K. S., Stoler, M., Kern, J. A., Tribble, C. G., and Laubach, V. E. (2000) Attenuation of lung reperfusion injury after transplantation using an inhibitor of nuclear factor-kappaB. *Am J Physiol* 279, L528-L536
38. Schwartz, M. D., Moore, E. E., Moore, F. A., Shenkar, R., Moine, P., Haenel, J. B., and Abraham, E. (1996) Nuclear factor-κB is activated in alveolar macrophages from patients with acute respiratory distress syndrome. *Crit Care Med* 24, 1285-1292

39. Moine, P., McIntyre, R., Schwartz, M. D., Kaneko, D., Shenkar, R., Le Tulzo, Y., Moore, E. E., and Abraham, E. NF-kappaB regulatory mechanisms in alveolar macrophages from patients with acute respiratory distress syndrome. (2000) *Shock* 13, 85-91
40. Bohrer, H., Qiu, F., Zimmerman, T., Zhang, Y., Jllmer, T., Mannel, D., Bottiger, B. W., Stern, D., Waldherr, R., Saeger, H.-D., Ziegler, R., Bierhaus, A., Martin, E., and Nawtroth, P. P. (1997) Role of NF-κB in the mortality of sepsis. *J Clin Invest* 100, 972-985
41. Karin, M. (1995) The regulation of AP-1 activity by mitogen-activated protein kinases. *J Biol Chem* 270, 16483-16486
42. Karin, M., Liu, Z., and Zandi, E. (1997) AP-1 function and regulation. *Curr Opin Cell Biol* 9, 240-246
43. Angel, P., and Karin, M. (1991) The role of Jun, Fos and the AP-1 complex in cell-proliferation and transformation. *Biochim Biophys Acta* 1072, 129-157
44. Hai, T., and Curran, T. (1991) Cross-family dimerization of transcription factors fos/jun and ATF/CREB alters DNA-binding specificity. *Proc Natl Acad Sci USA* 88, 3720-3724
45. Ziff, E. B. (1990) Transcription factors: a new family gathers at the cAMP response site. *Trends Genet* 6, 69-72
46. Davis, R. (2000) Signal transduction by the JNK group of MAP kinases. *Cell* 103, 239-252
47. Herlaar, E. and Brown, Z. (1999) p38 MAPK signaling cascade in inflammatory disease. *Mol Med Today* 5, 439-447
48. Ip, Y. T., and Davis, R. J. (1998) Signal transduction by the c-Jun N-terminal kinase (JNK)-from inflammation to development. *Curr Opin Cell Biol* 10, 205-219
49. Su, B., and Karin, M. (1996) Mitogen-activated protein kinase cascades and regulation of gene expression. *Curr Opin Immunol* 8, 402-411
50. Garrington, T. P., and Johnson, G. L. (1999) Organization and regulation of mitogen-activated protein kinase signaling pathways. *Curr Opin Cell Biol* 11, 211-218
51. Keyse, S. M. (1998) Protein phosphatases and the regulation of MAP kinase activity. *Semin Cell Dev Biol* 9, 143-152
52. Read, M. A., (1997) Tumor necrosis factor a-induced E-selectin expression is activated by the nuclear factor-κB and c-jun N-terminal kinase/p38 mitogen-kinase activated protein kinase pathways. *J Biol Chem.* 272, 2753-2761
53. Nick, J. A., Avdi N. J., Young, S. K., Lehman, L. A., McDonald, P. P., Frasch, S. C., Billstrom, M. A., Henson, P. M., Johnson, G. L., and Worthen, G. S. (1999) Selective activation and functional significance of p38α mitogen-activated protein kinase in lipopolysaccharide-stimulated neutrophils. *J Clin Invest* 103, 851-858
54. Bauer, G. J., Garcia, I. and Maier, R. V. (1997) Activation of stress-activated protein kinase is a mechanism of priming in macrophages. *Surg Forum* 48, 43-45
55. Ridley, S. H. (1997) Actions of IL-1 are selectively controlled by p38 mitogen-activated protein kinase. Regulation of prostaglandin H synthase-2, metalloproteinases, and IL-6 at different levels. *J Immunol* 158, 3165-3173
56. Jorres, R. A. and Magnussen, H. (1997) Oxidative stress in COPD. *Eur Respir Rev.* 7, 131-135
57. Patrick, D. A. (1997) Role of p38 MAP kinase in human neutrophil priming for superoxide production and elastase release. *Surg Forum* 48, 48-50
58. Badger, A. M., Bradbeer, J. N., Votta, B., Lee, J. C., Adams, J. L., and Griswold, D. E. (1996) Pharmacological profile of SB 203580, a selective inhibitor of cytokine suppressive binding protein/p38 kinase, in animal models of arthritis, bone resorption, endotoxin shock and immune function. *J Pharmacol Exp Ther* 279, 1453-1461
59. Carter, A. B., Monick, M. M., and Hunninghake, G. W. (1999) Both Erk and p38 kinases are necessary for cytokine gene transcription. *Am J Respir Cell Mol Biol* 20, 751-758
60. Nick, J. A., Young, S. K., Brown, K. K., Avdi N. J., Arndt, P. G., Suratt, B. T., Janes, M. S., Henson, P. M., and Worthen, G. S. (2000) Role of p38 mitogen-activated protein kinase in a murine model of pulmonary inflammation. *J Immunol* 164, 2151-2159

61. Underwood, D.C., Osborn, R. R., Kotzer, C. J., Adam, J. L., Lee, J. C., Webb, E. F., Carpenter, D. C., Bochnowicz, S., Thomas, H. C., Hay, D. W., and Griswold, D. E. (2000) SB 239063, a potent p38 MAP kinase inhibitor, reduces inflammatory cytokine production, airways eosinophil infiltration, and persistence. *J Pharmacol Exp Ther* 293, 218-288
62. Jackson, J. R., Bolognese, B., Hillegass, L., Kassis, S., Adams, J., Griswold, D. E., and Winkler, J. D. (1998) Pharmacological effects of SB 220025, a selective inhibitor of p38 mitogen-activated protein kinase, in angiogenesis and chronic inflammatory disease models. *J Pharmacol Exp Ther* 284:687-692, 1998
63. Niiro, H., Otsuka, T., Ogami, E., Yamaoka, K., Nagano, S., Akahoshi, M., Nakashima, H., Arinobu, Y., Izuhara, K., and Niho, Y. (1998) MAP kinase pathways as a route for regulatory mechanisms of IL-10 and IL-4 which inhibit COX-2 expression in human monocytes. *Biochem Biophys Res Commun* 250, 200-205
64. Tournier, C., Hess, P., Yang, D. D., Xu, J., Turner, T. K., Nimnual, A., Bar-Sagi, D., Jones, S. N., Flavell, R. A. and Davis, R. J. (2000) Requirement of JNK for stress-induced activation of the cytochrome c-mediated death pathway. *Science* 288, 870-874
65. Yujiri, T., Sather, S., Fanger, G. R., and Johnson, G. L. (1998) Role of MEKK-1 in cell survival and activation of JNK and ERK pathways defined by targeted gene disruption. *Science* 282, 1911-1914
66. Baud, V., Liu, Z.-G., Bennett, B., Suzuki, N., Xia, Y., and Karin, M. (1999) Signaling by proinflammatory cytokines: oligomerization of TRAF2 and TRAF6 is sufficient for JNK and IKK activation and target gene induction via an amino-terminal effector domain. *Genes Develop* 13, 1297-1308
67. Han, J., Huez, G., and Beutler, B. (1991) Interactive effects of the TNF promoter and 3'-untranslated regions. *J Immunol* 146, 1843-1848
68. Swantek, J. L., Cobb, M. H., and Geppert, T. D. (1997) Jun N-terminal kinase/stress-activated protein kinase (JNK/SAPK) is required for lipopolysaccharide stimulation of tumor necrosis factor alpha (TNF-α) translation: Glucocorticoids inhibit TNF-α translation by blocking JNK/SAPK. *Mol Cell Biol* 17, 6274-6282
69. Shanley, T. P, Vasi, N., Deneberg, A., and Wong, H. R. (2001) The serine/threonine phosphatase, PP2A: Endogenous regulator of inflammatory cell signaling. *J Immunol* 166, 966-972
70. Hambleton, J., Weinstein, S. L., Lem, L., and DeFranco, A. L. (1996) Activation of c-Jun N-terminal kinase in bacterial lipopolysaccharide-stimulated macrophages. *Proc Natl Acad Sci USA* 93, 2774-2778
71. Burack, W. R., and Shaw, A. S. (2000) Signal transduction: hanging of a scaffold. *Curr Opin Cell Biol* 12, 211-215
72. Chen, W., Monick, M. M., Carter, A. B., and Hunninghake, G. W. (2000) Activation of ERK2 by respiratory syncytial virus in A549 cells is linked to the production of interleukin-8. *Exp Lung Res* 26, 13-26
73. Procyk, K. J., Kovarik, P., von Gabain, A., and Baccarini, M. (1999) *Salmonella typhimurium* and lipopolysaccharide stimulate extracellularly regulated kinase activation in macrophages by a mechanism involving phosphatidylinositol 3-kinase and phospholipase D as novel intermediates. *Infect Immun* 67, 1011-1018
74. Meyer, C. F., Wang, X., Chang, C., Templeton, D., and Tan, T. H. (1996) Interaction between c-Rel and the mitogen-activated protein kinase kinase kinase-1 signaling cascade in mediating kappaB enhancer activation. *J Biol Chem* 271, 8971-8976
75. Janssen-Heininger, Y. M., Macara, I., and Mossman, B. T. (1999) Cooperativity between oxidants and tumor necrosis factor in the activation of nuclear factor (NF)-kappaB: requirement of Ras/mitogen-activated protein kinases in the activation of NF-kappaB by oxidants. *Am J Respir Cell Mol Biol* 20, 942-952

Chapter 2

INNATE IMMUNE MECHANISMS TRIGGERING LUNG INJURY

Robert M. Strieter, John A. Belperio, David Kelley, Ammar Sakkour, and Michael P. Keane
Department of Medicine, Division of Pulmonary and Critical Care Medicine, University of California, Los Angeles (UCLA) School of Medicine, Los Angeles, CA.

INTRODUCTION

The lungs comprise a unique interface between the body and the environment, presenting an alveolar surface area of approximately 75 square meters and only a minimal barrier of 4 to 8 μm between the alveolar airspace and the microvasculature. While this configuration is ideal for gas exchange, it also increases vulnerability to noxious stimuli and pathogens. Consequently, the lung must possess the capacity to generate a brisk innate host defense to both inhaled and hematogenous challenges. This response is characterized by acute inflammation and must provide prompt clearance of the offending agent while avoiding compromise of essential gas exchange. This acute pulmonary inflammatory response typically results in local increases in vascular permeability and a predominantly early neutrophilic influx followed by mononuclear cell infiltration. Once successful containment of the noxious agent has occurred, inflammation should then resolve with normal repair and tissue remodeling, and return to homeostasis. Because of the great capacity of the innate immune system to initiate acute inflammation, however, the lung may also be predisposed to tissue injury by excessive reactions generated by both local and distant mediators. In conditions such as acute lung injury (ALI), the over-exuberant tissue inflammation may result in severe, irreversible lung injury mediated primarily by elicited and activated leukocytes.

Many clinical entities, including trauma, pneumonia/sepsis, ischemia-reperfusion injury, as well as the acute respiratory distress syndrome (ARDS), are characterized by varying degrees of acute pulmonary inflammation and subsequent impairment of normal gas exchange function of the lung. The inflammatory response is initiated, propagated, and resolved by a complex yet coordinated intercellular interaction between immune and non-immune cells. For example, the host response to bacterial pneumonia is characterized histologically by proteinaceous exudate and massive neutrophil extravasation leading to consolidation of the lung. Once the inciting microbe is cleared, the inflammatory reaction resolves and normal repair and tissue remodeling occurs. This re-establishes normal lung function without the sequela of chronic inflammation and pulmonary fibrosis. In contrast, the acute inflammatory response associated with ARDS/ALI may culminate in severe lung injury, ultimately impairing lung function and impacting on host survival. While a variety of factors are involved in the innate response leading to acute pulmonary inflammation, cytokines constitute the largest and most pleiotropic group of mediators that regulate this response and will be the focus of this chapter. We will address recent advances in inflammation research that examine initiating events in cytokine-induced leukocyte recruitment into the lung during the pathogenesis of acute pulmonary inflammation.

Cellular Communication and Cytokine Networks

The fidelity of pulmonary inflammation is dependent upon intercellular communication. While this is often accomplished through direct cell-to-cell adhesive interaction via specific cellular adhesion molecules, cells also signal each other through soluble mediators, such as cytokines. These polypeptide molecules often have pleiotropic effects on a number of biological functions including proliferation, differentiation, recognition, and leukocyte recruitment. Their actions are mediated through paracrine and autocrine signaling via receptor-ligand interactions on specific cell populations. Under certain conditions, however, these molecules may behave as hormones. Cytokines display concentration-dependent effects, being expressed in low concentrations during normal homeostasis, with modest increases exerting local effects, and still greater increases resulting in systemic effects. Cytokine research investigating the biology of these proteins has rapidly expanded, and currently numerous cytokines have been isolated and characterized. Individual subpopulations of both immune and non-immune cells, endothelial cells, fibroblasts, and epithelial cells possess different capacities to elaborate and secrete specific cytokines in response to particular stimuli. Furthermore, cell populations vary in their expression of receptors for individual cytokines, and, as a result, differ in their capacity to respond to specific cytokine signals.

Investigations into the interactions between various cell populations have lead to the concept of cytokine networking. In this process, one population of cells may respond directly to specific stimuli (i.e., exogenous and/or endogenous) leading to the elaboration of a particular cytokine to exert distinct effects upon another population of cells. The targets respond by producing cytokines, which may serve as positive and negative feedback signals to the primary cell, or alternatively, initiate a cascade of events by affecting yet another array of target cells. Inflammatory effector cells, such as neutrophils and monocytes, may be locally recruited and activated in response to specific chemotactic signals resulting in further amplification of a cytokine cascade by non-immune resident cells (i.e., endothelial cells, fibroblasts, and epithelial cells). As many of the complexities of the innate immune cytokine cascade that mediate ALI have been elucidated, increasing evidence suggests that non-immune cells play crucial roles in the generation, maintenance, and resolution of both local and systemic inflammatory responses. Further understanding the role of cytokines in the innate response of acute pulmonary inflammation will lead to therapeutic intervention that will ultimately refine the reaction, eliminate the triggering agent, and preserve lung function without increased mortality of the host.

Initiating Events of the Innate Host Defense in ALI

Cytokines that are involved in the innate immune response of the lung are not constitutively expressed, and must be induced by specific signals that alert the host to invading microorganisms or early triggering events. Evolution has provided the mammalian host with two major forms of host defense: the innate and adaptive immune responses (1-7). The innate defense is the gatekeeper and sentry for the immediate host response against invading pathogens. The adaptive immune response is naïve, and develops in time through somatic generation of a diverse repertoire of receptors, prior to the development of an effective immune response to invading microbes (1-7). This difference in behavior suggests that the innate immune response has been genetically predetermined to recognize microorganism associated molecular patterns. The innate defense is designed to directly interact with, recognize, and immediately respond to the invading microorganism. In contrast, the adaptive immune response depends on two classes of specialized lymphocytes, T and B cells, with specific receptors that are somatically generated in response to antigen presentation by professional antigen presenting cells (i.e., dendritic cells, macrophages, and other B cells). This process allows for antigen-dependent clonal expansion of T and B cells resulting in learned, and long-term humoral and cell-mediated immune memory. This process does not occur immediately, however, and the delay in response could have a devastating impact on survival of the host. Therefore, the two immune responses are coordinate in their behavior, with innate immunity representing the most fundamental process of host defense.

Microorganisms are critical in initiating acute inflammation such as occurs in ALI. Microorganisms express highly conserved molecular patterns that are unique and distinct from the host. These include: viral double-stranded RNA; unmethylated CpG dinucleotides common in bacterial DNA but under-represented in vertebrate DNA; mannans of yeast; glycolipids of mycobacteria; lipoproteins of bacteria and parasites; lipoteichoic acids of Gram-positive bacteria; and lipopolysaccharide (LPS) of Gram-negative bacteria (3-8). The host has evolved specific pattern recognition receptors to detect these pathogen-associated molecules (3-5). While these receptors can be divided into secreted, endocytic, and signaling classes of receptors (3-5), the latter class of signaling receptors is critical in mediating the expression of a variety of cytokines that are subsequently necessary to amplify the acute inflammation of innate immunity.

The mammalian Toll-like receptors (TLR) are important signaling receptors in innate host defense, and have evolved from the *Drosophila* Toll gene (3-7, 9). Although the *Drosophila* Toll gene was identified in mediating dorsoventral polarization during embryogenesis of the fly (10), it is a transmembrane protein with a cytoplasmic domain that is homologous to the cytoplasmic domain of the mammalian interleukin-1 receptor (11). This finding supports the notion that both Toll and mammalian TLRs share similar signal-transduction pathways that ultimately involve the nuclear factor-κB (NF-κB) family of transcriptional factors (11, 12). NF-κB plays an important role as a "master switch" in the transactivation of a number of cytokines that are involved in the innate immune response and development of pulmonary inflammation (see Chapter 1).

Medzhitov and colleagues were the first to characterize a human Toll-like Receptor 4, TLR4, (13). The constitutively active mutant of TLR4 transfected into human cell lines induced the activation of NF-κB and the expression of pro-inflammatory cytokines genes for interleukin-1 (IL-1), interleukin-6 (IL-6), and interleukin-8 (IL-8) (13). In addition, TLR4 signal transduction and NF-κB transactivation resulted in the expression of IL-12 p40 and the molecules CD80 and CD86, which are co-stimulatory molecules necessary to bridge the innate to the adaptive immune response and activate naïve T cells in an antigen-dependent manner (13).

Subsequent work has identified TLR4 as the putative receptor for LPS signal-transduction on macrophages, dendritic cells, and B cells. In LPS resistant C3H/HeJ and C57BL/10ScCr mice, defective LPS signaling is related to mutations in TLR4 gene (15-17). LPS recognition and triggering of the innate host response, however, is more complex then direct interaction with TLR4. LPS first interacts with a serum protein, lipopolysaccharide-binding protein (LBP), that transfers LPS to CD14 anchored to the cell membrane by glycosylphosphoinositol (18). While CD14 lacks a trans-membrane and cytoplasmic domain for signaling coupling, it appears that the LPS/CD14 complex uses TLR4 as a co-receptor (9). Furthermore, MD-2 is a molecule that is constitutively associated with TLR4 and confers enhanced LPS responsiveness to TLR4 (19). These studies support the notion that LPS recognition by the host is dictated by a complex of at least three components, CD14, TLR4, and MD-2 (3-5).

Although another Toll-like receptor, TLR2, was initially thought to also signal-couple via LPS binding, studies have now shown that TLR2 -/- mice are able to respond to LPS. These mice, however, are unable to respond to other bacterial-associated molecular patterns (peptidoglycan and lipoproteins), suggesting that TLR2 is important for detecting microbial molecular patterns other than LPS (20, 21). The ability of these receptors to detect only certain pathogen-associated molecular patterns supports their high degree of specificity for detecting microorganisms and acting as sentinels in the initiation of acute inflammation by triggering expression of a variety of factors, including cytokines. The expression of cytokines provides further fidelity to the innate response as not all cells that participate in this response have receptors that sense microorganisms. The ability of cytokine networks to be activated after encountering a microorganism or other triggering events lead to autocrine, paracrine, and endocrine intercellular communication between immune and non-immune cells that ultimately amplifies the innate host defense, increases the inflammatory response, and elicits the recruitment of leukocytes that ultimately facilitates eradication of the microorganism or triggering stimuli.

Early Response Cytokines

The above mechanisms for pathogen recognition by the innate immune response are necessary in order to sense and to respond to microorganism molecular patterns. Early response cytokines are cytokines that are activated initially after TLR signal-coupling or in response to other triggering events. As such they amplify, engage, and activate additional cells, as well as signal the expression of more distal cytokines critical to the recruitment of leukocytes. Two of the most important early response cytokines in innate immunity and ALI are interleukin-1 (IL-1) and tumor necrosis factor-α (TNFα).

Although biochemically unrelated, TNFα and IL-1 demonstrate similar pleiotropic and overlapping effects on a variety cellular functions (22-38). These cytokines are primarily produced by mononuclear phagocytes, and because of their role for initiating further inflammatory responses, have been termed "early response cytokines." At sites of local inflammation, modest concentrations are essential, and serve to closely regulate cellular function. These early response cytokines dictate the events leading to further initiation, maintenance, and resolution of tissue injury via a cascade of cytokine activity. In contrast to the controlled events of local production of TNFα and IL-1, the exaggerated systemic release of these cytokines can result in a syndrome of multiorgan injury, including ALI, with increased host morbidity and mortality. Thus, TNFα and IL-1 have a broad spectrum of biologic activity that can influence the outcome of ALI.

Interleukin-1 Family of Cytokines

The interleukin-1 family of cytokines consists of two agonists, IL-1α and IL-1β, and one antagonist, interleukin-1 receptor antagonist (IL-1ra) (22-31). The IL-1 agonists are pleiotropic cytokines that exist as two distinct genes with protein isoforms with two isoelectric points, pI 5.0 and pI 7.0, for IL-1α and IL-1β, respectively (22-31). These two forms of IL-1 are also distinguished by whether they are found predominantly membrane associated (IL-1α) or secreted (IL-1β) (22-31). Both isoforms are produced by a variety of cells and bind to the type I IL-1 receptor on target cells with similar biologic function (22-31, 39, 40).

IL-1 ligand binding to the IL-1 type I receptor and the IL-1 receptor associated protein recruits an adapter molecule, MyD88, which in turn recruits IL-1 receptor associated kinase (IRAK). IRAK recruits the adapter molecule, tumor necrosis factor (TNF) receptor associated factor 6 (TRAF6), which recruits the NF-κB inducing kinase (NIK). NIK activates the IκB kinase complex (IKK) which phosphorylates IκB-α leading to ubiquitination and release of NF-κB for translocation to the nucleus and subsequent transactivation of a number of genes (i.e., cyclooxygenase, adhesion molecules, cytokines, inducible nitric oxide synthase, acute phase proteins, and chemokines) (12, 41, and Chapter 1). While IL-1 signaling can occur through other distinct pathways (i.e. p38/MAP kinase pathway and JNK pathways), the effects often synergize with those of NF-κB activation (42). Interestingly, the signal-coupling of IL-1 and the IL-1 type I receptor is identical to signal coupling of LPS on TLR4 (3-7, 42). These two divergent ligand-receptor pairs ultimately signal through the same cytoplasmic pathway leading to NF-κB activation, and transcription of several genes critical to the amplification of innate immunity. Redundancy in this system triggering NF-κB is not only important for amplification of the innate response, but also to initiate the transition to and development of adaptive immunity. This also demonstrates that an exogenous factor such as LPS may be the initial triggering event on specific cells that express the complex of CD14/TLR4/MD-2, however, endogenous ligands, such as IL-1 can further amplify this response. The presence of IL-1 receptors on essentially all immune and non-immune cells affords the ability of IL-1 to bind to the IL-1 type I receptor, activate, and engage all of these cells as participants of acute inflammation.

In addition to the IL-1 type I receptor, IL-1 also binds to an IL-1 type II or decoy receptor that does not signal (28, 29, 43-46). Binding of IL-1 to the IL-1 type II receptor may be a mechanism to sequester IL-1 from interacting with the IL-1 type I receptor (28, 29, 43-46). In addition, the IL-1 type II receptor is cleaved by metalloproteases on the cell surface releasing a soluble form of the IL-1 type II receptor (44). The presence of soluble IL-1 type II receptor can play a role in down-modulating IL-1 biology.

In contrast to the two IL-1 agonists, IL-1ra is the only known naturally occurring cytokine with specific antagonistic activity. IL-1ra consists of three isoforms, one secreted and two intracellular forms, derived from alternatively spliced mRNAs of the same gene (43). The discovery of IL-1ra has led to an appreciation of a dynamic balance between IL-1 agonists and IL-1ra in the maintenance of IL-1-dependent homeostasis and inflammation, and has necessitated investigations into the role of IL-1ra in disease (22-24, 26-29, 31, 47, 48). A variety of other cell populations, including polymorphonuclear cells, alveolar macrophages, fibroblasts, keratinocytes, and tumor cells, have subsequently been demonstrated to produce IL-1ra (22-24, 26-29, 31, 47-52). IL-1ra acts as a pure antagonist of either IL-1α or IL-1β, and when present in sufficient quantities can attenuate a variety of IL-1 actions in both *in vitro* and *in vivo* model systems (22-24, 26-29, 31, 47-53). The intracellular iso-forms of IL-1ra may also play a role in attenuating IL-1 biology under conditions of cellular injury or apoptosis when they are released from the cell (48). These studies have led to an appreciation that IL-1ra normally modulates IL-1-dependent activity and speculation that it may play a role in the resolution of the pulmonary inflammatory cascade necessary for the lung to return to homeostasis.

When LPS, IL-1 or TNFα are intratracheally injected, these inflammatory mediators induce an intra-alveolar inflammatory response composed of predominately of neutrophils, followed later by a mononuclear cell infiltrate (54). IL-1, however, is more potent than TNF in this response. In addition, LPS is capable of inducing both TNFα and IL-1 gene expression in the lung, leading to amplification of the inflammatory response. In fact, IL-1ra has been found to reduce the inflammatory response to LPS in the lungs (55). These findings suggest that IL-1ra has an important immunomodulating influence on IL-1, and its production by mononuclear phagocytes and other cells in the lung may impact on the pathogenesis of the innate response. Studies using genetic approaches, however, have led to findings that are less impressive for IL-1 in the innate immune response. For example, IL-1α-/- and IL-1β-/- animals display no phenotype at birth and appear similar to their wild-type littermates (56). To determine their response under conditions analogous to the innate host response, Horai and associates generated doubly deficient knockout animals (IL-1α/β -/- mice), as compared to IL-1α -/-, IL-1β -/-, and IL-ra -/- mice (56). When these mice were injected with a non-specific inducer of inflammation (i.e., turpentine), fever was suppressed in IL-1β -/- as well as IL-1α/β -/- mice, whereas IL-1α -/- mice displayed a normal febrile response. In contrast IL-1ra -/- mice showed an increased febrile response and this response was paralleled by increased levels of circulating glucocorticoids (56). In response to LPS, IL-1β -/- mice behave very similarly to IL-1β +/+ mice with regard to generation of IL-1α, IL-6, and TNF-α, and were equally sensitive to the lethal effects of LPS (57, 58). In response to influenza infection, however, IL-1β -/- mice demonstrated a higher mortality rate, as compared to IL-1β +/+ mice (59). The difference in above findings for the importance of IL-1 in mediating inflammation and participating in the innate host response may be related to the individual model systems or the development of redundancy that may have occurred during embryogenesis in the null mutant mice.

Tumor Necrosis Factor-α (TNFα)

TNFα is a primary mononuclear phagocyte-derived cytokine, which is recognized for its pleiotropic effects in mediating acute inflammation of the innate host response. TNFα is a member of a family of ligands that activate a unique family of receptors (60). TNF is produced primarily by monocytes and macrophages, and has many overlapping biologic activities with IL-1. TNFα was first described as a cytolytic agent that caused hemorrhagic necrosis of tumor cells *in vivo*, and also caused fever, cachexia, systemic shock, and the production of hepatic acute phase proteins (32, 33, 36, 38, 61).

In solution, TNFα is a homotrimer and binds to two different cell surface receptors, p55 and p75 (32, 33, 36, 38, 60, 61). The TNF-receptor family is comprised of transmembrane proteins with an extracellular domain that contains a recurring cysteine-rich motif and an intracellular domain that demonstrates more variability than the extracellular domain (60). The p55 receptor and the Fas receptor contain a 60 amino-acid domain known as the "death domain" which is essential for signal transduction of an apoptotic signal (60 and Chapter 15). The signal transduction of TNF merges with the same pathway of IL-1 and LPS at the level of NF-κB inducing kinase (NIK) and through mitogen-activated protein/ERK kinase 1/2 (MEK1/2) to activate NF-κB and mediate its pro-inflammatory function in innate immunity (41, 62, 63). Thus, the triumvirate of TLRs, IL-1RI, and the TNF receptor coupling are the recognition signals necessary to assure NF-κB nuclear translocation and transactivation of critical genes for perpetuation of acute inflammation of the innate immune response.

TNFα exhibits a variety of inflammatory effects that are important to innate host defense, including: induction of neutrophil- and mononuclear cell-endothelial cell adhesion and transendothelial migration via expression of adhesion molecules and chemokines; enhancement of a procoagulant environment by upregulating the expression of tissue factor and plasminogen activator inhibitor, with suppression of the protein C pathway; and acting as an early response cytokine in the promotion of the cytokine cascade (32, 33, 35, 36, 38, 60, 61). These pleiotropic properties are important in mediating acute inflammation and promoting ALI.

TNFα expression has been found to be increased in patients with pneumonia (64). TNFα has been found to be expressed early in animal models of pulmonary infection related to *Streptococcus pneumoniae*,

Klebsiella pneumoniae, *Pseudomonas aeruginosa*, *Legionella pneumophilia*, *Cryptococcus neoformans*, *Aspergillus fumigatus*, and *Pneumocystis carinii* (65-72). These pathogens represent a diverse array of extracellular and intracellular bacteria, and fungi that trigger the expression of TNFα. Moreover, when TNFα is depleted in these model systems there is evidence of markedly impaired clearance of these microorganisms (65-72). For example, Laichalk and colleagues have demonstrated that intratracheal inoculation of *Klebsiella pneumoniae* can result in a time-dependent expression of TNFα mRNA and protein within the lung (67). Furthermore, depletion of endogenous TNFα resulted in marked increase in bacteremia and accumulation of *Klebsiella pneumoniae* in the lung associated with an increased mortality (67). The mechanism responsible for the increased mortality in this model system, and others, appears to be related to a marked reduction in the recruitment of neutrophils to the lung during the pathogenesis of the pneumonia, and the failure to contain and clear the microorganism (67, 68, 71). Interestingly, TNFα does not directly mediate the migration of neutrophils and this effect is most likely related to the ability of TNFα to induce the expression of leukocyte-endothelial cell adhesion molecules and chemotactic factors (i.e., chemokines, see Chapter 3).

To further determine whether TNFα is critical to mediate an innate response to bacterial pneumonia, Standiford and associates augmented the expression of TNFα in the lungs of animals infected with *Klebsiella pneumoniae* using a strategy of recombinant adenoviral vector delivery of the murine TNFα cDNA (73). Over-expression of TNFα in the lung using this strategy resulted in significant infiltration of neutrophils, and improved survival of animals challenged concomitantly with intrapulmonary *Klebsiella pneumoniae*. This survival advantage was directly related to reduced bacteremia and burden of microorganisms in the lung (73). These studies indicate that TNFα is a critical endogenous component of innate host defense and acute pulmonary inflammation.

Synergy Between TNFα and IL-1

Although the pathogenesis of septic shock and the development of ALI are multifactorial, the role of TNFα and IL-1 in mediating septic shock and ALI has been clearly demonstrated in a number of studies. Waage and colleagues (74-76), examined sera from patients suffering from meningococcal septicemia with ALI. They found a significant correlation between serum TNFα levels and mortality. In a similar study of 55 patients with a clinical diagnosis of sepsis and purpura fulminans due to meningococcemia, serum levels of both TNFα and IL-1 correlated with mortality (77). In another study, patients were prospectively randomized to assess the efficacy of methylprednisolone administration in septic shock (78). Serum levels of

TNFα were detected in 33% of the patients with septic shock. TNFα levels were increased with equal frequency in patients with shock due to either Gram-positive or negative bacteria. The magnitude of TNFα measured also correlated with a higher incidence and severity of ALI and mortality.

In several animal studies, systemically administered TNFα can induce similar pathophysiological effects as either endotoxin or infusion of live Gram-negative bacteria. In response to TNFα, animals demonstrate metabolic acidosis, increased body temperature and circulating levels of catecholamines, consumptive coagulopathy, multiorgan dysfunction (renal, hepatic, gastrointestinal, and pulmonary), alterations in circulating leukocytes, and hypotension leading to shock (79). The concomitant administration of both TNFα and IL-1 is synergistic in mediating similar pathophysiological effects as TNFα alone (80). Interestingly, when *in vivo* protein synthesis was inhibited by actinomycin D and sublethal TNFα or IL-1 was administered to mice, a 100 percent mortality occurred within 8 to 12 hours, as compared to the absence of lethality without actinomycin D (81). These findings suggest that *de novo* protein synthesis is required to protect against the lethal effects of either TNFα or IL-1.

Inhibition of endogenously produced TNFα during bacteria-induced septic shock has been shown in animal models to significantly attenuate the pathogenesis of multiorgan injury and mortality. Tracey and associates (79) using a baboon model of septic shock, administered a monoclonal anti-human TNF antibody both prior to and after the injection of a LD100 dose of live *E. coli*. Only the monoclonal antibody administration prior to the lethal dose of *E. coli* decreased mortality. In addition, Hinshaw and colleagues (82) employing a similar model of *E. coli*-induced lethal septicemia in a baboon model, could delay the addition of monoclonal anti-TNF antibodies for up to 30 minutes after *E. coli* challenge and all animals survived. The endogenous expression and regulation of TNFα from murine models of endotoxemia has shown that TNFα is rapidly produced after a LD_{100} infusion of endotoxin (83). Peak levels of TNFα are seen as early as within one hour, with a rapid decline to relatively undetectable levels by 8 hours. Similar findings have been seen in human volunteer subjects injected with low doses of endotoxin (84). These results suggest that TNFα is under strict regulation.

TNFα and IL-1, in the context of septic shock, are potent mediators that trigger a cascade of events that can lead to ALI and multiorgan failure. Clinical studies have generally not shown a benefit of inhibition of either IL-1 or TNFα in clinical septic shock and ALI. Reasons why these approaches have failed are due to a variety of circumstances that include: criteria used for patient inclusion into the study; relative efficacy of the pharmacologic therapy against the target; timing of administration of the inhibitor; and failure of the study design to assess whether IL-1 or TNFα were present at the time of administration (85). Therefore, future studies may need to take these considerations into account prior to excluding a major role for IL-1 and/or TNFα in mediating the acute pulmonary inflammation.

Role of TNFα and IL-1 in Modulating Adhesion Molecule Expression.

One of the most critical roles for cytokines in ALI is their regulation of adhesion molecule expression. As mentioned, histopathology of lungs from patients with ALI are characterized by massive infiltration of neutrophils. The process by which neutrophils are recruited from the peripheral blood onto the pulmonary vascular endothelium is mediated, in large part, by TNFα and IL-1β via induced expression of a series of endothelial cell and leukocyte adhesion molecules (reviewed in reference 86). An initial phase of adhesion described as "rolling" is mediated by cytokine-upregulation of the selectin family of adhesion molecules on endothelium (e.g. E-selectin), which interacts with sialylated oligosaccharides constitutively expressed on neutrophils (87-89). The second phase of firm adhesion is a result of cytokine-activated β_2-integrin (e.g. CD11a,b,c/CD18) expression on neutrophils binding to a counter-receptor, intercellular adhesion molecule-1 (ICAM-1), expressed on endothelial cells (90). This efficient adhesive cascade results in the accumulation of neutrophils at the site of inflammation with the subsequent release of oxygen radical species, proteases, and other factors that impair endothelial barrier function and results in alveolar edema.

As an understanding of the role of adhesion molecule expression has unfolded, the goal of anti-adhesion molecule therapy has become an intriguing pursuit. While numerous pre-clinical animal trials have demonstrated that anti-adhesion molecule antibodies such as anti-ICAM-1 (91, 92), anti-E-selectin (93), anti-L-selectin (93, 94), and anti-P-selectin (95) can inhibit neutrophil accumulation and tissue injury in the lung, to date, no human trials have successfully implemented anti-adhesion molecule strategies. Furthermore, anti-adhesion molecule strategy is tempered by the appreciation that this leukocyte-adhesion molecule cascade is a necessary host response, as evidenced by individuals who suffer recurrent infections as a result of leukocyte adhesion deficiency (LAD) syndromes-1 and -2. The molecular basis of these defects are absent expression of the β-integrins (counter-receptor for ICAM-1) in LAD-1 and absence of sialyl-Lewis X (carbohydrate ligand for selectins) in LAD-2 (96). In light of this, inhibiting this cascade in the setting of an invading organism may be detrimental to host survival.

Summary

ALI remains a major cause of mortality in critical care medicine. The innate immune response to invading pathogens plays a key role in the initiation of this response. Pathogen pattern recognition receptors, such as the

Toll-like receptors on immunologically active cells, initiate a signal transduction cascade resulting in the expression of early cytokines (TNFα and IL-1). These inflammatory cytokines contribute to this pathophysiologic state via receptor-mediated signaling pathways that effect target cell responses including the up-regulation of adhesion molecule expression. The triggering of the leukocyte-endothelial cell adhesion cascade mediates leukocyte recruitment to the site of inflammation. The specificity of the leukocyte subtype is mediated by a family of chemotactic cytokines, or chemokines, as reviewed in the following chapter. It is hoped that continued application of molecular biologic techniques into the field of lung inflammation will enhance our understanding of this biologic response and further identify potential therapeutic targets.

REFERENCES

1. Delves, P.J., and Roitt, I.M. (2000) The immune system. Second of two parts. *N Engl J Med* 343, 108-117.
2. Delves, P.J., and Roitt., I.M. (2000) The immune system. First of two parts. *N Engl J Med.* 343, 37-49.
3. Medzhitov, R., and Janeway, C. Jr. (2000) Innate immune recognition: mechanisms and pathways. *Immunol Rev* 173, 89-97.
4. Medzhitov, R., and Janeway, C., Jr. (2000) Innate immunity. *N Engl J Med* 343, 338-344.
5. Medzhitov, R., and Janeway, C.A., Jr. (2000) How does the immune system distinguish self from nonself? *Semin Immunol* 12:185-188.
6. Brightbill, H.D., Libraty, D.H., Krutzik, S.R., Yang, R.B., Belisle, J. T., Bleharski, J.R., Maitland, M., Norgard, M.V., Plevy, S.E., Smale, S.T., Brennan, P. J., Bloom, B.R., Godowski, P. J., and Modlin, R. L. (1999) Host defense mechanisms triggered by microbial lipoproteins through toll-like receptors. *Science* 285, 732-736
7. Brightbill, H.D., and Modlin, R.L. (2000) Toll-like receptors: molecular mechanisms of the mammalian immune response. *Immunology* 101, 1-10.
8. Krieg, A.M., Love-Homan, L., Yi, A.K., and Harty, J.T. (1998) CpG DNA induces sustained IL-12 expression in vivo and resistance to Listeria monocytogenes challenge. *J Immunol* 161, 2428-2434
9. Beutler, B., and Poltorak, A. (2000) Positional cloning of Lps, and the general role of toll-like receptors in the innate immune response. *Eur Cytokine Netw* 11, 143-152
10. Hashimoto, C., Hudson, K.L., and Anderson, K.V. (1988) The Toll gene of Drosophila, required for dorsal-ventral embryonic polarity, appears to encode a transmembrane protein. *Cell* 52, 269-279
11. Gay, N.J., and Keith, F.J. (1991) Drosophila Toll and IL-1 receptor [letter]. *Nature* 351, 355-356
12. Ghosh, S., May, M.J., and Kopp, E.B. (1998) NF-kappa B and Rel proteins: evolutionarily conserved mediators of immune responses. *Annu Rev Immunol* 16, 225-260
13. Medzhitov, R., Preston-Hurlburt, P., and Janeway, C.A., Jr. (1997) A human homologue of the Drosophila Toll protein signals activation of adaptive immunity [see comments]. *Nature* 388, 394-397
14. Zlotnik, A., and Yoshie, O. (2000) Chemokines: a new classification system and their role in immunity. *Immunity* 12, 121-127
15. Poltorak, A., He, X, Smirnova, I., Liu, M.Y., Huffel, C.V., Du, X., Birdwell, D., Alejos, E., Silva, M., Galanos, C., Freudenberg, M., Ricciardi-Castagnoli, P., Layton, B., and Beutler, B. (1998) Defective LPS signaling in C3H/HeJ and C57BL/10ScCr mice: mutations in Tlr4 gene. *Science* 282, 2085-2088

16. Qureshi, S.T., Lariviere, L., Leveque, G., Clermont, S., Moore, K.J., Gros, P., and Malo, D. (1999) Endotoxin-tolerant mice have mutations in Toll-like receptor 4 (Tlr4) *J Exp Med* 189, 615-625
17. Hoshino, K., Takeuchi, O., Kawai, T., Sanjo, H., Ogawa, T., Takeda, Y., Takeda, K., and Akira, S. (1999) Cutting edge: Toll-like receptor 4 (TLR4)-deficient mice are hyporesponsive to lipopolysaccharide: evidence for TLR4 as the Lps gene product. *J Immunol* 162, 3749-3752
18. Wright, S.D., Tobias, P.S., Ulevitch, R.J., and Ramos, R.A. (1989) Lipopolysaccharide (LPS) binding protein opsonizes LPS-bearing particles for recognition by a novel receptor on macrophages. *J Exp Med* 170, 1231-1241
19. Shimazu, R., Akashi, S., Ogata, H., Nagai, Y., Fukudome, K., Miyake, K., and Kimoto M. (1999) MD-2, a molecule that confers lipopolysaccharide responsiveness on Toll- like receptor 4. *J Exp Med* 189, 1777-1782
20. Takeuchi, O., Kaufmann, A., Grote, K., Kawai, T., Hoshino, K., Morr, M., Muhlradt, P. F., and Akira. S., (2000) Cutting edge: preferentially the R-stereoisomer of the mycoplasmal lipopeptide macrophage-activating lipopeptide-2 activates immune cells through a toll-like receptor 2- and MyD88-dependent signaling pathway. *J Immunol* 164, 554-557
21. Takeuchi, O., Hoshino, K., Kawai, T., Sanjo, H., Takada, H., Ogawa, T., Takeda, K., and Akira, S. (1999) Differential roles of TLR2 and TLR4 in recognition of gram-negative and gram-positive bacterial cell wall components. *Immunity* 11, 443-451
22. Dinarello, C.A. (1988) Biology of interleukin 1. *FASEB J* 2, 108-115
23. Dinarello, C.A. (1985) An update of human interleukin 1: From molecular biology to clinical relevance. *J Clin Immunol* 5, 287-297
24. Dinarello, C.A., Cannon, J.G., Wolff, S. M., Bernheim, H.A., Beutler, B., Cerami, A., Figari, I.S., Palladino M.A., Jr, and O'Connor, J.V. (1986) Tumor necrosis factor (cachectin) is an endogenous pyrogen and induces production of interleukin 1. *J Exp Med* 163, 1433-1450
25. Dinarello, C.A. (1989) Interleukin-1 and its biologically related cytokines. *Advances in Immunology* 44, 153-205.
26. Dinarello, C.A. (1991) Interleukin-1 and interleukin-1 antagonism. *Blood* 77, 627-1635
27. Dinarello, C.A. (1996) Biologic basis for interleukin-1 in disease. *Blood* 87, 2095-2147
28. Dinarello, C.A. (1997) Interleukin-1. *Cytokine Growth Factor Rev* 8, 253-265
29. Dinarello, C.A. (1998) Interleukin-1 beta, interleukin-18, and the interleukin-1 beta converting enzyme. *Ann N Y Acad Sci* 856, 1-11
30. Arend, W.P. (1991) Interleukin-1 receptor antagonist- a new member of the interleukin-1 family. *J Clin Invest* 88, 1445-1451
31. Arend, W.P., Malyak, M., Guthridge, C.J., and Gabay, C. (1998) Interleukin-1 receptor antagonist: role in biology. *Annu Rev Immunol* 16, 27-55
32. Beutler, B., Krochin, N., Milsark, I.W., Luedke, C., and Cerami, A. (1986) Control of cachectin (tumor necrosis factor) synthesis: mechanisms of endotoxin resistance. *Science* 232, 977-979
33. Beutler, B., and Cerami, A. (1986) Cachectin and tumor necrosis factor as two sides of the same biological coin. *Nature* 320, 584-588
34. Beutler, B., and Cerami, A. (1989) The biology of cachectin/TNF-A primary mediator of the host response. *Ann Rev Immunol* 7, 625-650
35. Kunkel, S.L., Remick, D.G., Strieter, R.M., and Larrick, J.W. (1989) Mechanisms that regulate the production and effects of Tumor necrosis factor α. *Critical Reviews of Immunology* 9, 93-117
36. Le, J., and Vilcek, J. (1987) Tumor necrosis factor and interleukin 1: Cytokines with multiple overlapping biological activities. *Lab Invest* 56, 234-248
37. Larrick, J.W., and Kunkel, S.L. (1988) The role of tumor necrosis factor and interleukin-1 in the immunoinflammatory response. *Pharm Res* 5, 129-139
38. Cerami, A (1992) Inflammatory cytokines. *Clin Immunol Immunopathol* 62, S3-S10.

39. Vigers, G.P.A., Anderson, L.J., Caffes, P., and Brandhuber, B.J. (1997) Crystal structure of the type-I interleukin-1 receptor complexed with interleukin-1b. *Nature* 386, 190-194
40. Schreuder, H., Tardif, C., Trump-Kallmeyer, S., Soffientini, A., Sarubbi, E., Akeson, A., Bowlin, T., Yanofsky, S., and Barrett, R.W. (1997) A new cytokine-receptor binding mode revealed by the crystal structure of the IL-1 receptor with an antagonist. *Nature* 386, 194-200.
41. Murphy, J.E., Robert, C., and Kupper, T.S. (2000) Interleukin-1 and cutaneous inflammation: a crucial link between innate and acquired immunity. *J Invest Dermatol* 114, 602-608
42. Saklatvala, J., Dean, J., and Finch, A. (1999) Protein kinase cascades in intracellular signaling by interleukin-I and tumour necrosis factor. *Biochem Soc Symp* 64, 63-77
43. Mantovani, A., Muzio, M., Ghezzi, P., Colotta, C., and Introna, M. (1998) Regulation of inhibitory pathways of the interleukin-1 system. *Ann N Y Acad Sci* 840, 338-351
44. Orlando, S., Sironi, M., Bianchi, G., Drummond, A.H., Boraschi, D., Yabes, D., and Mantovani, A. (1997) Role of metalloproteases in the release of the IL-1 type II decoy receptor. *J Biol Chem* 272, 31764-31769
45. Mantovani, A. (1997) The interplay between primary and secondary cytokines. Cytokines involved in the regulation of monocyte recruitment. *Drugs* 54, 15-23
46. Colotta, F., Saccani, S., Giri, J.G., Dower, S.K., Sims, J.E., Introna, M., and Mantovani, A. (1996) Regulated expression and release of the IL-1 decoy receptor in human mononuclear phagocytes. *J Immunol* 156, 2534-2541
47. Arend, W.P., Smith, J.M.F., Janson, R.W., and Joslin, F.G. (1991) IL-1 receptor antagonist and IL-1β production in human monocytes are regulated differently. *J Immunol* 147, 1530-1536
48. Mantovani, A., Garlanda, C., Introna, M., and Vecchi. A. (1998) Regulation of endothelial cell function by pro- and anti-inflammatory cytokines. *Transplant Proc* 30, 4239-4243
49. Hannum, C.H., Wilcox, C.J., Arend, W.P., Joslin, F.G., Dripps, D.J., Heimdal, P.L., Armes, L.J., Sommer, A., Eisenberg, S.P., and Thompson. R.C. (1990) Interleukin-1 receptor antagonist activity of a human interleukin-1 inhibitor. *Science* 343, 336-340
50. Henderson, B., Thompson, R.C., Hardingham, T., and Lewthwaite, J. (1991) Inhibition of interleukin-1-induced synovitis and articular cartilage proteoglycan loss in the rabbit knee by recombinant human interleukin-1 receptor antagonist. *Cytokine* 3, 246-249
51. McIntyre, K.W., Stephan, G.J., Kolinsky, K.D., Benjamin, W.R., Plocinski, J.M., Kaffka, K.L. and Kilian, P.L. (1991) Inhibition of interleukin-1 binding and bioactivity in vitro and modulation of acute inflammation in vivo by IL-1 receptor antagonist and anti-IL-1 receptor monoclonal antibodies. *J Exp Med* 173, 931-939
52. Smith, D.R., Kunkel, S.L. Standiford, T.J., Chensue, S.W., Rolfe, M.W., Orringer, M.B., Whyte, R.I., Burdick, M.D., Danforth, J.M., Gilbert, A.R., and Strieter, R.M. (1993) The production of interleukin-1 receptor antagonist protein by human bronchogenic carcinoma. *Am J Path* 143, 794-803
53. Lukacs, N.W., Kunkel, S.L., Burdick, M.D., Lincoln, P.M., and Strieter, R.M. (1993) Interleukin-1 receptor antagonist blocks chemokine production in the mixed lymphocyte reaction. *Blood* 82, 3668-3674.
54. Ulich, T.R., L.R. Watson, S.M. Yin, K.Z. Guo, P. Wang, H. Thang, and J. del Castillo. The intratracheal administration of endotoxin and cytokines. (1991) Characterization of LPS-induced inflammatory infiltrate. *Am J Path* 138, 1485-1496
55. Ulich, T.R., Yin, S.M., Guo, K.Z., del Castillo, J., Eisenberg, S.P., and Thompson, R.C. (1991) The intratracheal administration of endotoxin and cytokines. III. The interleukin-1 (IL-1) receptor antagonist inhibits endotoxin- and IL-1- induced acute inflammation. *Am J Pathol* 138, 521-524
56. Horai, R., M. Asano, M., Sudo, K., Kanuka, H., Suzuki, M., Nishihara, M., Takahashi, M., and Iwakura, Y. (1998) Production of mice deficient in genes for interleukin (IL)-1alpha, IL- 1beta, IL-1alpha/beta, and IL-1 receptor antagonist shows that IL-1beta is crucial in turpentine-induced fever development and glucocorticoid secretion. *J Exp Med* 187, 1463-1475

57. Fantuzzi, G., Zheng, H., Faggioni, R., Benigni, F., Ghezzi, P., Sipe, J.D., Shaw, A.R., and Dinarello, C.A. (1996) Effect of endotoxin in IL-1 beta-deficient mice. *J Immunol* 157, 291-296.
58. Fantuzzi, G., and Dinarello, C.A. (1996) The inflammatory response in interleukin-1 beta-deficient mice: comparison with other cytokine-related knock-out mice. *J Leukoc Biol* 59, 489-493
59. Kozak, W., Zheng, H., Conn, C.A., Soszynski, D., van der Ploeg, L.H., and Kluger, M.J. (1995) Thermal and behavioral effects of lipopolysaccharide and influenza in interleukin-1 beta-deficient mice. *Am J Physiol* 269, R969-977
60. Bazzoni, F., and Beutler, B. (1996) The tumor necrosis factor ligand and receptor families. *N Engl J Med* 334, 1717-1725
61. Sherry, B., and Cerami, A. (1988) Cachectin/tumor necrosis factor exerts endocrine, paracrine, and autocrine control of the inflammatory responses. *J Cell Biol* 107, 1269-1277
62. Genersch, E., Hayes, K., Neuenfeld, Y., Haller, H., Reunanen, N., Westermarck, J., Hakkinen, L., Holmstrom, T.H., Elo, I., Eriksson, J.E., and Kahari, V.M. (2000) Sustained ERK phosphorylation is necessary but not sufficient for MMP-9 regulation in endothelial cells: involvement of Ras-dependent and -independent pathways. Enhancement of fibroblast collagenase (matrix metalloproteinase-1) gene expression by ceramide is mediated by extracellular signal-regulated and stress-activated protein kinase pathways. *J Cell Sci* 113, 319-4330
63. Reunanen, N., Westermarck, J., Hakkinen, L., Holmstrom, T.H., Elo, I., Eriksson, J.E., and Kahari, V.M. (1998) Enhancement of fibroblast collagenase (matrix metalloproteinase-1) gene expression by ceramide is mediated by extracellular signal-regulated and stress-activated protein kinase pathways. *J Biol Chem* 273, 5137-5145
64. Dehoux, M.S., Boutten, A., Ostinelli, J., Seta, N., Dombret, M.C., Crestani, B, Deschenes, M., Trouillet, J.L., and Aubier, M. (1994) Compartmentalized cytokine production within the human lung in unilateral pneumonia. *Am J Respir Crit Care Med* 150, 710-716
65. van der Poll, T., Keogh, C.V., Buurman, W.A., and Lowry, S.F. (1997) Passive immunization against tumor necrosis factor-alpha impairs host defense during pneumococcal pneumonia in mice. *Am J Respir Crit Care Med* 155, 603-608
66. Takashima, K., Tateda, K., Matsumoto, T., Iizawa, Y., Nakao, M., and Yamaguchi, K. (1997) Role of tumor necrosis factor alpha in pathogenesis of pneumococcal pneumonia in mice. *Infect Immun* 65, 257-260
67. Laichalk, L.L., Kunkel, S.L., Strieter, R.M., Danforth, J.M., Bailie, M.B., and T.J. Standiford. T.J. (1996) Tumor necrosis factor mediates lung antibacterial host defense in murine Klebsiella pneumonia. *Infect Immun* 64, 5211-5218
68. Gosselin, D., DeSanctis, J., Boule, M., Skamene, E., Matouk, C., and Radzioch, D. (1995) Role of tumor necrosis factor alpha in innate resistance to mouse pulmonary infection with Pseudomonas aeruginosa. *Infect Immun* 63, 3272-3278
69. Brieland, J.K., Remick, D.G., Freeman, P.T., Hurley, M.C., Fantone, J.C., and Engleberg. N.C. (1995) In vivo regulation of replicative Legionella pneumophila lung infection by endogenous tumor necrosis factor alpha and nitric oxide. *Infect Immun* 63, 3253-3258
70. Huffnagle, G.B., Toews, G.B., Burdick, M.D., Boyd, M.B., McAllister, K.S., McDonald, R.A., Kunkel, S.L., and Strieter, R.M. (1996) Afferent phase production of TNF-alpha is required for the development of protective T cell immunity to Cryptococcus neoformans. *J Immunol* 157, 4529-4536
71. Mehrad, B., Strieter, R.M. and Standiford, T.J. (1999) Role of TNF-alpha in pulmonary host defense in murine invasive aspergillosis. *J Immunol* 162, 1633-1640
72. Kolls, J.K., Beck, J.M., Nelson, S., Summer, W.R., and Shellito, J. (1993) Alveolar macrophage release of tumor necrosis factor during murine Pneumocystis carinii pneumonia. *Am J Respir Cell Mol Biol* 8, 370-376
73. Standiford, T.J., Wilkowski, J.M., Sisson, T.H., Hattori, N., Mehrad, B., Bucknell, K.A., and Moore, T.J. (1999) Intrapulmonary tumor necrosis factor gene therapy increases

bacterial clearance and survival in murine gram-negative pneumonia. *Hum Gene Ther* 10, 899-909
74. Waage, A., Halstensen, A., and Espevik, T. (1987) Association between tumor necrosis factor in serum and fatal outcome in patients with meningococcal disease. *Lancet* 8529, 355-357
75. Waage, A., and Bakke, O. (1988) Glucocorticoids suppress the production of tumor necrosis factor by lipopolysaccharide-stimulated human monocytes. *Immunology* 63, 299-302
76. Waage, A., Brandtzaeg, P., Espevik, T., and Halstensen, A. (1991) Current understanding of the pathogenesis of gram-negative shock. *Infect Dis Clin North Am* 5, 781-791
77. Girardin, E., Grau, G.E., Dayer, J.M., Roux-Lombard, P., and Lambert, P.H. (1988) Tumor necrosis factor and interleukin-1 in the serum of children wth severe infectious purpura. *N Eng J Med* 319, 397-400
78. Marks, J.D., Marks, C.B., Luce, J.M., Montgomery, A.B., Turner, J., Metz, C.A., and Murray, J.F. (1990) Plasma tumor necrosis in patients with septic shock: Mortality rate, incidence of adult respiratory distress syndrome. *Am Rev of Resp Dis* 141, 94-97
79. Tracey, K.J., Fong, Y., and Hesse, D.G. (1987) Anti-cachectin/TNF monoclonal antibodies prevent septic shock during lethal bacteremia in baboons. *Nature* 330, 662-666
80. Sipe, J.D. (1989) The molecular biology of interleukin 1 and the acute phase response. *Adv Intern Med* 34, 1-20
81. Shalaby, M.R., Halgunset, J., Haugen, O.A., Aarset, H., Aarden, L., Waage, A., Matsushima, K., Kvithyll, H., Boraschi, D., Lamvil, J., and Espevik, T. (1991) Cytokine-associated tissue injury and lethality in mice: A comparative study. *Clin Immunol Immunopath* 61, 69-82
82. Hinshaw, L.B., Tekamp-Olson, P., Chang, A.C.K., Lee, P.A., Taylor, F.B.J., Murray, C.K., Peer, G.T., Emergon, T.E.J., Poassey, R.B., and Juo, G.C. (1990) Survival of primates in LD100 septic shock following therapy with antibody to tumor necrosis factor (TNF-alpha). *Circ Shock* 30, 279-292
83. Remick, D.G., Strieter, R.M., III, J.P.L., Nguyen, D., Eskandari, M., and Kunkel, S.L. (1989) In vivo dynamics of murine tumor necrosis factor-α gene expression: Kinetics of dexamethasone-induced suppression. *Lab Invest* 60, 766-771
84. Michie, H.R., Mangue, K.R., Spriggs, D.R., Revhaug, A., O'Dwyer, S., Dinarello, C.A., Cerami, A., Wolff, S.M., and Wilmore, D.W. (1988) Detection of circulating tumor necrosis factor after endotoxin administration. *N Eng J of Med* 318, 1481-1484
85. Abraham, E. (1999) Why immunomodulatory therapies have not worked in sepsis. *Intensive Care Med* 25, 556-66
86. Lukacs, N.W., and Ward, P.A. (1996) Inflammatory mediators, cytokines, and adhesion molecules in pulmonary inflammation and injury. *Adv Immunol* 62, 257-304
87. Hogg, J.C., and Doerschuk, C.M. (1995) Leukocyte traffic in the lung. *Ann Rev Physiol* 57, 97-114
88. Imhof, B.A., and Dunon, D. (1995) Leukocyte migration and adhesion. *Adv Immunol.* 58, 345-416
89. Donnelly, S.C., Haslett,C., and Dransfield, I. (1994) Role of selectins in development of adult respiratory distress syndrome. *Lancet* 344, 215-219
90. Zimmerman, G.A., Prescott, S.M., and McIntyre, T.M. (1992) Endothelial cell interactions with granulocytes: tethering and signaling molecules. *Immunol Today* 13, 93-110
91. Kumasake, T., Quinlan, W.M., and Doyle, N.A., (1996) Role of intercellular adhesion molecule 1 (ICAM-1) in endotoxin-induced pneumonia evaluated using ICAM-1 antisense oligonucleotides, ICAM-1 monoclonal antibodies, and ICAM-1 mutant mice. *J Clin Invest* 97, 2362-2369
92. Mulligan, M.S., Wilson, G.P., Todd, R.F., Smith, C.W., Anderson, D.C., Varani, J., Issekutz, T.B., Myasaka, M., Tamatani, T., Rusche, J.R., Vaporciyan, A.A. and Ward, P.A. (1993) Role of ß1, ß2 integrins and ICAM-1 in lung injury after deposition of IgG and IgA immune complexes. *J Immunol* 150, 2407-2417

93. Ridings, P.C., Windsor, A.C., Jutila, M.A., Blocher, C.R., Fisher, B.J., Sholley, M.M., Sugerman, H.J., and Fowler A.A.. (1995) A dual-binding monoclonal antibody to E- and L-selectin attenuates sepsis-induced lung injury. *Am J Respir Crit Care Med* 151, 1995-2004
94. Mulligan, M.S., Miyasaka, M., Tamatani, T., Jones, M.L. and Ward, P.A. (1994) Requirements for L-selectin in neutrophil-mediated lung injury in rats. *J Immunol* 52, 832-840
95. Mulligan, M.S., Polley, M.J., Bayer, R.J., Nunn, M.F., Paulson , J.C,. and Ward, P.A. (1992) Neutrophil-dependent acute lung injury. Requirement for P-selectin (GMP-140). *J Clin Invest* 90, 1600-1607
96. Abbas, A.K., Lichtman, A.H., and Pober, J.S. (1994) Cytokines. In: Cellular and Molecular Immunology. Philadelphia, PA: W.B. Saunders Company, pp 417-418

Chapter 3

THE ROLE OF CHEMOKINES IN THE RECRUITMENT OF LEUKOCYTES DURING LUNG INFLAMMATION

Robert M. Strieter, John A. Belperio, David Kelley, Ammar Sakkour, and Michael P. Keane
Department of Medicine, Division of Pulmonary and Critical Care Medicine, University of California, Los Angeles (UCLA) School of Medicine, Los Angeles, CA.

INTRODUCTION

The recruitment of specific leukocyte subpopulations in response to lung injury is a fundamental mechanism of acute pulmonary inflammation. The elicitation of leukocytes is dependent upon a complex series of events, including reduced leukocyte deformability, endothelial cell activation and expression of endothelial cell-derived leukocyte adhesion molecules, leukocyte-endothelial cell adhesion, leukocyte activation and expression of leukocyte-derived adhesion molecules, leukocyte transendothelial migration, and leukocyte migration beyond the endothelial barrier along established chemotactic and haptotactic gradients. While the events of leukocyte extravasation may appear intuitive, it has taken over 150 years of research to elucidate the molecular and cellular steps involved in the process of leukocyte migration.

Chemotactic Factors

Historically, the first observations of leukocyte migration dates back to the initial observation of leukocyte adherence followed by trans-endothelial migration by Augustus Waller in 1846, who described extravasation of leukocytes in a frog tongue (1). This observation was followed by the first description of leukocyte migration in response to chemotactic signals (products of other leukocytes or killed bacteria), reported in the late nineteenth century (2). Although these studies were descriptive in nature, they were the first to establish leukocyte extravasation in response to a chemotactic signal.

The development of a chemotactic chamber in 1962 by Boyden was a historical event that allowed the quantitative analysis of leukocyte migration *in vitro* (3). This chamber separated leukocytes from a specific chemotactic signal using an interposing filter. Leukocytes that had migrated in response to a specific chemotactic stimulus could then be microscopically quantitated by counting leukocytes at either the "leading front" or adhered to the under surface of the filter. While molecules may behave *in vivo* as leukocyte chemotaxins, the use of these chambers allowed the assessment *in vitro* as to whether a molecule behaves as a direct or indirect leukocyte chemotaxin. Moreover, by modifying this technique, Zigmond and Hirsch could distinguish chemotaxis, a process of leukocyte migration in response to a concentration gradient, from chemokinesis, a property of random leukocyte motion (4).

In the late 1960's, investigators identified the first chemotactic molecules (5-7). These studies demonstrated that N-formyl-methionyl peptides from bacterial cell walls and the anaphylatoxin, C5a, were chemotactic for leukocytes. These findings were followed by the discovery that specific products of arachidonate metabolism were leukocyte chemotaxins. Both platelet activating factor (PAF) and leukotriene B_4 (LTB_4) were shown to have significant chemotactic activity for leukocytes at pM to nM concentrations (8, 9). These findings supported the premise that leukocyte recruitment is critical to the acute inflammation of innate immunity, and that a number of factors that possess potent and overlapping leukocyte chemotactic activity are necessary to assure continued leukocyte emigration at sites of inflammation.

While the above leukocyte chemotaxins are important in leukocyte extravasation, they appear to lack specificity for particular subsets of leukocytes. This nonspecificity is interesting in that the apparent nature of the stimulus and the subsequent spectrum of chemotactic factors produced, determines the subpopulation of leukocytes elicited during an inflammatory response. For example, during neutrophil extravasation associated with acute lung injury (ALI), neutrophil chemotaxins predominate over other leukocyte chemotactic factors. In contrast, a different set of leukocyte chemotaxins predominant when a specific antigenic stimulus in the lung leads to cell-mediated immunity, resulting in the production of chemotactic factors that recruit exclusively mononuclear cells leading to granulomatous inflammation. Thus, a diversity of chemotactic factors must exist with specific activity to target subsets of leukocytes and maintain leukocyte migration.

The salient feature of acute inflammation of innate immunity and the development of ALI is the extravasation of predominately neutrophils followed by mononuclear cells. These extravasating leukocytes contribute to the pathogenesis of inflammation and promote the eradication of the offending agent. In addition, the shear magnitude of increase in infiltrating cells, the activation of these cells, and the release of a variety of mediators, including additional cytokines that interact with resident non-leukocyte cellular populations, leads to further amplification of acute inflammation and lung injury. The maintenance of leukocyte recruitment during inflammation requires intercellular communication between infiltrating leukocytes and the endothelium, resident stromal cells, and parenchymal cells. These events are mediated via the recognition of the offending agent by the Toll-like receptors (TLRs), the generation of early response cytokines (e.g., IL-1β and TNFα), the expression of cell-surface adhesion molecules, and the production of chemotactic molecules, such as chemokines.

Chemokines

A new classification for chemokines has recently been reported (10), and will be used through out this chapter in conjunction with the older terminology. The human CXC, CC, C, and CX_3C chemokine families of chemotactic cytokines are four closely related polypeptide families that behave, in general, as potent chemotactic factors for neutrophils, eosinophils, basophils, monocytes, mast cells, dendritic cells, NK cells, and T and B lymphocytes (Tables 1 and 2).

Table 1. CC Chemokine/Receptor Family (reference 10)

Systematic Name	Human Ligand	Mouse Ligand	Chemokine Receptor(s)
CCL1	I-309	TCA-3, P500	CCR8
CCL2	MCP-1/MCAF	JE?	CCR2
CCL3	MIP-1α/LD78α	MIP-1α	CCR1, 5
CCL4	MIP-1β	MIP-1β	CCR5
CCL5	RANTES	RANTES	CCR1, 3, 5
(CCL6)	Unknown	C10, MRP-1	Unknown
CCL7	MCP-3	MARC?	CCR1, 2, 3
CCL8	MCP-2	MCP-2?	CCR3
(CCL9/10)	Unknown	MRP-2, CCF18 MIP-1γ	Unknown
CCL11	Eotaxin	Eotaxin	CCR3
(CCL12)	Unknown	MCP-5	CCR2
CCL13	MCP-4	Unknown	CCR2, 3
CCL14	HCC-1	Unknown	CCR1
CCL15	HCC-2/Lkn-1/MIP-δ	Unknown	CCR1, 3
CCL16	HCC-4/LEC	LCC-1	CCR1
CCL17	TARC	TARC	CCR4
CCL18	DC-CK1/PARC AMAC-1	Unknown	Unknown
CCL19	MIP-3β/ELC/exodus-3	MIP-3β/ELC/exodus-3	CCR7
CCL20	MIP-3α/LARC/exodus-1	MIP-3α/LARC/exodus-1	CCR6
CCL21	6Ckine/SLC/exodus-2	6Ckine/SLC/exodus-2/TCA-4	CCR7
CCL22	MDC/STCP-1	ABCD-1	CCR4
CCL23	MPIF-1	Unknown	CCR1
CCL24	MPIF-2/Eotaxin-2	Unknown	CCR3
CCL25	TECK	TECK	CCR9
CCL26	Eotaxin-3	Unknown	CCR3
CCL27	CTACK/ILC	ALP/CTACK/ILC ESkine	CCR10

Table 2. CXC Chemokine/Receptor Family (reference 10)

Systematic Name	Human Ligand	Mouse Ligand	Chemokine Receptor(s)
CXCL1	GROα/MGSA-α	GRO/KC?	CXCR2 > CXCR1
CXCL2	GROβ/MGSA-β	GRO/KC?	CXCR2
CXCL3	GROγ/MGSA-γ	GRO/KC?	CXCR2
CXCL4	PF4	PF4	Unknown
CXCL5	ENA-78	LIX?	CXCR2
CXCL6	GCP-2	Ckα-3	CXCR1, CXCR2
CXCL7	NAP-2	Unknown	CXCR2
CXCL8	IL-8	Unknown	CXCR1, CXCR2
CXCL9	Mig	Mig	CXCR3
CXCL10	IP-10	IP-10	CXCR3
CXCL11	I-TAC	Unknown	CXCR3
CXCL12	SDF-1α/β	SDF-1	CXCR4
CXCL13	BLC/BCA-1	BLC/BCA-1	CXCR5
CXCL14	BRAK/bolekine	BRAK	Unknown
(CXCL15)	Unknown	Lungkine	Unknown

The CXC chemokines can be further divided into two groups on the basis of a structure/function domain consisting of the presence or absence of three amino acid residues (Glu-Leu-Arg; 'ELR' motif) that precedes the first cysteine amino acid residue in the primary structure of these cytokines (10-18). The ELR^+ CXC chemokines are chemoattractants for neutrophils and act as potent angiogenic factors (14, 95-106). In contrast, the ELR^- CXC chemokines are chemoattractants for mononuclear leukocytes and are potent inhibitors of angiogenesis (14, 95-103). There is approximately 20% to 40% homology between the members of the four chemokine families (14, 95-103). Chemokines have been found to be produced by an array of cells including monocytes, alveolar macrophages, neutrophils, platelets, eosinophils, mast cells, T- and B-lymphocytes, NK cells, keratinocytes, mesangial cells, epithelial cells, hepatocytes, fibroblasts, smooth muscle cells, mesothelial cells, and endothelial cells. These cells can produce chemokines in response to a variety of factors that trigger the innate immune response, including microorganisms, IL-1β, TNFα, C5a, LTB4, and IFNs. The production of chemokines by both immune and non-immune cells supports the contention that these cytokines play a pivotal role in further orchestrating the inflammatory response of innate immunity and promote ALI.

An ideal leukocyte chemotactic factor is one that demonstrates longevity at the site of inflammation. For example, while C5a is rapidly inactivated to its des-arg form in the presence of serum/plasma peptidases and loses a significant portion of its neutrophil chemotactic activity, IL-8 (CXCL8) maintains 100% of its activity in the presence of serum (22). Thrombin and plasmin have been found to convert IL-8 (CXCL8) from the 77 to the 72 amino acid form, whereas, urokinase and tissue-type plasminogen activator are unable to induce NH_2-terminus truncation (20).

Moreover, IL-8 (CXCL8) incubated in the presence of neutrophil granule lysates and purified proteinase-3, can undergo significant conversion from the 77 to the 72 amino acid form that enhances it biological activity (22). The 72 amino acid form of IL-8 (CXCL8) binds to neutrophils 2-fold more than the 77 amino acid form, and is 2- to 3-fold more potent in inducing cytochalasin B-mediated neutrophil degranulation (20). Other studies demonstrated similar findings for NH_2-terminal truncations of ENA-78 (CXCL5) (23).

Although the above studies support the contention that the NH_2-terminus of CXC chemokines is critical for receptor binding and activation, the role of the heparin-binding domain in the COOH-terminus of these molecules may also play an important role in enhancing the response to IL-8 (CXCL8) (24). The combination of IL-8 (CXCL8) and heparin sulfate results in both a significant rise in cytosolic free Ca^{+2} and a 4-fold increase in neutrophil chemotaxis, as compared to IL-8 (CXCL8) alone (24). Interestingly, heparin in combination with IL-8 (CXCL8) enhances its ability to induce a rise in cytosolic free Ca^{+2}, but not in neutrophil chemotaxis. The affects of these glycosaminoglycans on IL-8 (CXCL8) activity is specific, as neither heparin sulfate nor heparin alters the affect of fMLP-induced neutrophil activation. These findings together with the importance of the 'ELR' motif in neutrophil binding and activation support the notion that both the NH_2- and COOH-terminal domains of ELR^+ CXC chemokines interplay in a cooperative fashion to optimize their ability to mediate neutrophil activation and chemotaxis. These findings also suggest that CXC chemokines can undergo further refinement at the local level of inflammation, enhancing their ability to further amplify recruitment of additional neutrophils.

NH_2-terminal processing of CC chemokines may also play a significant role in modifying their biological function for recruitment of mononuclear cells. The lymphocyte surface glycoprotein, CD26/dipeptidyl peptidase IV, is a membrane-associated peptidase that has a high degree of specificity for peptides with proline or alanine at the second position and cleaves off dipeptides at the NH_2-terminus (25). While NH_2-terminal truncation of the CXC chemokine granulocyte chemotactic protein-2 (GCP-2; CXCL6) by CD26 results in no change in chemotactic activity for neutrophils (25, 26), NH_2-terminal truncation of regulated on activation normal T-cell expressed and secreted chemokine (RANTES; CCL5), eotaxin (CCL11), and macrophage-derived chemokine (MDC; CCL22) by CD26 has been shown to markedly impair their ability to bind to their receptors and elicit chemotactic activity (25, 26). Interestingly, while CD26 NH_2-terminal truncation of RANTES (CCL5) results in reduced binding and activation of two of its CC chemokine receptors, CCR1 and CCR3, RANTES (CCL5) binding to its other receptor, CCR5, is actually preserved (26). This suggests that CD26 modification of RANTES (CCL5) increases its receptor selectivity and may significantly modify its biological behavior during both innate and adaptive immune responses. In contrast, NH_2-terminal processing of LD78β (CCL3), an isoform of macrophage inflammatory peptide-1α (MIP-1α; CCL3), by CD26 increases its chemotactic activity (27). This effect is mediated through both of its major CC chemokine receptors, CCR1 and CCR5 (27). These studies are important to local chemokine biology, as further extracellular processing of these potent leukocyte chemoattractants can play a major role in positively or negatively modifying their behavior in recruitment of leukocytes during inflammation.

Chemokine Receptors

Chemokine receptors belong to the largest known family of cell-surface receptors, the G-protein coupled receptors (GPCR) which mediate transmission of stimuli as diverse as hormones, peptides, glycopeptides, and chemokines (28). The functional unit consists of a seven transmembrane receptor coupled to the heterotrimeric G-protein. Currently, at least ten cellular CC chemokine receptors, five CXC chemokine receptors, one C chemokine receptor, and one CX_3C have been cloned, expressed, and identified to have specific ligand binding profiles (Tables 1, 2, and references 10, 15, 16, 29-31). The expression of these receptors on specific cells, in the context of the temporal expression of their respective chemokine ligands, plays an important role in mediating the initial and subsequent leukocyte infiltration during the evolution of the inflammatory response of innate immunity. For example, neutrophils expressing CXCR1 and CXCR2 will arrive early during innate immunity in response to ELR^+ CXC chemokines that are generated in response to microorganism recognition via TLRs, or in response to TNFα and/or IL-1β signaling. Mononuclear cells, especially monocytes, expressing CCR2, CCR1, and CCR5, will arrive later in response to MCP-1 (CCL2), MIP-1α (CCL3) and RANTES (CCL5). These elicited leukocytes play a critical role in the subsequent eradication of the offending agent (i.e. microorganism). In addition, specific chemokine ligand/receptor pairs are critical to the recruitment of immature dendritic cells to tissue, as well as subsequently mediating the migration of mature dendritic cells for antigen presentation to T and B cells in the secondary lymphoid tissue (35-40). This event is critical for the innate response to bridge and initiate the adaptive immunity in response to an offending agent .

CXC Chemokines in Pulmonary Inflammation

CXC chemokines have also been found to play a significant role in mediating neutrophil infiltration in the lung parenchyma and pleural space in response to endotoxin and bacterial challenge. Frevert and associates (41), have passively immunized rats with neutralizing KC (homologous to human GRO-α) antibodies prior to intratracheal LPS, and found a 71% reduction in neutrophil accumulation within the lung. Broaddus and associates (42-43), have found that passive immunization with neutralizing IL-8 antibodies blocked 77% of endotoxin-induced neutrophil influx in the pleura of rabbits. In the context of microorganism invasion, however, depletion of a CXC chemokine and reduction of infiltrating neutrophils may have a major negative impact on the host.

ELR$^+$ CXC chemokines have been implicated in mediating neutrophil sequestration in the lungs of patients with pneumonia. IL-8 (CXCL8) has been found in the bronchoalveolar lavage (BAL) fluid of patients with community acquired pneumonia and nosocomial pneumonia following trauma (44, 45). In animal models of pneumonia, ELR$^+$ CXC chemokines have been found in a number of model systems of pneumonia. For example, growth related gene (GRO-α; CXCL1), has been found in *Escherichia coli* pneumonia in rabbits (46). In addition, murine GRO-α (CXCL1) and GRO-β/γ (CXCL2/3), KC (CXCL1), and MIP-2 (CXCL2/3) have been found in murine models of *Klebsiella pneumoniae*, *Pseudomonas aeruginosa*, *Nocardia asteroides*, and *Aspergillus fumigatus* pneumonia, respectively (47-53). In a model of *Aspergillus fumigatus* pneumonia, neutralization of TNFα resulted in marked attenuation of the expression of murine GRO-α (KC; CXCL1) and GRO-β/γ (MIP-2; CXCL2/3), which was paralleled by a reduction in the infiltration of neutrophils and associated with increased mortality (49-50). In addition, Laichalk and associates (54), administered a TNFα agonist peptide consisting of the 11-amino-acid TNFα binding site (TNF70-80) to animals intratracheally inoculated with *Klebsiella pneumoniae* and found markedly increased levels of MIP-2 (CXCL2/3) associated with increased neutrophil infiltration.

To further establish the role of ELR$^+$ CXC chemokines in mediating the innate host defense and eradication of a variety of microorganisms in the lung, Greenberger and colleagues (47) demonstrated that depletion of MIP-2 (CXCL2/3) during the pathogenesis of murine *Klebsiella pneumoniae* pneumonia resulted in a marked reduction in the recruitment of neutrophils to the lung that was paralleled by increased bacteremia and reduced bacterial clearance in the lung. Since ELR$^+$ CXC chemokine ligands in the mouse use the CXC chemokine receptor, CXCR2, and several ELR$^+$ CXC chemokine ligands are expressed during murine models of pneumonia, this would suggest that targeting CXCR2 would delineate the importance of ELR$^+$ CXC chemokine ligand/CXCR2 biology during the pathogenesis of pneumonia and ALI. Standiford and associates, using specific neutralizing antibodies to CXCR2, demonstrated that blocking CXCR2 results in markedly reduced neutrophil infiltration in response to *Pseudomonas aeruginosa* (51), *Norcardia asteroides* (53), and *Aspergillus fumigatus* (50) pneumonias. The reduction in neutrophil elicitation was directly related to reduced clearance of the microorganisms and increased mortality in these model systems.

These studies have established the critical importance that ELR^+ CXC chemokine/CXCR biology plays in the acute inflammation of innate immune responses to a variety of microorganisms. Moreover, with the evolving clinical presence of multi-drug resistant microorganisms it is increasingly necessary to consider alternative means to eradicate these microbial pathogens. Tsai and associates (52) demonstrated that transgenic expression of murine GRO-α (KC; CXCL1) in the lung using a Clara cell-specific promoter, in the context of *Klebsiella pneumoniae* pneumonia, enhances host survival that is directly related to increased neutrophil recruitment and bacterial clearance in the lungs. This response was not accompanied by the increased expression of other pro-inflammatory cytokines, such as TNFα, IFN-γ, or IL-12. This study further indicated that the compartmentalized overexpression of an ELR^+ CXC chemokine could represent a novel approach to the treatment of antimicrobial resistant microorganisms. Furthermore, these studies demonstrate the importance of microorganism recognition, early response cytokine production (i.e., TNFα), and the subsequent generation of ELR^+ CXC chemokines associated with neutrophil elicitation and eradication of invading microbial pathogens.

Clinical studies attempting to correlate pulmonary IL-8 levels and the development and mortality of ALI have conflicted, however, most have suggested a strong correlation (55-60). Of particular interest is the findings of Donnelly and colleagues (59) that correlated early increases in BAL fluid IL-8 content in patients at-risk for subsequent development of ARDS, and importantly, also demonstrated that the alveolar macrophage is an important cellular source of IL-8 prior to neutrophil influx. High concentrations of IL-8 were found in the BAL fluid from trauma patients, some within 1 hour of injury and prior to any evidence of significant neutrophil influx. Patients who progressed to ALI had significantly greater BAL fluid levels of IL-8 than those patients who did not develop ALI. Interestingly, plasma levels of IL-8 were not significantly different in patients who developed ALI as compared to those that did not develop ALI.

Recent investigations have also shown that anoxia/hyperoxia, simulating an ischemia-reperfusion or hyperoxia environment, can lead to an induction of IL-8 gene expression with a significant increase in IL-8 production by mononuclear cells and endothelial cells (61, 62). IL-8 gene induction in hypoxic endothelial cells was associated with increased activation of the transcription factor NF-κB (61, 62). Of further clinical significance, endotoxin was found to further potentate this hyperoxic response. While these *in vitro* studies suggested that IL-8 may be a major neutrophil chemotaxin produced in the context of simulated ischemia-reperfusion, Sekido and associates (63) demonstrated that IL-8 significantly contributed to reperfusion lung injury using a rabbit model of lung ischemia-reperfusion injury. Reperfusion of the ischemic lung resulted in the production of IL-8 that correlated with maximal pulmonary neutrophil infiltration. Passive immunization of the animals with neutralizing antibodies to IL-8, prior to reperfusion of the ischemic lung, prevented neutrophil extravasation and tissue injury, suggesting a causal role for IL-8 in this model. In another model of ischemia-reperfusion injury demonstrating the importance of cytokine cascades between the liver and lung, Colletti and colleagues (64, 65) demonstrated that hepatic ischemia-reperfusion injury and the generation of TNFα can result in pulmonary-derived ENA-78. The production of ENA-78 in the lung was correlated with the presence of neutrophil-dependent lung injury, and passive immunization with neutralizing ENA-78 antibodies resulted in significant attenuation of lung injury. These studies support the notion that CXC chemokines are important in the elicitation of neutrophils in the lung under conditions of acute inflammation. Furthermore, under conditions of microorganism-induced pneumonia leading to ALI, the expression of CXC chemokines may be beneficial to both the eradication of the organism and host survival.

CC Chemokines in Pulmonary Inflammation

The CC chemokines, RANTES (CCL5), MIP-1α (CCL3), MIP-1β (CCL4), MCP-1 (CCL2), have also been implicated in mediating the innate host defense in animal models of *Influenza* A virus (66), *Paramyxovirus* pneumonia virus, *Aspergillus fumigatus*, and *Cryptococcus neoformans* pneumonias. The host response to *Influenza* A virus is characterized by an influx of mononuclear cells into the lungs that is associated with the increased expression of CC chemokine ligands (66). Dawson and colleagues have used a genetic approach (CCR5 and CCR2 knockout mice) to determine the role of CC chemokines in mediating the innate response to this virus (67). CCR5 -/- mice (note: CCR5 is the receptor for RANTES, MIP-1α, and MIP-1β) infected with *Influenza A* displayed increased mortality related to severe pneumonitis compared to wild-type animals. In contrast, CCR2 -/- mice (note: CCR2 is the receptor for MCP-1) infected with *Influenza* A were protected from the severe pneumonitis due to defective macrophage recruitment. The delay in macrophage accumulation in CCR2 -/- mice was correlated with high pulmonary viral titers (66). These studies support the potential of different roles that CC chemokine ligand/receptor biology plays during *Influenza A* infection. In addition, this study also demonstrates that macrophage recruitment during the innate response is critical to the development of adaptive immunity to this microbe.

Domachowske and associates (67) have examined the role of CC chemokine ligands (i.e. RANTES and MIP-1α) that bind to the CC chemokine receptor CCR1 in response to *Paramyxovirus* pneumonia virus infection in mice. This infection is associated with a predominant neutrophil and eosinophil infiltration into the lung that is accompanied by expression of CCR1 ligands (67). Using CCR1 -/- mice infected with *Paramyxovirus* pneumonia virus they found that the inflammatory response was minimal, the clearance of virus from lung tissue was reduced, and mortality was markedly increased. This suggested that the CC chemokine-dependent innate response limited the rate of virus replication *in vivo* and played an important role in reducing mortality.

The effect of CC chemokines in mediating the recruitment of mononuclear cells during the innate host defense is not limited to viral infections. Mehrad and colleagues (68), have shown that MIP-1α (CCL3) and the recruitment of mononuclear cells plays an important role in the eradication of invasive pulmonary aspergillosis. They demonstrated that in both immunocompetent and neutropenic mice, MIP-1α (CCL3) was induced in the lungs in response to intratracheal inoculation of *Aspergillus fumigatus*. Depletion of endogenous MIP-1α (CCL3) by passive immunization with neutralizing antibodies resulted in increased mortality in neutropenic mice, which was associated with a reduced mononuclear cell infiltration and markedly decreased clearance of lung fungal burden. Gao and associates (69) have confirmed this finding by assessing CCR1, the major CC chemokine receptor for MIP-1α (CCL3). CCR1 -/- mice exposed to *Aspergillus fumigatus* had a markedly increased mortality, as compared to CCR1 +/+ controls. These studies indicate that MIP-1α (CCL3) and the elicitation of mononuclear cells is a critical event in mediating host defense against *Aspergillus fumigatus* in the setting of neutropenia, and may be an important target in devising future therapeutic strategies against invasive pulmonary aspergillosis.

Cryptococcus neoformans is acquired via the respiratory tract and is a significant cause of fatal mycosis in immunocompromised patients. Both the innate and adaptive immune response are necessary to clear the microbe from the lung and prevent dissemination to the meninges. Huffnagle and associates (70, 71) have found that MCP-1 (CCL2) and MIP-1α (CCL3) play important roles in the eradication of *Cryptococcus neoformans* from the lung and prevent cryptococcal meningitis. In mice exposed to intratracheal *Cryptococcus neoformans*, both MCP-1 (CCL2) and MIP-1α (CCL3) expression directly correlates with the magnitude of infiltrating leukocytes. Depletion of endogenous MCP-1 (CCL2) with neutralizing antibodies markedly decreased the recruitment of both macrophages and CD4+ T cells, and inhibited cryptococcal clearance. Neutralization of MCP-1 (CCL2) also resulted in decreased BAL fluid levels of TNFα. Using the same model system, depletion of MIP-1α (CCL3) resulted in a significant reduction in total leukocytes and an increase in the burden of *Cryptococcus neoformans* in the lungs of these animals. Interestingly, depletion of MIP-1α (CCL3) did not decrease the levels of MCP-1 (CCL2), however, depletion of MCP-1

(CCL2) significantly reduced MIP-1α (CCL3) levels, demonstrating that induction of MIP-1α (CCL3) was largely dependent on MCP-1 (CCL2) production. Neutralization of MIP-1α (CCL3) also blocked the cellular recruitment phase of a recall response to cryptococcal antigen in the lungs of immunized mice. Thus, in both the context of active cryptococcal infection, or re-challenge with cryptococcal antigen, MIP-1α (CCL3) was required for maximal leukocyte recruitment into the lungs, most notably the recruitment of phagocytic effector cells (neutrophils and macrophages). These studies support the notion that CC chemokine ligand/receptor biology plays a critical role in innate host defense and development of pulmonary inflammation that is important for eradication of microorganisms.

Interplay of Early Response Cytokines, Adhesion Molecules, and CXC Chemokines in ALI

During the initiation phase of acute lung inflammation the movement of neutrophils from the pulmonary vascular compartment to interstitium and alveolar space is an early event in the propagation of further lung inflammation. Inflammatory stimuli from either side of the alveolar-capillary membrane may result in pulmonary microvascular alterations which lead to local increases in neutrophil sequestration and adhesion to the endothelium (72). Under the influence of chemokines, neutrophils in the microvascular compartment undergo directed migration along chemotactic and haptotactic gradients to the inflamed area. During recruitment, these neutrophils become activated, releasing various proteases, reactive oxygen metabolites, and cytokines, which amplify inflammation and contribute to further lung injury. As the acute inflammatory process changes from the initiation to maintenance and resolution stages, the cellular composition of the inflammatory lesions changes to a predominately mononuclear cell population. Thus, leukocyte elicitation is dynamic, with specific chemo-attractants expressed at specific temporal windows of the inflammatory response.

The cytokine networks existing between immune and non-immune cells of the alveolar-capillary membrane are necessary for intercellular communication during inflammation. The subsequent events mediated by these intercellular/cytokine interactions are crucial to the initiation and propagation of the inflammatory response that leads to pulmonary injury. Both TNFα and IL-1β are early response cytokines that are necessary not only for the initiation of acute inflammation, but are also required for persistence of the inflammatory response, that may evolve into chronic inflammation. The production of CXC chemokines by the major cellular components of the alveolar-capillary membrane, and their participation in the inflammatory response, may be critical for the orchestration of the directed migration of leukocytes into the lung. The alveolar-capillary membrane has traditionally been viewed simply as a structure for gas exchange, but an understanding of a more complex role has emerged with advances in molecular biological techniques and investigations of individual cell components. The alveolar-capillary membrane can now be viewed as a dynamic assembly of immune and non-immune cells that, through cytokine networking, can generate significant quantities of chemokines.

Importantly, the expression of CXC chemokines by the major cellular constituents of the lung is stimulus specific. Neutrophils, mononuclear phagocytes, and endothelial cells produce CXC chemokines in response to either LPS, TNFα, or IL-1β, but not to IL-6. Pulmonary fibroblasts and epithelial cells express CXC chemokines in response to specific host-derived signals, such as TNFα or IL-1β. These findings are significant since cells once thought of as "targets" of the inflammatory response, can actively participate as effector cells in the production of potent neutrophil chemoattractants. Thus, during acute inflammation, the production of TNFα and IL-1β can act in either an autocrine or paracrine fashion to stimulate contiguous cells, both immune and non-immune, to express CXC chemokines.

The paradigm for neutrophil extravasation is likely operative in the microvasculature of the lung, and consists of four or more steps. First, lung injury results in the activation of the microvascular endothelium in response to the local generation of TNFα or IL-1β, leading to expression of endothelial cell-derived ICAM-1. In conjunction, the neutrophil undergoes reduced deformability allowing it to more closely interact with the endothelium. The constitutive presence of neutrophil-derived L-selectin and other lectin moieties allows for the initial adhesive interaction of these cells with endothelial cell selectins leading to the "rolling" effect. Second, generation of CXC chemokines leads to the activation of neutrophils in the vascular compartment and expression of β2 integrins, while L-selectin is concomitantly shed. Third, the interaction of the neutrophil β2 integrins with their receptor/ligand, ICAM-1, results in the rapid arrest of neutrophils on the endothelium. Fourth, the subsequent events leading to neutrophil extravasation beyond the vascular compartment are dependent on chemotactic and haptotactic gradients (i.e., migration in response to an insoluble gradient), the continued expression of β2 integrins on neutrophils and ICAM-1 on non-immune cells, and the maintenance of a CXC chemokine-specific neutrophil chemotactic and haptotactic gradient. The participation of CXC chemokines in the inflammatory response appears to be critical for the orchestration of the directed migration of neutrophils into the lung. After arriving in the lung, these activated leukocytes can further respond to noxious or antigenic stimuli and induce pulmonary injury through the release of reactive oxygen metabolites, proteolytic enzymes, and additional cytokines.

Summary

The complete manifestation of ALI is dependent on the elicitation and activation of inflammatory cells into the lung leading to tissue injury. This same response, however, is necessary to control and eradicate infections of the pulmonary compartment. This is underscored by the important activation of TLRs, interaction of early response cytokines, adhesion molecules, and chemokines in the orchestration of the recruitment of leukocytes into the lung (Figure 1). The discovery of chemokine supergene families has greatly enhanced our understanding of the biology of leukocyte recruitment to the lung.

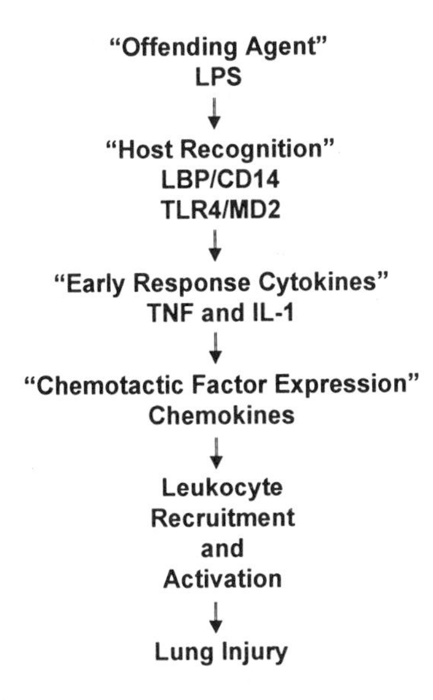

Figure 1. Activation of TLRs, interaction of early response cytokines and chemokines in the orchestration of the recruitment of leukocytes into the lung.

REFERENCES

1. Waller, A. (1846) Microscopical observations on the perforation of the capillaries by the corpuscles of the blood, and on the origin of mucous and pus-globules. *Philos Magazine* 29, 397
2. Massart, J., and Bordet, C. (1890) Recherches sur l'irritabilite des leucocytes et sur l'intervention de cette irritabilite dans la nutrition des cellules et dans l'inflammation. *J Med Chir Pharm Brux* 90, 169
3. Boyden, S. (1962) The chemotactic effect of mixtures of antibody and antigens on polymorphonuclear leukocytes. *J Exp Med* 115, 453

4. Zigmond, S.H., and Hirsch, J.G. (1973) Leukocyte locomotion and chemotaxis: New methods for evaluation and demonstration of cell-derived chemotactic factor. *J Exp Med* 137, 387
5. Ward, P.A., and Newman, L.J. (1969) A neutrophil chemotactic factor from human C'5. *J Immunol* 102, 93-99
6. Shin, H.S., Snyderman, R., Friedman, E., Mellors, A., and Mayer, M.M. (1968) Chemotactic and anaphylatoxic fragment cleaved from the fifth component of guinea pig complement. *Science* 162, 361-363
7. Becker, E.L., and Ward, P.A. (1969) Esterases of the polymorphonuclear leukocyte capable of hydrolyzing acetyl DL-phenyl-alanine beta-naphthyl ester. Relationship to the activatable esterase of chemotaxis. *J Exp Med* 129, 569-574
8. Lee, T.C., and Snyder, F. (1985) Function, metabolism and regulation of platelet activating factor and related ether lipids. *In* Phospholipids and cellular regulation. J.F. Kuo, editor. Boca Raton: *CRC Press Inc.*
9. Ford-Hutchinson, A.W., Bray, M.A., Doig, M.V., Shipley, M.E., and Smith, M.J. (1980) Leukotriene B, a potent chemokinetic and aggregating substance released from polymorphonuclear leukocytes. *Nature* 286, 264-265
10. Zlotnik, A., and Yoshie, O. (2000) Chemokines: a new classification system and their role in immunity. *Immunity* 12, 121-127
11. Strieter, R.M., Lukacs, N.W., Standiford, T.J. and Kunkel, S.L. (1993) Cytokines and lung inflammation. *Thorax* 48, 765-769
12. Strieter, R.M., Koch, A.E., Antony, V.B., Fick, R.B., Standiford, T.J., and Kunkel, S.L. (1994) The immunopathology of chemotactic cytokines: The role of interleukin-8 and monocyte chemoattractant protein-1. *J Lab Clin Med* 123, 183-196
13. Strieter, R.M., and Kunkel, S.L. (1994) Acute lung injury: The role of cytokines in the elicitation of neutrophils. *J Invest Med* 42, 640-665
14. Strieter, R.M., and Kunkel, S.L. (1994) Acute lung injury: the role of cytokines in the elicitation of neutrophils *J Invest Med* 42, 640-51
15. Strieter, R.M., and Kunkel, S.L. (1997) Chemokines in the lung. *In* Lung: Scientific Foundations, 2nd edition. R. Crystal, J. West, E. Weibel, and P. Barnes, editors. *New York: Raven Press*
16. Strieter, R.M., Kunkel, S.L., Keane, M.P., and Standiford, T.J. (1999) Chemokines in lung injury: Thomas A. Neff Lecture. *Chest* 116, 103S-110S
17. Luster, A.D. (1998) Chemokines-chemotactic cytokines that mediate inflammation. *N Engl J Med* 338, 436-445
18. Rollins, B.J. (1997) Chemokines. *Blood* 90, 909-928
19. Locati, M., and Murphy, P.M. (1999) Chemokines and chemokine receptors: biology and clinical relevance in inflammation and AIDS. *Annu Rev Med* 50, 425-440
20. Hebert, C.A., Vitangcol, R.V., and Baker J.B. (1991) Scanning mutagenesis of interleukin-8 identifies a cluster of residues required for receptor binding. *J Biol Chem* 266, 18989-18994
21. Hebert, C.A., and Baker, J.B. (1993) Interleukin-8: a review. *Cancer Invest* 11, 743-750
22. Clark-Lewis, I., Dewald, B., Geiser, T., Moser, B., and Baggiolini, M. (1993) Platelet factor 4 binds to interleukin 8 receptors and activates neutrophils when its N terminus is modified with Glu-Leu-Arg. *Proc Natl Acad Sci USA* 90, 3574-3577
23. Padrines, M., Wolf, M., Walz, A., and Baggiolini, M. (1994) Interleukin-8 processing by neutrophil elastase, cathepsin G, and proteinase-3. *FEBS Lett* 352, 231-235
24. Walz, A., Strieter, R.M., and Schnyder, S. (1993) Neutrophil-activating peptide ENA-78. *Adv Exp Med Biol* 351, 129-137
25. Webb, L.M.C., Ehrengruber, M.U., Clark-Lewis, I., Baggiolini, M. and Rot, A. (1993) Binding to heparan sulfate or heparin enhances neutrophil responses to interleukin 8. *Proc Natl Acad Sci USA* 90, 7158-7162
26. De Meester, I., Korom, S., Van Damme, J. and Scharpe, S. (1999) CD26, let it cut or cut it down. *Immunol Today* 20, 367-375

27. Proost, P., Struyf, S., Schols, D., Durinx, C., Wuyts, A., Lenaerts, J.P., DeClercq, E., De Meester, I., and Van Damme, J. (1998) Processing by CD26/dipeptidyl-peptidase IV reduces the chemotactic and anti-HIV-1 activity of stromal-cell-derived factor-1alpha. *FEBS Lett* 432, 73-76
28. Proost, P., Menten, P., Struyf, S., Schutyser, E., DeMeester, I., and Van Damme, J. (2000) Cleavage by CD26/dipeptidyl peptidase IV converts the chemokine LD78beta into a most efficient monocyte attractant and CCR1 agonist. *Blood* 96, 1674-1680
29. Iacovelli, L., Sallese, M., Mariggio, S., and de Blasi. A. (1999) Regulation of G-protein-coupled receptor kinase subtypes by calcium sensor proteins. *FASEB J* 13, 1-8
30. Broxmeyer, H.E., and Kim, C.H. (1999) Regulation of hematopoiesis in a sea of chemokine family members with a plethora of redundant activities. *Exp Hematol* 27, 1113-1123
31. Nibbs, R.J.B., Wylie, S.M., Pragnell, I.B., and Graham, G.J. (1997) Cloning and characterization of a novel murine beta chemokine receptor, D6. Comparison to three other related macrophage inflammatory protein- 1alpha receptors, CCR-1, CCR-3, and CCR-5. *J Biol Chem* 272, 12495-12504
32. Premack, B.A., and Schall, T.J. (1996) Chemokine receptors: gateways to inflammation and infection. *Nature Med* 2, 1174-1178
33. Imai, T., Chantry, D., Raport, C.J., Wood, C.L., Nishimura, M., Godiska, R., Yoshie, O., and Gray, P.W. (1998) Macrophage-derived chemokine is a functional ligand for the CC chemokine receptor 4. *J Biol Chem* 273, 1764-1768
34. Liao, F., Alderson, R., Su, J., Ullrich, S.J., Kreider, B.L., and Farber, J.M. (1997) STRL22 is a receptor for the CC chemokine MIP-3alpha. *Biochem Biophys Res Commun* 236, 212-217
35. Liao, F., Alkhatib, G., Peden, K.W., Sharma, G., Berger, E.A., and Farber, J.M. (1997) STRL33, A novel chemokine receptor-like protein, functions as a fusion cofactor for both macrophage-tropic and T cell line-tropic HIV-1. *J Exp Med* 185, 2015-2023
36. Sozzani, S., Luini, W., Borsatti, A., Polentarutti, N., Zhou, D., Piemonti, L., D'Amico, G., Power, C.A., Wells, T.N., Gobbi, M., Allavena, P., and Mantovani, A. (1997) Receptor expression and responsiveness of human dendritic cells to a defined set of CC and CXC chemokines. *J Immunol* 159, 1993-2000
37. Dieu, M.C., Vanbervliet, B., Vicari, A., Bridon, J.M., Oldham, E., Ait-Yahia, S., Briere, F., Zlotnik, A., Lebecque, S., and Caux, C. (1998) Selective recruitment of immature and mature dendritic cells by distinct chemokines expressed in different anatomic sites. *J Exp Med* 188, 373-386
38. Saeki, H., Wu, M.T., Olasz, E., and Hwang, S.T. (2000) A migratory population of skin-derived dendritic cells expresses CXCR5, responds to B lymphocyte chemoattractant in vitro, and co-localizes to B cell zones in lymph nodes in vivo. *Eur J Immunol* 30, 2808-2814
39. Cumberbatch, M., Dearman, R.J., Griffiths, C.E., and Kimber, I. (2000) Langerhans cell migration. *Clin Exp Dermatol* 25, 413-418
40. Dieu-Nosjean, M.C., Massacrier, C., Homey, B., Vanbervliet, B., Pin, J.J., Vicari, A., Lebecque, S., Dezutter-Dambuyant, C., Schmitt, D., Zlotnik, A., and Caux, C. (2000) Macrophage inflammatory protein 3alpha is expressed at inflamed epithelial surfaces and is the most potent chemokine known in attracting Langerhans cell precursors. *J Exp Med* 192, 705-718
41. Sozzani, S., Allavena, P., Vecchi, A., and Mantovani, A. (2000) Chemokines and dendritic cell traffic. *J Clin Immunol* 20, 151-160
42. Frevert, C.W., Huang, S., Danaee, H., Paulauskis, J.D., and Kobzik, L. (1995) Functional characterization of the rat chemokine KC and its importance in neutrophil recruitment in a rat model of pulmonary inflammation. *J Immunol* 154, 335-344
43. Boylan, A.M., Hebert, C.A., Sadick, M., Wong, W.L., Chuntharapai, A., Hoeffel, J.M., Hartiala, K.T., and Broaddus, V.C. (1994) Interleukin-8 is a major component of pleural liquid chemotactic acti .ty in a rabbit model of endotoxin pleurisy. *Am J Physiol* 267, L137-144

44. Broaddus, V.C., Boylan, A.M., Hoeffel, J.M., Kim, K.J., Sadick, M., Chuntharapai, A., and Hebert, C.A. (1994) Neutralization of IL-8 inhibits neutrophil influx in a rabbit model of endotoxin-induced pleurisy. *J Immunol* 152, 2960-2967
45. Boutten, A., Dehoux, M.S., Seta, N., Ostinelli, J., Venembre, P., Crestani, B., Dombret, M.C., Durand, G., and Aubier, M. (1996) Compartmentalized IL-8 and elastase release within the human lung in unilateral pneumonia. *Am J Respir Crit Care Med* 153, 336-342
46. Rodriguez, J.L., Miller, C.G., DeForge, L.E., Kelty, L., Shanley, C.J., Bartlett, R.H., and Remick, D.G. (1992) Local production of interleukin-8 is associated with nosocomial pneumonia. *J Trauma* 33, 74-81
47. Johnson, M.C., 2nd, Kajikawa, O., Goodman, R.B., Wong, V.A., Mongovin, S.M., Wong, W.B., Fox-Dewhurst, and R., Martin. T.R. (1996) Molecular expression of the alpha-chemokine rabbit GRO in Escherichia coli and characterization of its production by lung cells in vitro and in vivo. *J Biol Chem* 271, 10853-10858
48. Greenberger, M.J., Strieter, R.M., Kunkel, S.L., Danforth, J.M., Laichalk, L.L., McGillicuddy, D.C., and Standiford, T.J. (1996) Neutralization of macrophage inflammatory protein-2 attenuates neutrophil recruitment and bacterial clearance in murine Klebsiella pneumonia. *J Infect Dis* 173, 159-165
49. Standiford, T.J., Kunkel, S.L., Greenberger, M.J., Laichalk, L.L., and Strieter, R.M. (1996) Expression and regulation of chemokines in bacterial pneumonia. *J Leukoc Biol* 59, 24-28
50. Mehrad, B., and Standiford, T.J. (1999) Role of cytokines in pulmonary antimicrobial host defense. *Immunol Res* 20, 15-27
51. Mehrad, B., Strieter, R.M., Moore, T.A., Tsai, W.C., Lira, S.A., and Standiford, T.J. (1999) CXC chemokine receptor-2 ligands are necessary components of neutrophil-mediated host defense in invasive pulmonary aspergillosis. *J Immunol* 163, 6086-6094
52. Tsai, W.C., Strieter, R.M., Mehrad, B., Newstead, M.W., Zeng, X., and Standiford, T.J. (2000) CXC chemokine receptor CXCR2 is essential for protective innate host response in murine Pseudomonas aeruginosa pneumonia. *Infect Immunol* 68, 4289-4296
53. Tsai, W.C., Strieter, R.M., Wilkowski, J.M., Bucknell, K.A., Burdick, M.D., Lira, S.A., and Standiford, T.J. (1998) Lung-specific transgenic expression of KC enhances resistance to Klebsiella pneumoniae in mice. *J Immunol* 161, 2435-2440
54. Moore, T.A., Newstead, M.W., Strieter, R.M., Mehrad, B., Beaman, B.L., and Standiford, T.J. (2000) Bacterial clearance and survival are dependent on CXC chemokine receptor-2 ligands in a murine model of pulmonary Nocardia asteroides infection. *J Immunol* 164, 908-915
55. Laichalk, L.L.,Bucknell, K.A., Huffnagle, G.B., Wilkowski, J.M., Moore, T.A., Romanelli, R.J., and Standiford. T.J. (1998) Intrapulmonary delivery of tumor necrosis factor agonist peptide augments host defense in murine gram-negative bacterial pneumonia. *Infect Immun* 66, 2822-2826
56. Jorens, P.G., VanDame, J., DeBecker, W., Bossaert, L., DeJongh, R.F., Herman, A.G., and Rampart, M. (1992) Interleukin-8 in the bronchoalveolar lavage fluid from patients with the adult respiratory distress syndrome (ARDS) and patients at risk for ARDS. *Cytokine* 4, 592-597
57. Miller, E.J., Cohen, A.B., Nago, S., Griffith, D., Maunder, R.J., Martin, T.R., Weiner-Kronish, J.P., Sticherling, M., Christophers, E., and Matthay, M.A. (1992) Elevated levels of NAP-1/Interleukin-8 are present in the airspaces of patients with the adult respiratory distress syndrome and are associated with increased mortality. *Am Rev Resp Dis* 146, 427-432
58. Hack, C.E., Hart, M., Strack-vanSchijndel, R.J.M., Eerenberg, A.J.M., Nuijens, J.H., Thijs, L.G., and Aarden, L.A., (1992) Interleukin-8 in sepsis: relation to shock and inflammatory mediators. *Infect Immun* 60, 2835-2842
59. Chollet-Martin, S., Montravers, P., Gilbert, C., Elbim, C., Desmonts, J.M., Fagon, J.Y., and Gougerot-Pocidalo, M.A. (1993) High levels of interleukin-8 in the blood and

alveolar spaces of patients with pneumonia and adult respiratory distress syndrome. *Infect. Immun* 61, 4553-4559

60. Donnelly, S.C., Strieter, R.M., Kunkel, S.L., Walz, A., Robertson, C.R., Carter, D.C., Grant, I.S., Pollok, A.J., and Haslett, C. (1993) Interleukin-8 and development of adult respiratory distress syndrome in at-risk patient groups. *Lancet* 341, 643-647
61. Goodman, R.B., Strieter, R.M., Martin, D.P., Steinberg, K.P., Milberg, J.A., Maunder, R.J., Kunkel, S.L., Walz, A., Hudson, L.D., and Martin, T.R. (1996) Inflammatory cytokines in patients with persistence of the acute respiratory distress syndrome. *Am J Respir Crit Care Med* 154, 602-611
62. Metinko, A.P., Kunkel, S.L., Standiford, T.J., and Strieter, R.M. (1992) Anoxia-hyperoxia induces monocyte derived interleukin-8. *J Clin Invest* 90, 791-798
63. Karakurum, M., Shreeniwas, R., Chen, J., Pinsky, D., Yan, S-D., Anderson, M., Sunouchi, K., Major, J., Hamilton, T., Kuwabara, K., Rot, A., Nowygrod, R., Stern, D. (1994) Hypoxia induction of interleukin-8 gene expression in human endothelial cells. *J Clin Invest* 93, 1564-1570
64. Sekido, N., Mukaida, N., Harada, A., Nakanishi, I., Watanabe, Y., and Matsushima, K. (1993) Prevention of lung reperfusion injury in rabbits by a monoclonal antibody against interleukin-8. *Nature* 365, 654-657
65. Colletti, L.M., Remick, D.G., Burtch, G.D., Kunkel, S.L., Strieter, R.M., and Campbell, D.A., Jr. (1990) Role ot tumor necrosis factor-alpha in the pathophysiologic alterations after hepatic ischemia/reperfusion injury in the rat. *J Clin Invest* 85, 1936-1943
66. Colletti, L.M., Kunkel, S.L., Walz, A., Burdick, M.D., Kunkel, R.G., Wilke, C.A., and Strieter, R.M. (1995) Chemokine expression during hepatic ischemia/reperfusion-induced lung injury in the rat. The role of epithelial neutrophil activating protein. *J Clin Invest* 95, 134-141
67. Dawson, T.C., Beck, M.A., Kuziel, W.A., Henderson, F., and Maeda, N. (2000) Contrasting effects of CCR5 and CCR2 deficiency in the pulmonary inflammatory response to influenza A virus. *Am J Pathol* 156, 1951-1959
68. Domachowske, J.B., Bonville, C.A., Gao, J.L., Murphy, P.M., Easton, A.J., and Rosenberg, H.F. (2000) The chemokine macrophage-inflammatory protein-1 alpha and its receptor CCR1 control pulmonary inflammation and antiviral host defense in paramyxovirus infection. *J Immunol* 165, 2677-2682
69. Mehrad, B., Moore, T.A., and Standiford, T.J. (2000) Macrophage inflammatory protein-1 alpha is a critical mediator of host defense against invasive pulmonary aspergillosis in neutropenic hosts. *J Immunol* 165, 962-968
70. Gao, J.L., Wynn, T.A., Chang, Y., Lee, E.J., Broxmeyer, H.E., Cooper, S., Tiffany, H., H. Westphal, H., Kwon-Chung, J. and Murphy, P.M. (1997) Impaired host defense, hematopoiesis, granulomatous inflammation and type 1-type 2 cytokine balance in mice lacking CC chemokine receptor 1. *J Exp Med* 185, 1959-1968
71. Huffnagle, G.B., R.M. Strieter, T.J. Standiford, R.A. McDonald, M.D. Burdick, S.L. Kunkel, and G.B.Toews. (1995) The role of monocyte chemotactic protein-1 (MCP-1) in the recruitment of monocytes and CD4+ T cells during a pulmonary Cryptococcus neoformans infection. *J Immunol* 155, 4790-4797
72. Huffnagle, G.B., Strieter, R.M., McNeil, L.K., McDonald, R.A., Burdick, M.D., Kunkel, S.L., and Toews, G.B. (1997) Macrophage inflammatory protein-1alpha (MIP-1alpha) is required for the efferent phase of pulmonary cell-mediated immunity to a Cryptococcus neoformans infection. *J Immunol* 159, 318-327
73. Doerschuk, C.M., Mizgerd, J.P., Kubo, H., Qin, L., and Kumasaka, T. (1999) Adhesion molecules and cellular biomechanical changes in acute lung injury: Giles F. Filley Lecture. *Chest.* 116, 37S-43S

Chapter 4

REGULATION OF LUNG IMMUNITY: SIGNIFICANCE OF THE CYTOKINE ENVIRONMENT

Nicholas W. Lukacs, Theodore Standiford, Cory Hogaboam, and Steven L. Kunkel
University of Michigan Medical School, Department of Pathology and Division of Pulmonary and Critical Care Medicine

INTRODUCTION

The disruption of lung structure and pulmonary function can be a devastating event leading to acute deteriorating health and chronic respiratory illness (1-4). Regulation of lung function is maintained by intricate mechanisms that promote a balance between host defenses against injurious events and reactive inflammatory responses. An excessive inflammatory response can result in acute lung injury (ALI), leading to organ dysfunction. The development and regulation of inflammation and immune responses within the lung are dependent upon several complex interactions including the nature of the infectious agent, structural cell production of chemotactic factors, activation of recruited leukocytes, and release of specific cytokines. The cytokine responses that occur in the lung dictate the nature of ensuing activation events and the type of immune response that subsequently occurs. For example, if an infectious or noxious foreign agent initiates a response that predominantly induces a Th1 type response characterized by high levels of IL-12 and IFN-γ, an environment may be initiated that would augment subsequent anti-pathogen immune responses within the lung. In contrast, if a foreign agent induces a predominant Th2 type response, then subsequent responses may be skewed. The Th2 response may alter the ability to clear intracellular pathogens and increase the host's susceptibility to developing allergic responses. This chapter will address the influence of early cytokine production on the nature of the subsequent immune responses.

Regulation of ALI and Cytokine Responses in Sepsis

The induction of septic shock is commonly associated with bacterial infections and an intense systemic inflammatory response that leads to severe physiologic and immunologic dysfunction of multiple organs (5-10). The lung is one of the most common target organs in septic patients, often necessitating the use of mechanical ventilatory support. Within the lungs of septic patients, local production of inflammatory cytokines and chemokines initiate an intense and sustained inflammatory response resulting in the lung injury associated with the acute respiratory distress syndrome (ARDS), tissue remodeling, and eventual respiratory failure. A number of inflammatory cytokines and chemokines have been identified in patients that develop ARDS, including TNFα, IL-1β, MIP-1α, and IL-8. Recent evidence in animal models, however, suggest that the deleterious effects induced by the inflammatory cytokines can be counterbalanced by endogenous anti-inflammatory cytokines and chemokines produced locally within the lung (11-15). These anti-inflammatory cytokines can be broadly characterized as type 2 cytokines.

A number of investigators have established that a significant increase of interleukin-10 (IL-10) production, both systemically in the circulation and locally within the lung, is associated with sepsis and ARDS (13, 16, 17). It remains controversial, however, as to whether these responses correlate to the development of or the protection from septic responses. There is no doubt that IL-10 is a potent anti-inflammatory cytokine that can down-regulate TNFα and other inflammatory stimuli, as well as most chemokines (see Chapter 17). However, recent studies in animal models of sepsis employing cecal ligation and puncture (CLP) have demonstrated that IL-10 produced locally within the lung can lead to increased susceptibility to severe bacterial infections (14, 18). The occurrence of secondary infections during sepsis is often an initiator of or contributor to the progression of ARDS. Thus, although IL-10 may have a beneficial systemic effect by regulating over-production of proinflammatory cytokines, IL-10 locally in the lung may decrease the immune response mounted against secondary infections.

A second regulatory type 2 cytokine that has recently gained notice is interluekin-13 (IL-13). This cytokine shares similar features of IL-10 as reflected by its ability to regulate inflammatory cytokine production from mononuclear phagocytic cells. Although no IL-13 has been detected in the plasma of patients with sepsis or in volunteers receiving endotoxin (19), studies in animal models have demonstrated that IL-13 may have protective effects in septic-like responses (19-22). In a recent report using a CLP model of sepsis, IL-13 was found only in the tissues of the septic animals with no detectable levels in the serum (23). Thus, IL-13 appears to play a local role at the organ level. The administration of exogenous recombinant IL-13 to animals undergoing lethal septic responses protected them from death. In these same studies, the neutralization of IL-13 caused a significant increase in lethality associated with increased levels of inflammatory and chemotactic cytokines.

Unlike IL-10, IL-13 has the ability to differentially induce a number of chemokines from various cell populations (24-28). For example, although IL-13 can downregulate the production of neutrophil chemotactic CXC and CC chemokines, such as IL-8 and MIP-1α, it is known to induce the production of other CC chemokines, such as MDC, C10, and MCP-1. The preferential induction of certain chemokines by IL-13 suggests that these chemokines may play a specific role in protection against the septic response. A number of studies have demonstrated that IL-13-inducible chemokines play a key role in the regulation of bacterial clearance and protection from the adverse effects of septic responses (23, 30, 31). Thus, IL-13 not only modulates inflammatory cytokines that may lead to unchecked pulmonary inflammation, but also augments bacterial clearance guarding the host from opportunistic infections.

The novel observations that specific chemokines have the ability to augment anti-bacterial immune responses suggested that these molecules might have therapeutic potential. The administration of exogenous MCP-1 to mice undergoing LPS-induced septic responses attenuated the lethal response (32). In addition, the administration of anti-MCP-1 antibody increased mortality in mice subjected to a sublethal dose of endotoxin. In both cases, the effects appeared to be correlated to the levels of TNFα and IL-12, suggesting that MCP-1 may modulate inflammatory cytokine production. Other studies have also demonstrated that MCP-1 can increase bacterial clearance via the upregulation of macrophage bactericidal activity (33). Studies further examining the role of the IL-13-inducible chemokines (C10 and MDC) have shown that these chemokines have a significant effect on the septic response (31, 34, 35). In a model of CLP-induced bacterial peritonitis, the effect of these two CC chemokines was primarily on bacterial clearance with little effect on inflammatory cytokines, such as IL-12 and TNFα. The ability of these IL-13-induced chemokines to facilitate more efficient clearance of bacteria also modulated production of both the systemic inflammatory cytokines and lung neutrophil chemokines. Thus, controlling the response more efficiently at the site of bacterial contamination led to a lower incidence of both systemic inflammation and remote organ dysfunction. These studies demonstrate how cytokines, such as IL-13 and IL-13-inducible chemokines, with a traditional role in antigen-specific immune responses, can impact on the innate immune response and control the deleterious outcome of a disease.

Chemokines: A Family of Cytokines with Diverse Functions

Historically, chemokines have been viewed as leukocyte chemoattractants that regulate cellular movement from the circulatory system into inflamed tissue. However, as investigators continue to examine the function of chemokines in both disease and homeostatic circumstances, one finds a complex regulation of function and interaction among multiple cell types. Chemokine receptors identified on various structural cells, mediate functions ranging from chemotaxis to growth regulation to activation. As reviewed in the chapter by Strieter and colleagues (see Chapter 3), chemokines have been primarily divided into two main families based upon their sequence homology and the position of the first two cysteine residues (36, 37). This group of chemoattractants has grown considerably over the years as groups have utilized microarray technology with expressed sequence tab databases (38). As this continues, however, the function of all of these members must be characterized in order to determine their individual role in the inflamed and non-inflamed environment. Two other distinct families are the C family and the CXXXC family. Both of these families have only a single member, lymphotactin and neurotactin (fractalkine), respectively.

Although a number of chemokines have been identified as lymphocyte activating factors, the exact role that each chemokine plays during an immune response remains incompletely defined. When examining T lymphocytes skewed toward a type 1, or type 2, response *in vitro* it appears that there is preferential expression of certain chemokine receptors. T lymphocytes that have a type 1 phenotype (characterized by IFN-γ production) appear to preferentially express CXCR3 and CCR5. T-cells that are skewed toward a type 2 phenotype preferentially express CCR3, CCR4, CCR8, and CXCR4 (39-46). It is interesting to speculate as to whether these receptors are the cause, or result of, lymphocyte differentiation. It is conceivable, however, that these receptors would dictate the ligands required to allow these cells to migrate into inflamed tissues. Indeed, if chemokine production is examined during specific types of immune responses, distinct chemokine phenotypes can be observed. During type 1 immune responses, there appears to be a predominant dependence on MIP-1, RANTES, and interferon-inducible protein-10 (IP-10). In contrast, during type 2 immune responses one can observe the preferential expression of eotaxin, MDC, TARC, and TCA3 (40, 43, 47-53). A number of the studies that described these observations established the importance of these molecules during the immune response by neutralization experiments. These latter chemokines are induced by the type 2 cytokines, IL-4 and IL-13, thus correlating the type 2 immune response to the chemokines and the chemokine receptors that are expressed on Th2 lymphocytes. Thus, the local production of certain chemokines could dictate the type of lymphocyte recruited to the lung. Because these correlations in animal models are not exclusive, one must be cautious about extrapolating the significance of these findings to human disease states. It remains intriguing to speculate on how to therapeutically target specific chemokines or chemokine receptors during unique diseases that demonstrate a predicted cytokine phenotype.

The functions of chemokines have been expanding over the past several years and it is now evident that they modify the immune response either directly by activating the antigen presenting cell (APC) and T lymphocyte, or indirectly by recruiting the proper cell populations. Initial studies found that CC family chemokines (RANTES, MIP-1α and MIP-1β) could effectively initiate an antigen specific response *in vivo* when used in place of an adjuvant such as Freund's complete adjuvant (54). These early studies led to the hypothesis that chemokines could effectively skew immune responses toward either a type 1 (IFN–γ producing) or a type 2 (IL-4 producing) response (53, 55-59). It now appears that chemokines not only have the ability to recruit specific subsets of lymphocytes, but also can aid in determining the type of immune response that occurs (Table 1). These and other aspects may have a significant effect on the development of immunity within the lung.

Table 1. The role of specific chemokines in skewing the immune response

Chemokine	Receptors	Functions
RANTES	CCR1, CCR5	Recruits monocytes and lymphocytes; Skews towards type Th1 immune response
MIP-1α	CCR1, CCR5	Directs a Th1 immune response; Upregulates dendritic cell-derived IL-12 production
MCP-1	CCR2	Recruits monocytes, lymphocytes and basophils; Directs a Th2 immune response; Decreased IL-12 and increases IL-10
TCA3	CCR8	Recruits neutrophils and Th2 lymphocytes; Directs a Th2 immune response
IP-10, MIG, ITAC	CXCR3	Recruits monocytes and lymphocytes; Skews the immune response to Th1 type
SDF-1α	CXCR4	Recruits naïve and Th2 lymphocytes; May skew the immune response to Th2 type

The Role of Chemokines in Anti-bacterial and Anti-mycobacterial Immune Responses

The induction of immune responses during bacterial infections in the lung relies on an effective activation of innate immune responses that can clear the organisms before they colonize the lung. Examinations of bronchoalveolar lavage (BAL) samples from patients with pneumococcal and pseudomonal pneumonias have shown high levels of IL-8 (8, 60-64). These early studies posed the hypothesis as to whether the presence of this potent neutrophil chemoattractant was beneficial or detrimental to the clearance of the microorganism and health of the lung. A number of *in vivo* studies have addressed the importance of chemokines in initiating and maintaining an effective anti-bacterial response. In a model of *Klebsiella pneumoniae* infection in the lungs of mice, the influx of neutrophils mediated by MIP-2 (functional homologue of IL-8), were found to be required for the early clearance of the bacteria. Interestingly, the neutralization of MIP-2 had little effect on the eventual survival of the mice (65). Similar findings were observed in a rabbit model of *E. coli* infection that examined the role of another IL-8 homologue, GRO-α (66). These data suggested that perhaps multiple CXC chemokines expressed during bacterial pneumonia could compensate for the neutrophil recruitment and anti-bacterial effect even in the absence another CXC chemokine (67). In support of this concept, when KC (another CXCR2 ligand with IL-8 homology) was overexpressed during *Klebsiella* infection in mice, a significant effect on bacterial clearance and enhanced survival was observed (68). In other studies in which the primary murine neutrophil chemokine receptor CXCR2 was blocked, it was observed that this pathway was the most critical for neutrophil recruitment and bacterial clearance (67, 69). Interestingly, because of the redundancy of the CXCR2 ligand system, blocking only one of the chemokines was not as effective as blocking the receptor itself. These studies demonstrated how chemokines, produced locally by macrophage and epithelial cell populations in response to the bacterial infection, could initiate the protective innate immune responses.

These same chemokines have also been identified as being important in other models of pulmonary infections, including *Pseudomonas*, *Nocardia*, and *Aspergillus*. Understanding the regulation of these critical chemokines during pulmonary infections is also important for possibly altering or enhancing bacterial clearance to facilitate disease resolution. For example, neutralization of IL-10 allowed increased CXC chemokine production, associated with increased neutrophil accumulation, and enhanced bacterial clearance and survival (70, 71), indicating an important regulatory mechanism that may be targeted. Altogether, these data demonstrate the central role for these chemokine interactions in anti-bacterial defenses of the lung.

The inability of the pulmonary innate and acquired immune systems to properly dispose of mycobacterial infections has historically had a devastating effect on large populations. More recently the occurrence of new mycobacterial species that are resistant to current therapeutic strategies has led to the re-emergence of this important disease. The ability of this organism to evade the local pulmonary immune response has been targeted by the research community with several important findings resulting from the use of specific animal models. The successful clearance of any intracellular pathogen has been associated with a strong Th1 type immune response, characterized by high levels of IL-12 and IFN-γ. This allows the proper activation of macrophages leading to the intracellular killing of the infecting organism (72-74). The early production of IL-12 by infected macrophages appears to be a key trigger in this response leading to the direct activation of IFN-γ production from infiltrating NK cells and antigen-specific T lymphocytes. However, during infection of macrophages, induction of IL-12 may be directly and/or indirectly regulated by the response to the mycobacterial organism. Although *in vitro* data suggest that macrophage/dendritic cells infected with mycobacteria produce IL-12 (75, 76, 112), clinical and experimental data suggest that there is a shift in the immune response away from type 1 towards a type 2 response (12-14, 72, 77-79). This shift allows the organism to become established within the lung and impairs the clearance of the mycobacteria.

A critical aspect of bacterial clearance is the recruitment of the correct cell populations, which is in turn dependent upon the upregulation of specific chemokines. As indicated earlier, there appear to be chemokines that are differentially associated with type 1 and type 2 immune responses. Using animal models of mycobacterial antigen-induced pulmonary inflammation, the specific role of chemokines during type 1 immune responses was examined. Examination of granuloma formation in the lungs of mycobacterium-sensitized animals has identified specific chemokines that mediate the formation of the type 1 granulomatous response. In particular, RANTES appears to be a critical cytokine to recruit primarily mononuclear cell populations (56, 79). More interesting, administration of exogenous RANTES increases type 1 immune responses and can down-regulate type 2 immune responses, suggesting that this chemokine may play a regulatory role in recruitment of specific immune cells. Interestingly, it was recently demonstrated that *T. gondii* induces CCR5-dependent, IFN-γ production that is independent of IL-12 (80). Thus, RANTES appears to be associated with the immune response to intracellular mycobacteria.

The primary sources of RANTES in the lung are epithelial cells and macrophages. Thus, the production of RANTES during the initial phases of mycobacteria infection may help to skew the immune response toward a proper Th1 type response to enhance bacterial clearance. Interestingly, mycobacteria infection of epithelial cell populations downregulates RANTES production and may be one mechanism to explain how this organism evades the anti-bacterial immune response (81). In contrast to RANTES production, mycobacteria infection can directly induce MCP-1 production in epithelial cells (82-84), and likewise, MCP-1 is upregulated in severely infected patients with tuberculosis (85). In this context, one might predict that MCP-1 could be involved in a strategy to alter the proper anti-bacterial immune response. A number of studies have now demonstrated that there is an inverse relationship between IL-12 and MCP-1 during an immune specific response (32, 57, 86, 87). Neutralization of MCP-1 during Th2 type granulomatous responses not only decreased the development of the type 2 response but also increased the ability of isolated granuloma macrophages to produce IL-12. In addition, the overexpression of MCP-1 during the development of a type 1 mycobacteria antigen response downregulates the Th1 type cytokines and reduces the IFN-γ-driven pulmonary granulomatous responses.

Additional data using MCP-1 overexpressing and null mutant animals provided further evidence that MCP-1 may alter anti-bacterial immune responses. The overexpression of MCP-1 in transgenic mice decreased the ability of these animals to clear intercellular pathogens (88). In contrast, in MCP-1 -/- mice type 2 granulomatous responses were downregulated as reflected by decreases in IL-4 and IL-5 expression, while type 1 anti-mycobacteria responses remained intact (89). A final piece of evidence suggesting that MCP-1 can skew the immune response is derived from *in vitro* analysis of T cell activation. In these studies, co-culture of antigen-specific lymphocytes with MCP-1 and antigen caused a significant increase in Th2 type (IL-4) and a decrease in Th1 type (IFN-γ) cytokines (59, 90). Thus, although MCP-1 is an activator of macrophage anti-bacterial responses, it appears that MCP-1 is also involved in skewing the immune response toward a Th2 type response by regulating critical T cell-derived factors. Altogether, these studies suggest that chemokines produced during the initial or innate immune response may impact on the development of pathogen-specific acquired immunity. Furthermore, pathogens such as mycobacteria may directly influence the production of specific chemokines during an anti-pathogen response as a means toward establishing a successful infection.

Host Responses To Viral Infections Can Modulate the Pulmonary Immune Environment

How the host responds to respiratory viral infections can dictate subsequent responses within the lung environment. If the individual responds to a viral infection with a battery of Th1 type cytokines, including IL-12 and IFN-γ, a cellular immune response is promoted and the virus is quickly eradicated (91-95). In contrast, if a Th2 type response is triggered, with IL-4 and IL-13 production, the anti-viral defense may be insufficient and the appropriate cellular immune responses may be attenuated. Recent evidence suggests that respiratory viral infections are a common cause of asthma exacerbations and the response to viral infections may dictate the severity of subsequent asthmatic responses. In particular, respiratory syncytial virus (RSV) is known to cause asthma exacerbations. In many children less than 2 years of age, RSV infection can significantly alter airway function leading to long-term airway hyperreactivity with subsequent development of asthma later in life. The mechanisms that promote these long-term pulmonary problems are not clear; however, it appears that there may be a genetic predisposition toward the development of a specific immunologic response. The evidence of varied clinical responses to RSV infection supports this contention; i.e. not all children respond in a detrimental way to RSV and most clear the virus without sequelae. This suggests that a particular disease phenotype may be associated with unique mediator responses to RSV in the lung.

Preferential cytokine profiles during viral infections can be observed using inbred mouse strains (96). Recent studies from our own laboratory indicate that certain strains of mice that differ at the H-2 MHC locus have different cytokine and physiological responses to a primary RSV infection (Table 2). In particular, there appears to be a correlation between airway hyperreactivity (AHR) and IL-13 production, and an inverse correlation between IL-12 and development of AHR. These results tend to follow the responses previously observed in RSV vaccine studies in genetically varied mice. In addition, in human RSV disease there appears to be a correlation between low IL-12 production and severity of the pathophysiologic changes and intensity of the pulmonary inflammation in infants hospitalized for RSV infections. Therefore, the cytokine environment in the lung appears to correlate to disease severity and pathophysiologic outcome. This has been verified in an animal model of RSV-induced AHR in mice that had increased IL-13 production (97). When IL-13 was neutralized there was a significant decrease in the development of AHR, and an associated increase in IL-12 along with a decrease in viral antigen in the lungs. Thus, inappropriate production of IL-13 during RSV infection may lead to a more severe allergic pulmonary response.

Table 2. Primary RSV responses and cytokine environment in lungs of mice are dictated by the MHC genotype.

Mouse strain (MHC)	IL-12	IL-13	Development of AHR*	Exacerbation of allergen response
C57BL/6 (H-2b)	+++	+/–	+	Downregulates
Balb/c (H-2d)	+/–	++	++++	Upregulates
DBA/J (H-2d)	–	+++	+++	Upregulates
CBA/J (H2k)	+	+	++	No change

*AHR, airway hyperreactivity

Asthma is one of the fastest growing diseases in developed countries, especially in young children (98, 99). Asthma-like responses can be induced and/or exacerbated by viral infections that impact lung function through a combination of viral- and inflammation-induced damage (100-106). Recent evidence indicates that cytokines produced during an allergic or infectious response dictate how the pulmonary response will proceed (104, 107). Studies that have coupled RSV infection with allergic responses have clearly indicated that the exacerbation of allergen-induced AHR is associated with RSV-induced IL-5 (104, 108, 109). These latter results correlated to findings from RSV-infected patients with severe respiratory distress who demonstrated increased IL-5 levels (110). In contrast, using a short-term model of RSV infection (4 days), in conjunction with allergen challenge, no increase in Th2 cytokines was observed; however, there was an increase in AHR (111). In additional studies, using a mouse model of long-term RSV infection and allergen sensitization, RSV-induced IL-13 appeared to play a role in the exacerbation of the asthmatic responses (*N.W Lukacs, unpublished data*). In fact, the cytokine environment (IL-12 versus IL-13) developed during the primary RSV infection dictates whether that strain of mouse will have an exacerbated or regulated asthmatic response (Table 2). Additional studies have demonstrated that the combination of various pulmonary viral infections in murine models of asthma can alter the subsequent airway responses (104, 112-114). Thus, the cytokine response mounted within the lungs during viral infection can dictate the type of immune response (type 1 versus type 2) and the severity of subsequent allergic responses (Figure 1). This concept can also be observed in bacterial infections and CpG oligomers that induce a predominant Th1 response augmenting IL-12 from dendritic cells leading to increased IFN-γ production (115-120).

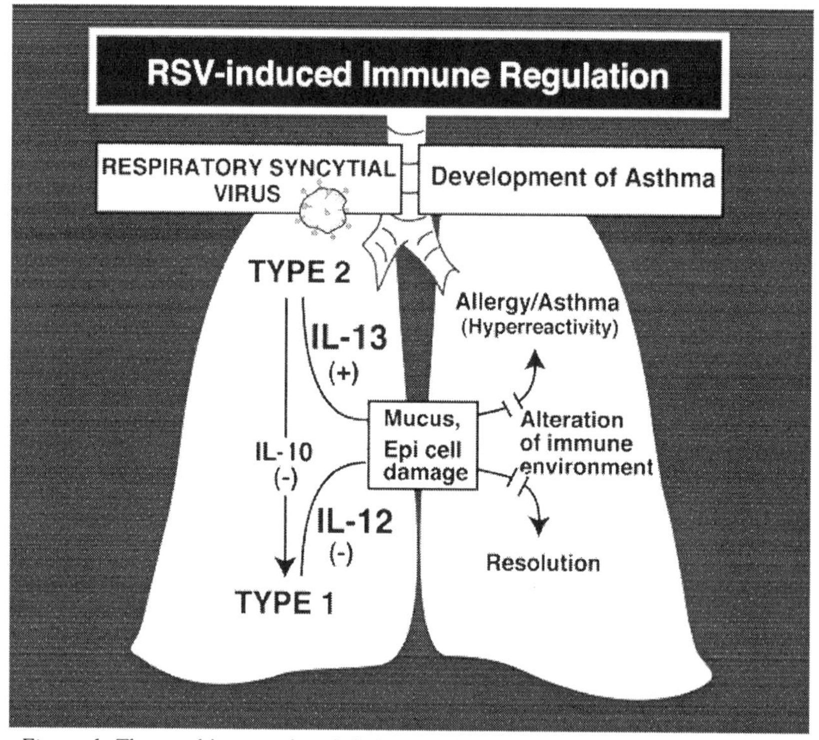

Figure 1. The cytokines produced during a RSV infection can alter the pulmonary environment and dictate the phenotype of subsequent responses to potential allergens. The nature of these responses are likely determined by a number of interacting factors including genetic, environmental, and the infectious agent itself.

Summary

The effective host defense against infections within tissue is reliant on a vigorous inflammatory response including production of cytokines and chemokines leading to the recruitment and activation of neutrophil and monocyte/macrophage populations. These early innate cell responses likely dictate several aspects of the lung responses, including the clearance of the invading pathogen, the intensity of lung pathophysiologic changes and the development of specific acquired immune responses. The coordination of cytokine production along with cell recruitment and activation becomes critical for the activation of the proper immune responses. Understanding how these interrelated responses develop will aid in determining the viable targets for therapeutic intervention in acute and chronic pulmonary diseases.

REFERENCES

1. Croce, M.A. (2000) Diagnosis of acute respiratory distress syndrome and differentiation from ventilator-associated pneumonia. *Am J Surg* 179, 26S-29S
2. Gant, V., and Parton, S. (2000) Community-acquired pneumonia. *Curr Opin Pulm Med* 6, 226-233
3. Reed, R.L., 2nd. (2000) Contemporary issues with bacterial infection in the intensive care unit. *Surg Clin North Am* 80, 895-909
4. Schneider, R.F. (1999) Bacterial pneumonia. *Semin Respir Infect* 14, 327-332
5. Brown, S.D. (1998) ARDS. History, definitions, and physiology. *Respir Care Clin N Am* 4, 567-582
6. Hudson, L.D., and Steinberg, K.P. (1999) Epidemiology of acute lung injury and ARDS *Chest* 116, 74S-82S
7. Matuschak, G.M. (1996) Lung-liver interactions in sepsis and multiple organ failure syndrome. *Clin Chest Med* 17, 83-98
8. Monton, C., and Torres, A. (1998) Lung inflammatory response in pneumonia. *Monaldi Arch Chest Dis* 53, 56-63
9. Sachdeva, R.C., and Guntupalli, K.K. (1997) Acute respiratory distress syndrome. *Crit Care Clin* 13, 503-521
10. Sessler, C.N., Bloomfield, G.L., and Fowler, 3rd, A.A. (1996) Current concepts of sepsis and acute lung injury. *Clin Chest Med* 17 213-235
11. Matsukawa, A., Hogaboam, C.M., Lukacs, N.W., Lincoln, P.M., Strieter, R.M., and Kunkel, S.L. (2000) Endogenous MCP-1 influences systemic cytokine balance in a murine model of acute septic peritonitis. *Exp Mol Pathol* 68, 77-84
12. Neidhardt, R., Keel, M., Steckholzer, U., Safret, A., Ungethuem, U., Trentz, O., and Ertel, W. (1997) Relationship of interleukin-10 plasma levels to severity of injury and clinical outcome in injured patients. *J Trauma* 42, 863-870
13. Sherry, R.M., Cue, J.I., Goddard, J.K., Parramore, J.B., and DiPiro, J.T. (1996) Interleukin-10 is associated with the development of sepsis in trauma patients. *J Trauma* 40, 613-616
14. Steinhauser, M.L., Hogaboam, C.M., Kunkel, S.L., Lukacs, N.W., Strieter, R.M., and Standiford, T.J. (1999) IL-10 is a major mediator of sepsis-induced impairment in lung antibacterial host defense. *J Immunol* 162, 392-399
15. Walley, K.R., Lukacs, N.W., Standiford, T.J., Strieter, R.M., and Kunkel, S.L. (1996) Balance of inflammatory cytokines related to severity and mortality of murine sepsis. *Infect Immun* 64, 4733-4838
16. Lo, C.J., Fu, M., and Cryer, H.G. (1998) Interleukin 10 inhibits alveolar macrophage production of inflammatory mediators involved in adult respiratory distress syndrome. *J Surg Res* 79, 179-184
17. Parsons, P.E., Moss, M., Vannice, J.L., Moore, E.E., Moore, F.A., and Repine, J.E. (1997) Circulating IL-1ra and IL-10 levels are increased but do not predict the development of acute respiratory distress syndrome in at-risk patients. *Am J Respir Crit Care Med* 155, 1469-1473
18. Reddy, R.C., Chen, G.H., Newstead, M.W., Moore, T., Zeng, X., Tateda, K., Standiford, T.J. (2001) Alveolar macrophage deactivation in murine septic peritonitis: role of interleukin 10. *Infect Immun* 69, 1394-1401

19. van der Poll, T., de Waal Malefyt, R., Coyle, S.M., and Lowry, S.F. (1997) Antiinflammatory cytokine responses during clinical sepsis and experimental endotoxemia: sequential measurements of plasma soluble interleukin (IL)-1 receptor type II, IL-10, and IL-13. *J Infect Dis* 175, 118-122
20. Baumhofer, J.M., Beinhauer, B.G., Wang, J.E., Brandmeier, H., Geissler, K., Losert, U., Philip, R., Aversa, G., and Rogy, M.A. (1998) Gene transfer with IL-4 and IL-13 improves survival in lethal endotoxemia in the mouse and ameliorates peritoneal macrophages immune competence. *Eur J Immunol* 28, 610-615
21. Muchamuel, T., Menon, S., Pisacane, P., Howard, M.C., and Cockayne, D.A. (1997) IL-13 protects mice from lipopolysaccharide-induced lethal endotoxemia: correlation with down-modulation of TNF-alpha, IFN-gamma, and IL-12 production. *J Immunol* 158, 2898-2903
22. Nicoletti, F., Mancuso, G., Cusumano, V., Di Marco, R., Zaccone, P., Bendtzen, K., and Teti, G. (1997) Prevention of endotoxin-induced lethality in neonatal mice by interleukin-13. *Eur J Immunol* 27, 1580-1583
23. Matsukawa, A., Hogaboam, C.M., Lukacs, N.W., Lincoln, P.M., Evanoff, H.L., Strieter, R.M., and Kunkel, S.L. (2000) Expression and contribution of endogenous IL-13 in an experimental model of sepsis. *J Immunol* 164, 2738-2744
24. Andrew, D.P., Chang, M.S., McNinch, J., Wathen, S.T., Rihanek, M., Tseng, J., Spellberg, J.P., and Elias, 3rd, C.G. (1998) STCP-1 (MDC) CC chemokine acts specifically on chronically activated Th2 lymphocytes and is produced by monocytes on stimulation with Th2 cytokines IL-4 and IL-13. *J Immunol* 161, 5027-5038
25. Goebeler, M., Schnarr, B., Toksoy, A., Kunz, M., Brocker, E.B., Duschl, A., and Gillitzer, R. (1997) Interleukin-13 selectively induces monocyte chemoattractant protein-1 synthesis and secretion by human endothelial cells. Involvement of IL- 4R alpha and Stat6 phosphorylation. *Immunology* 91, 450-457
26. Li, L., Xia, Y., Nguyen, A., Y. Lai, H., Feng, L., Mosmann, T.R., and Lo, D. (1999) Effects of Th2 cytokines on chemokine expression in the lung: IL-13 potently induces eotaxin expression by airway epithelial cells. *J Immunol* 162, 2477-2487
27. Orlofsky, A., Wu, Y., and Prystowsky, M.B. (2000) Divergent regulation of the murine CC chemokine C10 by Th(1) and Th(2) cytokines. *Cytokine* 12, 220-228
28. Pype, J.L., Dupont, L.J., Menten, P., Van Coillie, E., Opdenakker, G., Van Damme, J., Chung, K.F., Demedts, M.G., and Verleden, G.M. (1999) Expression of monocyte chemotactic protein (MCP)-1, MCP-2, and MCP-3 by human airway smooth-muscle cells. Modulation by corticosteroids and T- helper 2 cytokines. *Am J Respir Cell Mol Biol* 21, 528-536
29. DiPiro, J.T. (1997) Cytokine networks with infection: mycobacterial infections, leishmaniasis, human immunodeficiency virus infection, and sepsis. *Pharmacotherapy* 17, 205-223
30. Matsukawa, A., Hogaboam, C.M., Lukacs, N.W., Lincoln, P.M., Strieter, R.M., and Kunkel, S.L. (1999) Endogenous monocyte chemoattractant protein-1 (MCP-1) protects mice in a model of acute septic peritonitis: cross-talk between MCP-1 and leukotriene B4. *J Immunol* 163, 6148-6154
31. Steinhauser, M.L., Hogaboam, C.M., Matsukawa, A., Lukacs, N.W., Strieter, R.M., and Kunkel, S.L. (2000) Chemokine C10 promotes disease resolution and survival in an experimental model of bacterial sepsis. *Infect Immun* 68, 6108-6114
32. Zisman, D.A., Kunkel, S.L., Strieter, R.M., Tsai, W.C., Bucknell, K., Wilkowski, J., and Standiford, T.J. (1997) MCP-1 protects mice in lethal endotoxemia. *J Clin Invest* 99, 2832-2836
33. Nakano, Y., Kasahara, T., Mukaida, N., Ko, Y.C., Nakano, M., and Matsushima, K. (1994) Protection against lethal bacterial infection in mice by monocyte-chemotactic and -activating factor. *Infect Immun* 62, 377-383
34. Chvatchko, Y., Hoogewerf, A.J., Meyer, A., Alouani, S., Juillard, P., Buser, R., Conquet, F., Proudfoot, A.E., Wells, T.N., and Power, C.A. (2000) A key role for CC

chemokine receptor 4 in lipopolysaccharide-induced endotoxic shock. *J Exp Med* 191, 1755-1764

35. Matsukawa, A., Hogaboam, C.M., Lukacs, N.W., Lincoln, P.M., Evanoff, H.L., and Kunkel, S.L. (2000) Pivotal role of the CC chemokine, macrophage-derived chemokine, in the innate immune response. *J Immunol* 164, 5362-5368
36. Baggiolini, M., Dewald, B., and Moser, B. (1997) Human chemokines: an update. *Annu Rev Immunol* 15, 675-705
37. Proost, P., Wuyts, A., and van Damme, J. (1996) The role of chemokines in inflammation. *Int J Clin Lab Res* 26, 211-223
38. Zlotnik, A., Morales, J., and Hedrick, J.A. (1999) Recent advances in chemokines and chemokine receptors. *Crit Rev Immunol* 19, 1-47
39. Annunziato, F., Cosmi, L., Galli, G., Beltrame, C., Romagnani, P., Manetti, R., Romagnani, S., Maggi, E. (1999) Assessment of chemokine receptor expression by human Th1 and Th2 cells in vitro and in vivo. *J Leukoc Biol* 65, 691-699
40. Bonecchi, R., Bianchi, G., Bordignon, P.P., D'Ambrosio, D., Lang, R., Borsatti, A., Sozzani, S., Allavena, P., Gray, P.A., Mantovani, A., and Sinigaglia, F. (1998) Differential expression of chemokine receptors and chemotactic responsiveness of type 1 T helper cells (Th1s) and Th2s. *J Exp Med* 187, 129-134
41. D'Ambrosio, D., Iellem, A., Bonecchi, R., Mazzeo, D., Sozzani, S., Mantovani, A., and Sinigaglia, F. (1998) Selective up-regulation of chemokine receptors CCR4 and CCR8 upon activation of polarized human type 2 Th cells. *J Immunol* 161, 5111-5115
42. O'Garra, A., McEvoy, L.M., and Zlotnik, A. (1998) T-cell subsets: chemokine receptors guide the way. *Curr Biol* 8, R646-649
43. Rossi, D., and Zlotnik, A. (2000) The biology of chemokines and their receptors. *Annu Rev Immunol* 18, 217-242
44. Sallusto, F., Kremmer, E., Palermo, B., Hoy, A., Ponath, P., Qin, S., Forster, R., Lipp,M., and Lanzavecchia, A. (1999) Switch in chemokine receptor expression upon TCR stimulation reveals novel homing potential for recently activated T cells. *Eur J Immunol* 29, 2037-2045
45. Sallusto, F., Lanzavecchia, A., and Mackay, C.R. (1998) Chemokines and chemokine receptors in T-cell priming and Th1/Th2- mediated responses. *Immunol Today* 19, 568-574
46. Sallusto, F., Lenig, D., Mackay, C.R., and Lanzavecchia, A. (1998) Flexible programs of chemokine receptor expression on human polarized T helper 1 and 2 lymphocytes. *J Exp Med* 187, 875-883
47. Kunkel, S.L. 1996. Th1- and Th2-type cytokines regulate chemokine expression. B*iol Signals* 5, 197-202
48. Li, L., Xia, Y., Nguyen, A., Feng, L., and Lo, D. (1998) Th2-induced eotaxin expression and eosinophilia coexist with Th1 responses at the effector stage of lung inflammation. *J Immunol* 161, 3128-3135
49. Sauty, A., Dziejman, M., Taha, R.A., Iarossi, A.S. Neote, K., Garcia-Zepeda, E.A., Hamid, Q., and Luster, A.D. (1999) The T cell-specific CXC chemokines IP-10, Mig, and I-TAC are expressed by activated human bronchial epithelial cells. *J Immunol* 162, 3549-3558
50. Schrum, S., Probst, P., Fleischer, B., and Zipfel, P.F. (1996) Synthesis of the CC-chemokines MIP-1alpha, MIP-1beta, and RANTES is associated with a type 1 immune response. *J Immunol* 157, 3598-3604
51. Siveke, J.T., and Hamann, A. (1998) T helper 1 and T helper 2 cells respond differentially to chemokines. *J Immunol* 160, 550-554
52. Teran, L. M., Mochizuki, M., Bartels, J., Valencia, E.L., Nakajima, T., Hirai, K., and Schroder, J.M. (1999) Th1- and Th2-type cytokines regulate the expression and production of eotaxin and RANTES by human lung fibroblasts. *Am J Respir Cell Mol Biol* 20, 777-786

53. Zhang, S., Lukacs, N.W., Lawless, V.A., Kunkel, S.L., and Kaplan, M.H. (2000) Cutting edge: differential expression of chemokines in Th1 and Th2 cells is dependent on Stat6 but not Stat4. *J Immunol* 165, 10-14
54. Taub, D.D., Ortaldo, J.R., Turcovski-Corrales, S.M., Key, M.L., Longo, D.L., and Murphy, W.J. (1996) Beta chemokines costimulate lymphocyte cytolysis, proliferation, and lymphokine production. *J Leukoc Biol* 59, 81-89.
55. Boring, L., Gosling, J., Chensue, S.W., Kunkel, S.L., Farese, Jr., R.V., Broxmeyer, H.E., and Charo, I.F. (1997) Impaired monocyte migration and reduced type 1 (Th1) cytokine responses in C-C chemokine receptor 2 knockout mice. *J Clin Invest* 100, 2552-2561
56. Chensue, S.W., Warmington, K.S., Allenspach, E.J., Lu, B., Gerard, C., Kunkel, S.L., and Lukacs, N.W. (1999) Differential expression and cross-regulatory function of RANTES during mycobacterial (type 1) and schistosomal (type 2) antigen-elicited granulomatous inflammation. *J Immunol* 163, 165-173
57. Chensue, S.W., Warmington, K.S., Ruth, J.H., Sanghi, P.S., Lincoln, P., and Kunkel, S.L. (1996) Role of monocyte chemoattractant protein-1 (MCP-1) in Th1 (mycobacterial) and Th2 (schistosomal) antigen-induced granuloma formation: relationship to local inflammation, Th cell expression, and IL-12 production. *J Immunol* 157, 4602-4608
58. Karpus, W.J., and Lukacs, N.W. (1996) The role of chemokines in oral tolerance. Abrogation of nonresponsiveness by treatment with antimonocyte chemotactic protein-1. *Ann N Y Acad Sci* 778, 133-144
59. Lukacs, N.W., Chensue, S.W., Karpus, W.J., Lincoln, P., Keefer, C., Strieter, R.M., and Kunkel, S.L. (1997) C-C chemokines differentially alter interleukin-4 production from lymphocytes. *Am J Pathol* 150, 1861-1868
60. Boutten, A., Dehoux, M.S., Seta, N., Ostinelli, J., Venembre, P., Crestani, B., Dombret, M.C., Durand, G., and Aubier, M. (1996) Compartmentalized IL-8 and elastase release within the human lung in unilateral pneumonia. *Am J Respir Crit Care Med* 153, 336-342
61. Chollet-Martin, S., Montravers, P., Gibert, C., Elbim, C., Desmonts, J.M., Fagon, J.Y., and Gougerot-Pocidalo, M.A. (1993) High levels of interleukin-8 in the blood and alveolar spaces of patients with pneumonia and adult respiratory distress syndrome. *Infect Immun* 61, 4553-4559
62. Grunewald, T., Schuler-Maue, W., and Ruf, B. (1993) Interleukin-8 and granulocyte colony-stimulating factor in bronchoalveolar lavage fluid and plasma of human immunodeficiency virus-infected patients with Pneumocystis carinii pneumonia, bacterial pneumonia, or tuberculosis [letter]. *J Infect Dis* 168, 1077-1078
63. Ponglertnapagorn, P., Oishi, K., Iwagaki, A., Sonoda, F., Watanabe, K., Nagatake, T., Matsushima, K., and Matsumoto, K. (1996) Airway interleukin-8 in elderly patients with bacterial lower respiratory tract infections. *Microbiol Immunol* 40, 177-182
64. Rodriguez, J.L., Miller, C.G., DeForge, L.E., Kelty, L., Shanley, C.J., Bartlett, R.H., and Remick, D.G. (1992) Local production of interleukin-8 is associated with nosocomial pneumonia. *J Trauma* 33, 74-81
65. Greenberger, M.J., Strieter, R.M., Kunkel, S.L., Danforth, J.M., Laichalk, L.L., McGillicuddy, D.C., and Standiford, T.J. (1996) Neutralization of macrophage inflammatory protein-2 attenuates neutrophil recruitment and bacterial clearance in murine Klebsiella pneumonia. *J Infect Dis* 173, 159-165
66. Johnson, M.C., 2nd, Kajikawa, O., Goodman, R.B., Wong, V.A., Mongovin, S.M., Wong, W.B., Fox-Dewhurst, R., and Martin, T.R. (1996) Molecular expression of the alpha-chemokine rabbit GRO in Escherichia coli and characterization of its production by lung cells in vitro and in vivo. *J Biol Chem* 271, 10853-10858
67. Moore, T. A., Newstead, M.W., Strieter, R.M., Mehrad, B., Beaman, B.L., and Standiford, T.J. (2000) Bacterial clearance and survival are dependent on CXC chemokine receptor-2 ligands in a murine model of pulmonary Nocardia asteroides infection. *J Immunol*164, 908-915

68. Tsai, W.C., Strieter, R.M., Wilkowski, J.M., Bucknell, K.A., Burdick, M.D., Lira, S.A., and Standiford, T.J. (1998) Lung-specific transgenic expression of KC enhances resistance to Klebsiella pneumoniae in mice. *J Immunol* 161, 2435-2440
69. Tsai, W.C., Strieter, R.M., Mehrad, B., Newstead, M.W., Zeng, X., and Standiford, T.J. (2000) CXC chemokine receptor CXCR2 is essential for protective innate host response in murine Pseudomonas aeruginosa pneumonia. *Infect Immun* 68, 4289-4296
70. Greenberger, M.J., Strieter, R.M., Kunkel, S.L., Danforth, J.M., Goodman, R.E., and Standiford, T.J. (1995) Neutralization of IL-10 increases survival in a murine model of Klebsiella pneumonia. *J Immunol* 155, 722-729
71. Shanley, T.P., Vasi, N., and Denenberg, A. (2000) Regulation of chemokine expression by IL-10 in lung inflammation. *Cytokine* 12, 1054-1064
72. Hartmann, P., and Plum, G. (1999) Immunological defense mechanisms in tuberculosis and MAC-infection. *Diagn Microbiol Infect Dis* 34, 147-152
73. Infante-Duarte, C., and Kamradt, T. (1999) Thl/Th2 balance in infection. *Springer Semin Immunopathol* 21, 317-338
74. Munk, M. E., and Emoto, M. (1995) Functions of T-cell subsets and cytokines in mycobacterial infections. *Eur Respir J Suppl* 20, 668S-675S
75. Atkinson, S., Valadas, E., Smith, S.M., Lukey, P.T., and Dockrell, H.M. (2000) Monocyte-derived macrophage cytokine responses induced by M. bovis BCG. *Tuber Lung Dis* 80, 197-207
76. Wang, J., Wakeham, J., Harkness, R., and Xing, Z. (1999) Macrophages are a significant source of type 1 cytokines during mycobacterial infection. *J Clin Invest* 103, 1023-1029
77. Chensue, S.W., Warmington, K., Ruth, J., Lincoln, P., Kuo, M.C., and Kunkel, S.L. (1994) Cytokine responses during mycobacterial and schistosomal antigen- induced pulmonary granuloma formation. Production of Thl and Th2 cytokines and relative contribution of tumor necrosis factor. *Am J Pathol* 145, 1105-1113
78. Chensue, S.W., Warmington, K., Ruth, J.H., and Kunkel, S.L. (1997) Effect of slow release IL-12 and IL-10 on inflammation, local macrophage function and the regional lymphoid response during mycobacterial (Thl) and schistosomal (Th2) antigen-elicited pulmonary granuloma formation. *Inflamm Res* 46, 86-92
79. Chensue, S.W., Warmington, K., Ruth, J.H., Lukacs, N., and Kunkel, S.L. (1997) Mycobacterial and schistosomal antigen-elicited granuloma formation in IFN-gamma and IL-4 knockout mice: analysis of local and regional cytokine and chemokine networks. *J Immunol* 159, 3565-3573
80. Zou, W., Borvak, J., Marches, F., Wei, S., Galanaud, P., Emilie, D., and Curiel, T.J. (2000) Macrophage-derived dendritic cells have strong Thl-polarizing potential mediated by beta-chemokines rather than IL-12. *J Immunol* 165, 4388-4396.
81. Sangari, F.J., Petrofsky, M., and Bermudez, L.E. (1999) Mycobacterium avium infection of epithelial cells results in inhibition or delay in the release of interleukin-8 and RANTES. *Infect Immun* 67, 5069-5075
82. Lin, Y., Zhang, M., and Barnes, P.F. (1998) Chemokine production by a human alveolar epithelial cell line in response to Mycobacterium tuberculosis. *Infect Immun* 66, 1121-1126
83. Rao, S.P., Hayashi, T., and Catanzaro, A. (2000) Release of monocyte chemoattractant protein (MCP)-1 by a human alveolar epithelial cell line in response to mycobacterium avium. *FEMS Immunol Med Microbiol* 29, 1-7
84. Sadek, M.I., Sada, E., Toossi, Z., Schwander, S.K., and Rich, E.A. (1998) Chemokines induced by infection of mononuclear phagocytes with mycobacteria and present in lung alveoli during active pulmonary tuberculosis. *Am J Respir Cell Mol Biol* 19, 513-521
85. Lin, Y., Gong, J., Zhang, M., Xue, W. and Barnes, P.F. (1998) Production of monocyte chemoattractant protein 1 in tuberculosis patients. *Infect Immun* 66, 2319-2322
86. Karpus, W.J., Kennedy, K.J., Kunkel, S.L., and Lukacs, N.W. (1998) Monocyte chemotactic protein 1 regulates oral tolerance induction by inhibition of T helper cell 1-related cytokines. *Inflamm Res* 187, 733-741

87. Matsukawa, A., Hogaboam, C.M., Lukacs, N.W., Lincoln, P.M., Strieter, R.M., and Kunkel, S.L. (2000) Endogenous MCP-1 influences systemic cytokine balance in a murine model of acute septic peritonitis. *Exp Mol Pathol* 68, 77-84
88. Rutledge, B.J., Rayburn, H., Rosenberg, R., North, R.J., Gladue, R.P., Corless, C.L., and Rollins, B.J. (1995) High level monocyte chemoattractant protein-1 expression in transgenic mice increases their susceptibility to intracellular pathogens. *J Immunol* 155, 4838-4843
89. Lu, B., Rutledge, B.J., Gu, L., Fiorillo, J., Lukacs, N.W., Kunkel, S.L., North, R., Gerard, C., and Rollins, B.J. (1998) Abnormalities in monocyte recruitment and cytokine expression in monocyte chemoattractant protein 1-deficient mice. *Inflamm Res* 187, 601-608
90. Karpus, W.J., Lukacs, N.W., Kennedy, K.J., Smith, W.S., Hurst, S.D., and Barrett, T.A. (1997) Differential CC chemokine-induced enhancement of T helper cell cytokine production. *J Immunol* 158, 4129-4136
91. Always, W.H., Record, F.M., and Openshaw, P.J. (1993) Phenotypic and functional characterization of T cell lines specific for individual respiratory syncytial virus proteins. *J Immunol* 150, 5211-5218
92. Busse, W.W., Gern, J.E., and Dick, E.C. (1997) The role of respiratory viruses in asthma. *Ciba Found Symp* 206, 208-13.
93. Erb, K.J., and Le Gros, G. (1996) The role of Th2 type CD4+ T cells and Th2 type CD8+ T cells in asthma. *Immunol Cell Biol* 74, 206-208
94. Kay, A.B. (1997) T cells as orchestrators of the asthmatic response. *Ciba Found Symp* 206, 56-67
95. Wimalasundera, S.S., Katz, D. R., and Chain, B. M. (1997) Characterization of the T cell response to human rhinovirus in children: implications for understanding the immunopathology of the common cold. *J Infect Dis* 176, 755-9
96. Hussell, T., Georgiou, A., Sparer, T.E., Matthews, S., Pala, P., and Openshaw, P.J. (1998) Host genetic determinants of vaccine-induced eosinophilia during respiratory syncytial virus infection. *J Immunol* 161, 6215-6222
97. Tekkanat, K.K., Maassab, H.F., Cho, D.S., Lai, J.J., John, A., Berlin, A., Kaplan, M.H., and Lukacs, N.W. (2001) IL-13-Induced Airway Hyperreactivity During Respiratory Syncytial Virus Infection Is STAT6 Dependent. *J Immunol* 166, 3542-3548
98. Mielck, A., Reitmeir, P., and Wjst, M. (1996) Severity of childhood asthma by socioeconomic status. *Int J Epidemiol* 25, 388-393
99. Schuh, S., D. Johnson, D., Stephens, D., Callahan, S., and Canny, G. (1997) Hospitalization patterns in severe acute asthma in children. *Pediatr Pulmonol* 23, 184-192
100. Corne, J.M., and Holgate, S.T. (1997) Mechanisms of virus induced exacerbations of asthma. *Thorax* 52, 380-389
101. Grunberg, K., Timmers, M.C., Smits, H.H., de Klerk, E.P., Dick, E.C., Spaan, W.J., Hiemstra, P.S., and Sterk, P.J. (1997) Effect of experimental rhinovirus 16 colds on airway hyperresponsiveness to histamine and interleukin-8 in nasal lavage in asthmatic subjects in vivo [see comments]. *Clin Exp Allergy* 27, 36-45
102. Martinez, F.D. (1997) Definition of pediatric asthma and associated risk factors. *Pediatr Pulmonol Suppl* 15, 9-12
103. Rooney, J.C., and Williams, H.E. (1971) The relationship between proved viral bronchiolitis and subsequent wheezing. *J Pediatr* 79, 744-747
104. Schwarze, J., Hamelmann, E., Bradley, K.L., Takeda, K., and Gelfand, E.W. (1997) Respiratory syncytial virus infection results in airway hyperresponsiveness and enhanced airway sensitization to allergen. *J Clin Invest* 100, 226-233
105. Sly, P.D., and Hibbert, M.E. (1989) Childhood asthma following hospitalization with acute viral bronchiolitis in infancy. *Pediatr Pulmonol* 7, 153-158
106. Teichtahl, H., Buckmaster, N., and Pertnikovs, E. (1997) The incidence of respiratory tract infection in adults requiring hospitalization for asthma. *Chest* 112, 591-596

107. van Schaik, S.M., Tristram, D.A., Nagpal, I.S., Hintz, K.M., Welliver, 2nd, R.C., and Welliver, R.C. (1999) Increased production of IFN-gamma and cysteinyl leukotrienes in virus- induced wheezing. *J Allergy Clin Immunol* 103, 630-636
108. Schwarze, J., Cieslewicz, G., Hamelmann, E., Joetham, A., Shultz, L.D., Lamers, M.C., and Gelfand, E.W. (1999) IL-5 and eosinophils are essential for the development of airway hyperresponsiveness following acute respiratory syncytial virus infection. *J Immunol* 162, 2997-3004
109. Schwarze, J., Makela, M., Cieslewicz, G., Dakhama, A., Lahn, M., Ikemura, T., Joetham, A., and Gelfand, E.W. (1999) Transfer of the enhancing effect of respiratory syncytial virus infection on subsequent allergic airway sensitization by T lymphocytes. *J Immunol* 163, 5729-5734
110. Oymar, K., Elsayed, S., and Bjerknes, R. (1996) Serum eosinophil cationic protein and interleukin-5 in children with bronchial asthma and acute bronchiolitis. *Pediatr Allergy Immunol* 7, 180-186
111. Peebles, R.S., Sheller, J.R., Collins, R.D., A. K. Jarzecka, A.K., Mitchell, D.B., Parker, R.A., and Graham, B.S. (2001) Respiratory syncytial virus infection does not increase allergen-induced type 2 cytokine production, yet increases airway hyperresponsiveness in mice. *J Med Virol* 63, 178-188
112. Matsuse, H., Behera, A.K., Kumar, M., Rabb, H., Lockey, R.F., and Mohapatra, S.S. (2000) Recurrent respiratory syncytial virus infections in allergen-sensitized mice lead to persistent airway inflammation and hyperresponsiveness. *J Immunol* 164, 6583-6592
113. Openshaw, P. J., and Hewitt, C. (2000) Protective and harmful effects of viral infections in childhood on wheezing disorders and asthma. *Am J Respir Crit Care Med* 162, S40-S43
114. Tsitoura, D.C., Kim, S., Dabbagh, K., Berry, G., Lewis, D.B., and Umetsu, D.T. (2000) Respiratory infection with influenza A virus interferes with the induction of tolerance to aeroallergens. *J Immunol* 165, 3484-3491
115. Chu, R.S., Targoni, O.S., Krieg, A.M., Lehmann, P.V., and Harding, C.V. (1997) CpG oligodeoxynucleotides act as adjuvants that switch on T helper 1 (Th1) immunity. *Inflamm Res* 186, 1623-1631
116. Jakob, T., Walker, P.S., Krieg, A.M., Udey, M.C., and Vogel, J.C. (1998) Activation of cutaneous dendritic cells by CpG-containing oligodeoxynucleotides: a role for dendritic cells in the augmentation of Th1 responses by immunostimulatory DNA. *J Immunol* 161, 3042-3049
117. Kline, J.N., Waldschmidt, T.J., Businga, T.R., Lemish, J.E., Weinstock, J.V., Thorne, P.S., and Krieg, A.M. (1998) Modulation of airway inflammation by CpG oligodeoxynucleotides in a murine model of asthma. *J Immunol* 160, 2555-2559
118. Krieg, A.M., Love-Homan, L., Yi, A.K., and Harty, J.T. (1998) CpG DNA induces sustained IL-12 expression in vivo and resistance to Listeria monocytogenes challenge. *J Immunol* 161, 2428-2434
119. Ramachandra, L., Chu, R.S., Askew, D., Noss, E.H., Canaday, D.H., Potter, N.S., A. Johnsen, A., Krieg, A.M., Nedrud, J.G., Boom, W.H., and Harding, C.V. (1999) Phagocytic antigen processing and effects of microbial products on antigen processing and T-cell responses. *Immunol Rev* 168, 217-239
120. Weiner, G.J., Liu, H.M., Wooldridge, J.E., Dahle, C.E., and Krieg, A.M. (1997) Immunostimulatory oligodeoxynucleotides containing the CpG motif are effective as immune adjuvants in tumor antigen immunization. *Proc Natl Acad Sci USA* 94, 10833-10837

Chapter 5

ROLE OF COMPLEMENT IN ACUTE LUNG INJURY

Markus S. Huber-Lang and Peter A. Ward
Department of Pathology, University of Michigan Medical School, Ann Arbor, MI

INTRODUCTION

Because the respiratory tract is continuously exposed to pathogens and/or particles in the inhaled air, appropriate immune defense mechanisms have evolved to prevent lung inflammatory injury. The tight interplay between the cellular defense system and the complement system is linked by the lung inflammatory response. Complement-enhanced generation of chemokines and complement-related recruitment of leukocytes have been demonstrated to play important roles in the progression of the inflammatory response, which often leads to progressive lung injury and irreversible scarring with consequent impairment of respiratory function. Complement is also involved in the triggering events that lead to repair and remodeling of the lung parenchyma with the ultimate resolution of the inflammatory process. This review focuses on the impact of complement activation products, especially C3a, C5a, and C5b-9 membrane attack complex (MAC), on the lung inflammatory response and addresses the implications for novel therapeutic approaches to acute lung injury (ALI).

Systemic Activation of Complement

The importance of complement activation products as major participants in local and systemic inflammatory conditions has been shown in many clinical studies (1-3). Increased plasma levels of complement activation products have been found after severe physical and thermal trauma, hemorrhagic shock, sepsis, and sepsis-related multi-organ failure syndrome (4, 5). The complement activation products C3a, C5a, and C5b-9 are known to play crucial roles in the pathophysiology of ALI and its most severe manifestation, the adult respiratory distress syndrome (ARDS).

Whereas increased levels of C3a and C5a were found in patients with ARDS, or in patients who subsequently developed ARDS, no evidence of systemic complement activation was found in patients who did not develop ARDS (6-8). Furthermore, an inverse correlation was established between plasma levels of C3a and C5a, and survival of patients with ARDS (9-11).

Three different pathways of complement activation have been described (Figure 1). Complement activation can be triggered by target-bound antibody (classical pathway), by polysaccharide structures of microbes (MBLectin pathway), or by "foreign" surface structures (alternative pathway). All three pathways merge at the pivotal activation of C3, and subsequently, of C5 with release of complement activation products C3a and C5a. In the common terminal pathway, further complement factors (C6, C7, C8, and C9) are activated in a non-proteolytic manner and assembled into the MAC, leading to membrane lysis of microorganisms (12).

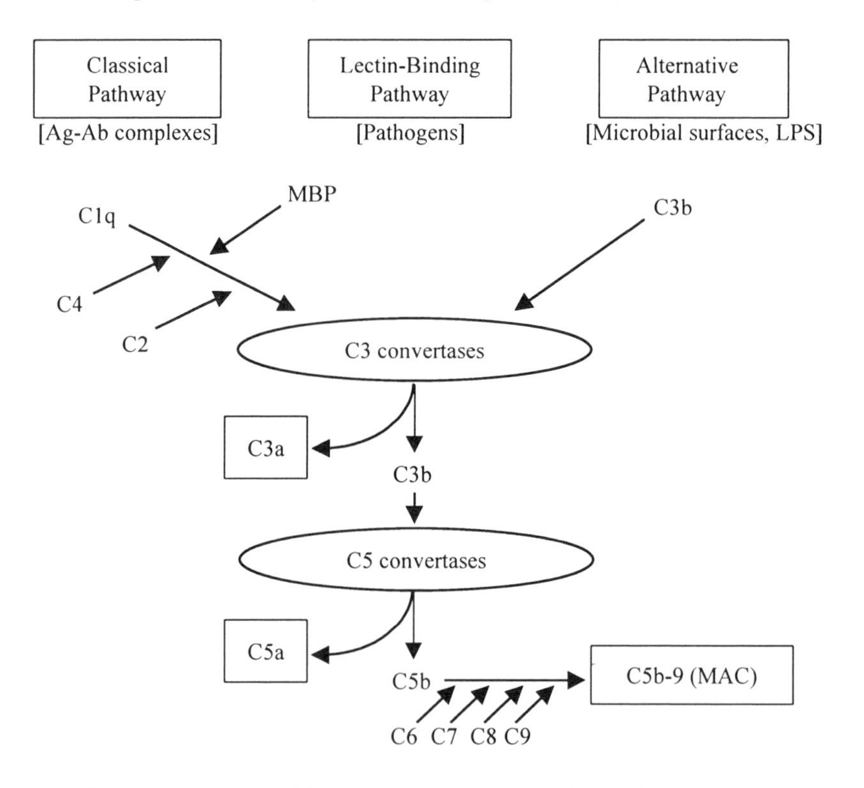

Figure 1. Activation of the complement cascade by three different pathways.

Experimentally, various animal models have been utilized to examine the effects of complement activation on the development of ALI. *In vivo* infusion of zymosan-activated plasma, as a source of activated complement components (especially C3a and C5a), has been shown to induce neutropenia, but there is little evidence of lung injury unless the activated plasma is repetitively infused (13). Purified cobra venom factor (CVF) is a C3b homologue and a powerful activator of the alternative pathway. Extensive systemic complement activation by intravenous infusion of CVF has been demonstrated to cause pulmonary capillary injury and neutrophil accumulation in the lung (14). Within minutes after intravenous CVF administration, serum levels of C5a reached a maximum. This was accompanied by increased expression of both the endothelial cell adhesion molecule, P-selectin, and the blood neutrophil β_2-integrin, CD11b/CD18, resulting in neutrophil adhesion to activated endothelium (15, 16). Complement-dependent recruitment of neutrophils can result in oxidant-mediated damage of endothelial cells and surrounding tissues, leading to increased vascular permeability with edema, hemorrhage, and inflammatory cell infiltration, characteristic histopathological signs of ALI. Repetitive intravenous infusion of C5a has been shown to induce similar effects. Thus, ALI can be experimentally simulated by systemic activation of complement, with excessive generation of the potent pro-inflammatory mediators, C3a and C5a.

Production of Complement Factors in the Lung

Plasma complement proteins are mainly produced in the liver (> 90%). A few components have their origin predominantly outside the liver: C1 in the intestinal epithelium and monocytes/macrophages, and C7 mainly in granulocytes. There is increasing evidence, however, that virtually all complement components required for activation of the classical and alternative pathway can be synthesized extrahepatically in the lung by different lung cells (Table 1). Lung macrophages have been reported to produce C2, C3, C4, C5, and factor B *in vitro* (17). Lung epithelial type II cells and lung fibroblasts are known to synthesize and secrete not only numerous complement components, but also complement regulatory proteins such as factor B, H, I, and C1s inhibitor (18).

Table 1. Complement production by different lung cells

Cell Type	Complement Factors
Lung epithelial type II – cells	C1, 4, 3, 5, 6, 7, 8, 9
Lung fibroblasts	C1, 4, 3, 5, 6, 8, 9
Alveolar macrophages	C2, C3, C4, C5

Furthermore, it has been demonstrated that different cytokines (IL-1α, IL-1β, IFN-γ, etc.) can modulate C3 and C5 production by human type II respiratory epithelial cells independently. In human monocytes/macrophages, IFNγ has been shown to selectively stimulate the synthesis of C2, factor B, and C1 inhibitor, but to decrease C3 production, indicating a complex linkage between the chemokine/cytokine network and locally produced complement components (19). Thus, during acute lung inflammation, there may be at least two sources of complement components (especially C3 and

C5). One source of complement may be derived from leakage of plasma proteins when vascular permeability has developed during ALI, while a second source may be locally produced complement from alveolar epithelial cells, fibroblasts, or lung macrophages. The contribution of locally produced complement activation products in the development of ALI, however, has not been clearly defined.

Local Activation of Complement

The well-described mechanisms of systemic complement activation, via the three different pathways, involve formation of C3 and C5 convertases. In both convertases the catalytically active center is known to reside in serine protease domains (20). Besides these convertases, non-complement-derived proteases such as bacteria-derived arginine-specific cysteine protease, proteases generated by macrophages (21), and some proteins of the coagulation cascade (e.g. kallikrein) have been demonstrated to cleave complement components, such as C3 and C5, to generate biologically active anaphylatoxins (22, 23).

Clinical and experimental investigations have described increased levels of C5a in bronchoalveolar lavage (BAL) fluids obtained in the context of bacterial or aspiration pneumonia (24, 25), immune complex or LPS-induced ALI (2, 26), and chronic lung disease (27). When C5 was instilled intratracheally into rabbits and hamsters, various C5 fragments were detectable in BAL fluids, and an intense pulmonary inflammatory reaction with extravasation of neutrophils was observed (28, 29). These data support the presence of a local mechanism for cleavage and activation of complement components within the lung.

After intratracheal application of C5a or its biological degradation product $C5a_{desarg}$, extensive pulmonary accumulation of neutrophils developed, which underscores the important chemotactic role of C5a in mediating neutrophil influx into the alveolar space (30). Furthermore, treatment with neutralizing anti-C5a antibody greatly reduced immune complex-induced acute lung inflammation as reflected by decreases in both lung vascular permeability and intra-pulmonary accumulation of neutrophils (31). In this same study, a compartmentalized role for C5a in mediating immune complex-induced inflammation was hypothesized, as airway instillation of anti-C5a, but not intravenous infusion of anti-C5a, was protective. Accordingly, a significant amount of C5 activation products (C5a) may be generated in the airway compartment independent of components from the vascular complement system (Figure 2).

The published data suggest that lung macrophages can be activated by a variety of agonists (e.g. bacterial products, TNFα, etc.) to express a serine protease on their cell surface that could cleave C5 to C5a, which in turn acts as a potent chemoattractant for blood neutrophils. C5a could also function as an autocrine activator of lung macrophages to induce the release of additional chemotactically active cytokines, as has been suggested by experimental data (31).

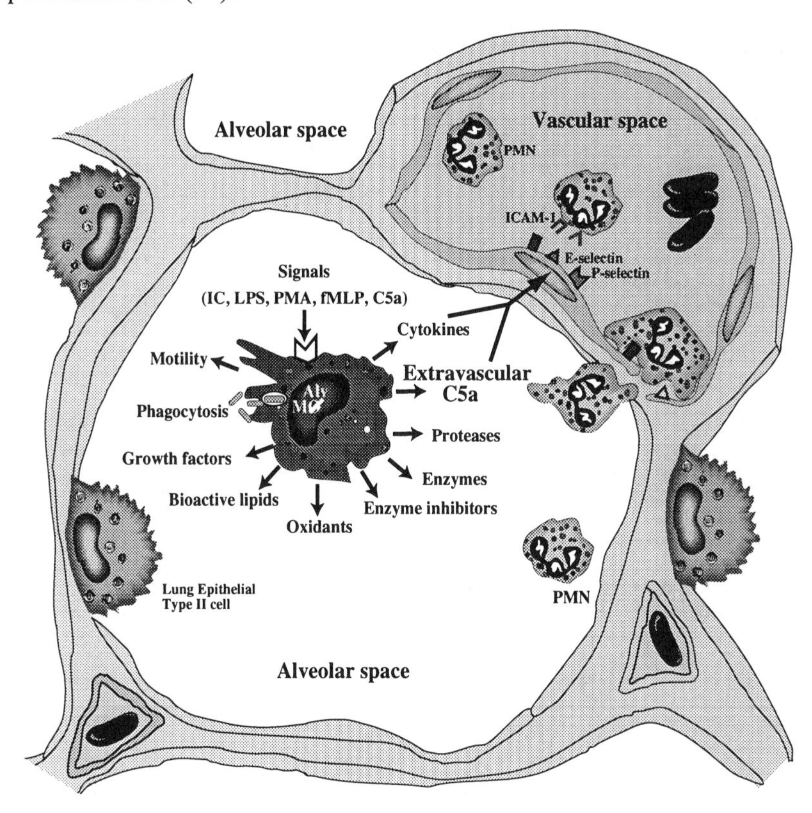

Figure 2. Generation of the complement activation product, C5a by alveolar macrophage and production of multiple mediators by activated macrophages.

Inflammatory Function of Complement Activation Products

The importance of anaphylatoxins as major participants in inflammatory processes is suggested by many studies (see above). C3a, C5a and MAC have been demonstrated to be required for the full development of injury in various lung inflammation models. C5a, known for its powerful chemotactic activity for neutrophils, and MAC have also been shown to activate endothelial cells, causing surface expression of the adhesion molecules, P-selectin (32, 33), and its counter-receptors on neutrophils (15, 16). Other endothelial adhesion molecules, such as E-selectin and intercellular adhesion molecule-1 (ICAM-1) are upregulated by sub-lytic concentrations of C5b-9 (MAC) in presence of low TNF levels (33). In addition, MAC is known to induce endothelial cell production of inflammatory mediators, such as interleukin-8 (IL-8) and monocyte chemoattractant protein-1 (MCP-1) (34).

Recruitment of circulating neutrophils to the site of inflammation occurs due to interaction with upregulated endothelial adhesion molecules and the network of lung tissue derived chemokines such as cytokine-induced neutrophil chemoattractant [CINC] and macrophage inflammatory protein-1α (MIP-1α) (see Chapters 2 and 3). Lung endothelial cells, in concert with recruited phagocytic cells, respond to activated complement components (C3a, C5a, and MAC) with generation of oxygen radical species such as superoxide anion (32). Such oxygen radical species not only result in killing of microorganisms, but also in additional damage to lung tissue (see Chapters 6, 7, and 9). Upon activation by complement, phagocytic cells release other important proinflammatory mediators. For example, C3a and C5a have been shown to induce or enhance the release of prostaglandins and leukotrienes from phagocytic cells. Thus, the combined activation of neutrophils, endothelial cells, and phagocytic cells by complement activation products appears to be required for the full production of lung microvascular injury (Table 2).

Table 2. Inflammatory function of complement activation products

Activation Product	Target Cells	Functions
C3a	Basophils, Mast-cells	Histamine release
	Eosinophils	Chemotactic
C5a	Neutrophils, monocytes	Chemotactic activity; Oxidative release
	Basophils, Mast-cells	Histamine release
	Smooth muscle cells	Contraction
	Endothelial cells	Secretion of vWF; upregulation of P-selectin, oxidative burst
	Platelets	release of serotonin
	Alveolar macrophage	Enhanced production of chemokines
	Lung epithelial type II cell	Chemokine release
C5b-9 (MAC)	Microorganisms	Cytotoxic activity
	Macrophage	Activation
	Endothelial cell	Upregulation of adhesion molecules (P-selectin, E-selectin, ICAM-1) Chemokine expression; secret vWF (at low TNF levels)

vWf = von Willebrand Factor
ICAM = intracellular adhesion molecule

In addition, C3a and C5a have been demonstrated to cause contraction of smooth muscle (35), a common biological activity of pro-inflammatory peptides causing increased airway resistance. Both anaphylatoxins are also able to activate mast cells and basophils to release phlogistic factors (e.g. histamine), causing leakage of soluble proteins from the vascular compartment into the airspace, which then may damage the integrity of the blood-air barrier. Furthermore, activated complement components (C3a and C5a) are known to activate platelets (36) and to interact with the kinin-coagulation cascade, resulting in enhancement of the inflammatory response by induction of a dense network of proinflammatory mediators.

Therapeutic Interventions in the Complement System

The target of complement blockade has important theoretical and clinical implications. To date, only consumptive complement depletion with CVF, which drastically reduces plasma levels of C3, has been effective in experimental settings, though this approach is not realistically applicable in the clinical setting. Recently, alternative therapeutic approaches involving blockade of the complement system have been developed. Human recombinant soluble complement receptor 1 (sCR1) interferes with formation of C3 and C5 convertases, reducing both C3a and C5a anaphylatoxin generation as well as the formation of C3b and MAC. As demonstrated in a variety of experimental models, complement blockade with sCR1 has provided effective anti-inflammatory activity (37). A theoretical drawback of therapy with sCR1 may be interference in the generation of protective opsonic complement products, such as C3b or the protective lytic MAC complex. Clinical trials to date, however, suggest that this is not a significant clinical problem.

The importance of complement activity for lung mucosal defenses against infection has been demonstrated in mice deficient in the C5a receptor (C5aR-/-). These animals had impaired clearance of intrapulmonary-instilled *Pseudomonas aeruginosa* in comparison to their wild type littermates (C5aR +/+), despite having a marked increase in neutrophil influx (38). Conversely, excessive generation of C5a with a dysregulated inflammatory response has also been shown experimentally and clinically to be damaging to the lung (6-8). *In vivo* blockade of C5 by whole antibody or antibody fragments has been shown to block inflammatory responses in a variety of tissues (39, 40). Preliminary clinical trials suggest that C5 blockade in patients undergoing cardiopulmonary bypass results in improved clinical parameters (41).

Given these caveats, the goal of a therapeutic strategy aimed at modulating the complement system must include moderate complement suppression that is highly selective and incorporates the following two concepts: 1) complement blockade should target a component that does not result in the formation of large amounts of immune complexes, as would be the case with antibody to C3; and 2) the complement blockade target should be highly selective without affecting other important complement components, such as C3b, which is required for optimal phagocytosis of bacteria, or C5, which is involved in formation of MAC and is necessary for lysis of Gram-negative bacteria.

In the IgG immune complex model of ALI in rats, complement depletion by treatment with purified CVF markedly reduced the permeability index by 49 % (2). In addition, airway administration of anti-C5a antibody was also highly protective by reducing the pulmonary vascular permeability by 73% (2). These data confirm that in IgG immune complex-induced lung inflammation, C5a plays a key role in the pathophysiology of ALI. Antibody against C5a does not interfere with the hemolytic activity of serum and therefore does not block the generation of MAC, and may fulfill the requirements outlined above. When administered intra-tracheally, anti-C5a greatly reduced pulmonary vascular permeability and diminished the recruitment of neutrophils in BAL fluids, implicating an effective and promising therapeutic approach against acute lung inflammation (2). Thus, selective blockade of C5a, or its corresponding receptor, may present novel therapeutic approaches to the treatment of ALI in humans.

Summary and Conclusion

The complement system plays a central role in the innate immunity of the host defense system, but also contributes to the development and progression of ALI. Many, if not all, of the complement components of the classical and alternative pathways can be generated within the lung by different cells. Furthermore, proteases derived from lung cells and from infiltrated inflammatory cells may cleave C3 and C5 to generate C3a and C5a. In concert with the complement activation products C3a and C5b-9 (MAC), C5a enhances production of various proinflammatory mediators by alveolar macrophages and upregulates endothelial adhesion molecules. These mediators and adhesion molecules facilitate the recruitment of circulating blood neutrophils into the site of inflammation resulting in tissue damage by the mechanisms described above. Inhibition of the biological activity of C5a by antibodies, or by use of C5aR antagonists, has demonstrated substantial protective effects in experimental models of ALI. The data support the promising therapeutic applications of complement inhibition in cases of dysregulated inflammatory injury.

ACKNOWLEDGEMENTS

The work represented in this review was funded by NIH grant HL-31963 and GM-29507. We are grateful to Beverly Schumann and Peggy Otto for their excellent assistance in preparing the manuscript.

REFERENCES

1. Weinburg, P.F., Matthay, M.A., Webster, R.O., Roskos, R.V., Goldstein, I.M., and Murray, J. F. (1984) Biologically active products of complement and acute lung injury in patients with the sepsis syndrome. *Am Rev Respir Dis* 130, 791-796
2. Mulligan, M.S., Schmid, E., Beck-Schimmer, B., Till, G.O., Friedl, H.P., Brauer, R.B., Hugli, T.E., Miyasaka, M., Warner, R.L., Johnson, K.J., and Ward, P.A. (1996) Requirement and role of C5a in acute lung inflammatory injury in rats. *J Clin Invest* 98, 503-512
3. Schmid, E., Piccolo, M.-T.S., Friedl, H.P., Warner, R.L., Mulligan, M.S., Hugli, T.E., Till, G. O., and Ward, P.A. (1997) Requirement for C5a in lung vascular injury following thermal trauma to rat skin. *Shock* 8, 119-124
4. Goya, T., Morisaki, T., and Torisu, M. (1994) Immunologic assessment of host defense impairment in patients with septic multiple organ failure: Relationship between complement activation and changes in neutrophil function. *Surgery* 115, 145-155
5. Stöve, S., Welte, T., Wagner, T.O.F., Kola, A., Klos, A., Bautsch, W., and Köhl, J. (1996) Circulating complement proteins in patients with sepsis or systemic inflammatory response syndrome. *Clin Diagn Lab Immunol* 3, 175-183
6. Hammerschmidt, D.E., Weaver, L.J., Hudson, L.D., Craddock, P.R., and Jacob, H.S. (1980) Association of complement activation and elevated plasma-C5a with adult respiratory distress syndrome. *Lancet* 1(8175), 947-949
7. Solomkin, J.S., Cotta, L.A., Satoh, P.S., Hurst, J.M., and Nelson, R.D. (1985) Complement activation and clearance in acute illness and injury: evidence for C5a as a cell-directed mediator of the adult respiratory distress syndrome in man. *J Surgery* 97, 668-678
8. Zillow, G., Sturm, J.A., Rother, U., and Kirschfink, M. (1990) Complement activation and the prognostic value of C3a in patients at risk of adult respiratory distress syndrome. *Clin Exp Immunol* 79, 151-157
9. Kapur, M.M., Jain, P., and Gidh, M. (1986) The effect of trauma on serum C3 activation and its correlation with injury severity score in man. *J Trauma* 26, 464-466
10. Nakae, H., Endo, S., Inada, K., and Yoshida, M. (1996) Chronological changes in the complement system in sepsis. *Jpn J Surg* 26, 225-229
11. Hecke, F., Schmidt, U., Kola, A., Bautsch, W., Klos, A., and Köhl, J. (1997) Analysis of complement proteins in polytrauma patients-correlation with injury severity, sepsis and outcome. *Shock* 7, 74-83
12. Prodinger, W.M., Würzner, R., Erdei, A., and Dierich, M.P. (1999) Complement. Paul, W.E. (ed.) Fundamental Immunology, Lippincott-Raven Publisher, Philadelphia 967-995
13. Perkowski, S.Z., Havill, A.M., Flynn, J.T., and Gee M.H. (1983) Role of intrapulmonary release of eicosanoids and superoxide anion as mediators of pulmonary dysfunction and endothelial injury in sheep with intermittent complement activation. *Circ Res* 53, 574-583
14. Till, G.O., Johnson, K.J., Kunkel, R., and Ward, P.A. (1987) Activation of C5 by cobra venom factor is required in neutrophil-mediated lung injury in the rat. *Am J Pathol* 129, 44-53

15. Mulligan, M.S., Yeh, C.G., Rudolph, A.R., and Ward, P.A. (1992) Protective effects of soluble CR1 in complement- and neutrophil-mediated tissue injury. *J Immunol* 148, 1479-1485
16. Mulligan, M.S., Smith, C.W., Anderson, D.C., Todd III, R.F., Miyasaka, M., Tamatani, T., Issekutz, T.B., and Ward, P.A. (1993) Role of leukocyte adhesion molecules in complement-induced lung injury. *J Immunol* 150, 2401-2406
17. Hetland, G., Johnson, E., and Aasebo, U. (1986) Human alveolar macrophages synthesize the functional alternative pathway of complement and active C5 and C9 *in vitro*. *Scand J Immunol* 24, 603-608
18. Strunk, R.C., Eidlen, D.M., and Mason, R.J. (1988) Pulmonary alveolar type II epithelial cells synthesize and secrete proteins of the classical and alternative complement pathways. *J Clin Invest* 81, 1419-1426
19. Rothman, B.L., Despins, A.W., and Kreutzer, D.L. (1990) Cytokine regulation of C3 and C5 production by the human type II pneumocyte cell line, A549. *J Immunol* 145, 592-598
20. Narayana, S.V.L., Babu, Y.S., and Volankis, J.E. (2000) Inhibition of complement serine proteases as a therapeutic strategy. In: Lambris, J. D., and Holers, V. M. (eds): Therapeutic interventions in the complement system. Human Press, Totowa, NJ 113-153
21. Snyderman, R., Shin, H.S., and Dannenberg, A.M. (1972) Macrophage proteinase and inflammation: The production of chemotactic activity from the fifth component of complement by macrophage proteinase. *J Immunol* 109, 896-898
22. Ward, P.A., and Newman, L.J. (1969) Neutrophil chemotactic factor from human C'5. *J Immunol* 102, 93-99
23. Wingrove, J.A., DiScipio, R.G., Chen, Z., Potempa, J., Travis, J., and Hugli, T.E. (1992). Activation of complement components C3 and C5 by a cysteine proteinase (gingipain-1) from porphyromonas (bacteroides) gingivalis. *J Bio Chem* 276, 18902-18907
24. Hopkins, H., Stull, T., von Essen, S.G., Robbins, R.A., and Rennard, S.I. (1989) Neutrophil chemotactic factors in bacterial pneumonia. *Chest* 95, 1021-1027
25. Ishii, Y., Kobayashi, J., and Kitamura, S. (1989) Chemotactic factor generation and cell accumulation in acute lung injury induced by endotracheal acid instillation. *Prostaglandins Leukot Essent Fatty Acids* 37, 65-70
26. Ward, P.A. (1996) Role of complement in lung inflammatory injury. *Am J Path* 149, 1081-1086
27. Groneck, P., Oppermann, M. and Speer, C.P. (1993) Levels of complement anaphylatoxin C5a in pulmonary effluent fluids of infants at risk of chronic lung disease and effects of dexamethasone treatment. *Ped Res* 34, 586-590
28. Henson, P.M., McCarthy, K., Larsen, G.L., Webster, R.O., Giclas, P.C., Dreisin, R.B., King, T. E., and Shaw, J.O. (1979) Complement fragments, alveolar macrophages, and alveolitis. *Am J Pathol* 97, 93-110
29. Desai, U., Kreutzer, D.L., Showell, H.T., Arroyave, C.V., and Ward, P.A. (1979) Acute lung inflammatory pulmonary reactions induced by chemotactic factor. *Am J Pathol* 96, 71-83
30. Desai, U., Dickey, B., Varani, J., and Kreutzer, D.L. (1984) Demonstration of C5 cleaving activity in bronchoalveolar fluids and cells: a mechanism of acute and chronic alveolitis. *J Exp Pathol* 1, 201-216
31. Czermak, B.J., Sarma, V., Bless, N.M., Schmal, H., Friedl, H.P., and Ward, P.A. (1999) *In vitro* and i*n vivo* dependency of chemokine generation on C5a and TNFα. *J Immunol* 162, 2321-2325
32. Foreman, K.E., Vaporciyan, A.A., Bonish, B.K., Jones, M.L., Johnson, K.J., Glovsky, M.M., Eddy, S.M., and Ward, P.A. (1994) C5a-induced expression of P-selectin in endothelial cells. *J Clin Invest* 94, 1147-1155
33. Kilgore, K.S., Shen, J.P., Miller, B.F., Ward, P.A., and Warren, J.S. (1995) Enhancement by the complement membrane attack complex of TNFα-induced endothelial cell expression of E-selectin and ICAM-1. *J Bio Chem* 155, 1434-1441

34. Kilgore, K.S., Schmid, E., Shanley, T.P., Flory, C.M., Maheswari, V., Tramontini, T.L., Cohen, H., Ward, P.A., Friedl, H.P., and Warren, J. S. (1997) Sublytic concentrations of the membrane attack complex of complement induce endothelial interleukin-8 and monocyte chemoattractant protein-1 through nuclear factor-kappa B activation. *Am J Pathol* 150, 2019-2031
35. Frank, M.M., and Fries, L.F. (1991) The role of complement in inflammation and phagocytosis. *Immunol Today* 12, 322-326
36. Meuer, S., Ecker, U., Hadding, U., Bitter-Suermann, D. (1981) Platelet-serotonin release by C3a and C5a: two independent pathways of activation. *J Immunol* 126, 1506-1509
37. Mulligan, M.S., Polley, M.J., Bayer, R.J., Nunn, M.F., Paulson, J.C., and Ward, P.A. (1992) Neutrophil-dependent acute lung injury: requirement for P-selectin (GMP-140). *J Clin Invest* 90, 1600-1607
38. Hoepken, U.E., Lu, B., Gerard, N.P., and Gerard, C. (1996) The C5a chemoattractant receptor mediates mucosal defense to infection. *Nature* 383, 86-89
39. Vakeva, A.P., Agah, A., Rollins, S.A., Matis, L.A., Li, L., Stahl, G.L. (1998) Myocardial infarction and apoptosis after myocardial ischemia and reperfusion: role of the terminal complement components and inhibition by anti-C5 therapy. *Circulation* 97, 2259-2267
40. Kyriakides, C., Austen, W.G., Wang, Y., Favuzza, J., Moore, F.D., Hechtman, H.B. (2000) Soluble P-selectin moderates complement dependent injury. *J Trauma* 48, 32-38
41. Fitch, J.C., Rollins, S., Matis, L., Alford, B., Aranki, S., Collard, C.D., Dewar, M., Elefteriades, J., Hines, R., Kopf, G., Kraker, P., Li, L., O'Hara, R., Rinder, C., Rinder, H., Shaw, R., Smith, B., Stahl, G., Shernan, S.K. (1999) Pharmacology and biological efficacy of a recombinant, humanized, single-chain antibody C5 complement inhibitor in patients undergoing coronary artery bypass graft surgery with cardiopulmonary bypass. *Circulation* 100, 2499-2506

Chapter 6

OXIDANT STRESS IN ACUTE LUNG INJURY

Jerry J. Zimmerman
Division of Critical Care Medicine, Children's Hospital and Regional Medical Center, Seattle, WA

INTRODUCTION

Fridovich has noted that, "The aerobic lifestyle offers many advantages but is fraught with danger (1)." The lungs represent the first aspect of a pathway of oxygen delivery from the environment to the mitochondria where efficient ATP production is facilitated via respiration. At the interface between the atmosphere and the circulation, the lungs are poised not only for gas exchange, but also as a target of both airway and blood-borne insults. Thus, a variety of prodromes can initiate acute lung injury (ALI) which is consistently manifested and amplified by inflammatory host auto-injury. A key aspect of this pulmonary inflammatory response is mediated by reactive oxygen and nitrogen species. This review will focus on reactive oxygen and nitrogen species chemistry, cellular pathophysiology in relation to ALI, and therapeutic interventions designed to modulate the contribution of these chemicals to pulmonary inflammation and injury.

Sources of reactive oxygen/nitrogen species

There are four main sources of the parent oxyradical species, superoxide anion: mitochondrial bleed, eicosanoid metabolism, the respiratory burst of leukocytes, and xanthine oxidase (Figure 1).

Normally there occurs a 1 to 2% univalent electron bleed from the mitochondrial electron transport chain primarily in the region of flavoprotein NADPH dehydrogenase and the ubiquinone-cytochrome b (2). Hyperoxia, a common clinical scenario, increases this univalent electron bleed production of superoxide anion (3). Other insults that result in increased mitochondrial superoxide anion production may injure mitochondrial DNA, lipids, and proteins despite the abundance of mitochondrial free radical scavengers.

Activation of eicosanoid metabolism, initially through activation of phospholipase A_2, not only enhances production of various reactive lipid peroxides along the lipoxygenase and cyclooxygenase metabolic pathways, but also results in direct production of superoxide anion during the conversion of prostaglandin G to prostaglandin H (4).

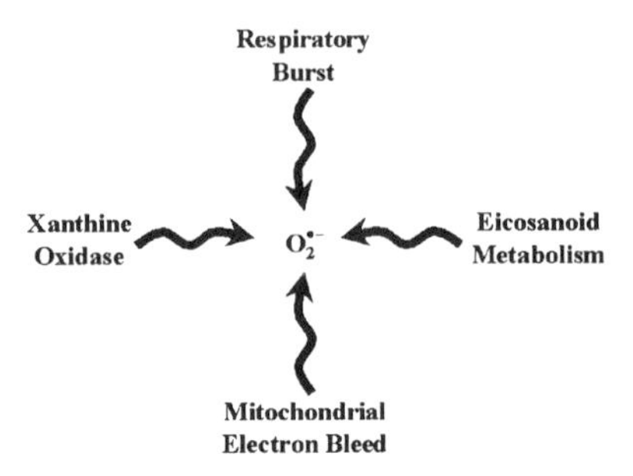

Figure 1. Sources of superoxide anion

Upon activation, various phagocytes may produce a virtual explosion of superoxide anion and metabolites via NADPH oxidase (5, 6). Congenital absence of neutrophil NADPH oxidase is phenotypically expressed as chronic granulomatous disease, the clinical manifestation of absent or inadequate respiratory burst production of superoxide anion. More recently it has become apparent that similar NADPH oxidases are found in various cell types and may function as a universal oxygen sensing mechanism that may be relevant in hypoxic pulmonary vasoconstriction, vascular smooth muscle proliferation, chemo-receptor activation, and erythropoietin gene expression (7).

Xanthine oxidase represents another key mechanism of superoxide production in the setting of ischemia-reperfusion. During ischemia cellular hypoxia results in complete hydrolysis of all high-energy phosphate bonds, leading to accumulation of xanthine and hypoxanthine. Concurrently, xanthine dehydrogenase, which normally utilizes NAD as an electron acceptor during purine catabolism, is converted via proteolytic and oxidative changes to xanthine oxidase, which can now utilize molecular oxygen as the electron acceptor. With reestablishment of perfusion a surfeit of oxygen and purine substrate is now available to xanthine oxidase resulting in a burst of superoxide as well as hydrogen peroxide production (8). Xanthine dehydrogenase can be converted to xanthine oxidase in endothelial cells via the action of neutrophil elastase, a common protease mediator in inflammation (9). Interestingly, normal pulmonary tissue contains no xanthine oxidase or xanthine dehydrogenase (10). However, it is known that in the setting of ischemia-reperfusion (e.g. hypovolemic shock) splanchnic-derived xanthine oxidase may translocate to the lungs and mediate distal oxidant injury (11,12). For example, during cardiac surgery circulating xanthine oxidase concentrations double after release of the aortic cross clamp

(13). Xanthine oxidase-derived oxidants may also promote neutrophil sequestration (14). Finally, xanthine oxidase levels are regulated by alterations in oxygen tension and this effect does not appear to be dependent upon either new mRNA or protein synthesis (15).

Other sources of superoxide production in the lung include cytochrome P_{450}, myoglobin, hemoglobin, and endothelial NADH oxido-reductase (16). Nitric oxide (NO) represents the other primary parent oxyradical species (17). Enhanced NO production, in the setting of ALI, by both alveolar epithelial cells and alveolar macrophages is known to occur as a result of increased production of a variety of inflammatory mediators, which lead to increased activation of transcription factors such as NF-κB, AP-1, and STAT-1, which are known to function synergistically in augmenting transcription of inducible NO synthase (18-20).

Reactive Oxygen and Nitrogen Species Chemistry

Superoxide and NO are capable of undergoing various chemical reactions depending on the relative concentrations of each species (Figure 2). Dismutation of superoxide anion by superoxide dismutase and combination of superoxide anion with NO to form peroxynitrite both occur through reactions characterized by extremely high-rate constants (21, 22, and Chapter 7). Particularly in the lung, the level of oxygen may regulate NO levels as the Michaelis constant for oxygen for constitutive NO synthase would indicate that increased oxygen tension would increase NO production and provide a mechanism for local vasodilation and hence regulation of ventilation-perfusion (23). Reactive nitrogen species generally mediate nitrosative reactions (production of nitrotyrosine) or oxidative reactions (peroxynitrite-mediated oxidation) (24). It is important to note that although NO can combine with superoxide to generate the highly oxidative species peroxynitrite or nitrogen dioxide (25), NO may also attenuate oxidative injury as a free-radical reaction terminator (e.g. NO + lipid peroxide) (26). Nitrate and nitrite are end products of NO metabolism and provide a relative indicator of reactive oxygen/nitrogen species flux.

Myeloperoxidase, a key neutrophil oxidant enzyme produces hypochlorous acid via the reaction of hydrogen peroxide (H_2O_2) with chloride anion (27). Formation of chlorotyrosine represents an important biologic marker of this reaction and hence the role of neutrophil myeloperoxidase in oxidant stress (28). Hypochlorous acid is known to mediate a variety of types of oxidant stress in addition to its role as a key bacteriocidal agent. Hypochlorous acid is particularly adept in terms of sulfhydryl and methionine oxidation, while mediating less DNA and lipid peroxidation as compared to hydrogen peroxide (29). A key role for complexed iron is the production of hydroxyl radical ($HO^{\bullet}$) via the Haber-Wiess reaction. Superoxide-dependent release of free iron from ferritin has been proposed to facilitate this process (30). Hydroxyl radical can initiate lipid peroxidation directly and is also thought to be involved in the activation of phospholipase A_2,which stimulates not only eicosanoid metabolism but, as noted above, additional production of reactive lipid peroxide species.

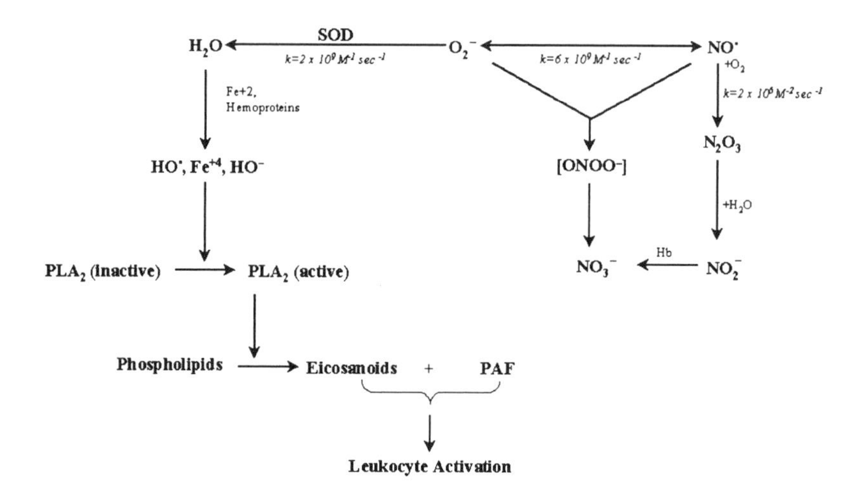

Figure 2. Superoxide and NO reactions (Adapted from: Grisham MB, *Transplant Proc* 1995; 27:2842).

Macromolecular Oxidative Injury

Lipid peroxidation reflects a key aspect of oxidant stress as all cell and organelle membranes are composed of polyunsaturated fatty acids, which represent the targets for lipid oxidant stress. Lipid peroxidation is initiated by abstraction of allylic methylene hydrogens by other lipid peroxide species or hydroxyl radical. As the degree of unsaturation in the fatty acid increases, the number of potential sites of extractable allylic hydrogens increases (31). Following a single initiating event, lipid peroxidation may generate self-sustaining autocatalytic chain reactions.

Important by-products of lipid peroxidation are β-scission-derived aldehydes such as 4-hydroxy-2-nonenal and malondialdehyde. These species are very reactive and significantly extend the oxidative injury potential of lipid peroxidation. For example, β-scission-derived aldehydes are known to induce cell death via changes in mitochondrial permeability (32). Lipid peroxides may themselves initiate both protein and nucleic acid oxidation (33). Finally, oxidatively fragmented phosphatidyl choline species are known to activate human neutrophils through a receptor for platelet activating factor, thus providing an explanation for neutrophil activation in the setting of lipid peroxidation (34).

F_2 isoprostanes are now recognized as a valuable index of lipid peroxidation and oxidative stress in general. These species are formed by non-enzymatic oxidation of arachadonic acid. Because these species are stable in blood, urine, and even exhaled breath, they are valuable indicators of lipid peroxidation reactions *in vivo*. Furthermore, these species are now recognized to exert their own biologic activity including increasing airway resistance and enhancing of plasma exudation, two scenarios common in ALI (35). Similarly, products of linoleic acid oxidation have also been promoted as markers of lipid peroxidation. In particular 9- and 13-hydroxy-octadecadienoic acid (9-, 13-HODE) can be formed by the action of 15-lipoxygenase and cytochrome P_{450}, but the majority of the former species is formed by non-enzymatic lipid oxidation processes (36). As with the F_2 isoprostane, the HODE-related species also exhibit important biologic activity. For example, leukotoxin (9,10-epoxy-12-octadecenoate) has been isolated from lung lavages of rats breathing pure oxygen and from patients with ALI. This lipid peroxidation species can mediate pulmonary neutrophil sequestration and inflammatory lung edema, activate NO synthase, and inhibit mitochondrial respiration (37).

Reactive oxygen and nitrogen species also mediate various oxidative insults towards proteins. For example, human neutrophils and the myeloperoxidase-hydrogen peroxide enzyme system are known to oxidize amino acids to various reactive aldehydes (38). This same enzyme system can also mediate oxidation of adjacent protein tyrosines to generate dityrosine and associated protein cross-links (39). Similar to lipid peroxide-mediated protein oxidation, tyrosyl radicals generated by the action of myeloperoxidase can initiate lipid peroxidation (40). Although virtually any amino acid can be oxidatively modified, cysteine sulfhydryls and methionine sulfur ethers are particularly susceptible to oxidation. Similarly, ion sulfur moieties represent additional common targets of oxidant stress. In this regard, NO concentration and cellular redox status in general are known to regulate endothelial xanthine oxidase (41). Interruption of glycolysis by oxidative inhibition of glyceraldehyde dehydrogenase or interruption of the Kreb's cycle at aconitase are additional well-known examples of oxidant stress on individual proteins. Another example of oxidant-mediated alteration of proteins involves peroxynitrite and surfactant proteins. Peroxynitrite is known to injure pulmonary surfactant by mediating surfactant phospholipid-associated lipid peroxidation, resulting in increased minimal surface tension, and by altering surfactant protein A with the formation of 3-nitrotyrosine (42). Evidence of protein oxidation may be assessed by various metabolites of tyrosine including dityrosine as noted above, nitrotyrosine, and chlorotyrosine. In addition, protein carbonyls also reflect oxidative stress in critically ill patients and the extent of protein carbonyl formation appears to correlate with circulating neutrophil myeloperoxidase (43).

Probably the most sinister macromolecular reaction of oxygen/nitrogen species relates to nucleic acid injury (44). For example, peroxynitrite-mediated DNA strand breakage is known to activate poly ADP-ribose polymerase resulting in cellular energy depletion (45). Oxidative insult to DNA would be expected to be associated with teratogenesis, mutagenesis, carcinogenesis, and altered cellular repair following an oxidative insult (44). Interestingly, oxidant stress secondary to TNFα, or hyperoxia also increases the transcription of antioxidant enzymes such as manganese superoxide dismutase. Similarly, hyperoxia in the lung increases the expression of messenger RNA for surfactant proteins A and C (46). Specific nucleic acid oxidation products (e.g. 8-hydroxyguanine) can be utilized to assess oxidative insult towards DNA (47). Either gas chromatography/mass spectroscopy, or ELISA can be utilized to quantitate 8-hydroxyguanine.

Cellular Oxidant-Mediated Alterations

The effects of reactive oxygen/nitrogen species on pulmonary epithelial and endothelial cell function have been studied extensively. Reactive oxygen/nitrogen species stress is known to increase cell membrane fluidity and protein/phospholipid disorganization resulting in increased endothelial permeability (48,49). Ion transport pathways across the cellular membrane are altered by oxidant stress, hence affecting signal transduction. Moreover, excessive calcium influx following oxidant cell membrane injury may overwhelm intracellular calcium binding defenses and initiate cell necrosis (50). Excess peroxynitrite stress can inhibit mitochondrial electron transport directly (51). Endogenous xanthine oxidase-derived oxygen metabolites have been demonstrated to inhibit surfactant production by inhibiting incorporation of palmitic acid into phosphatidyl choline (52). Similarly, non-lethal insults by hydrogen peroxide towards type II alveolar cells results in decreased phosphatidyl choline synthesis secondary to inhibition of glycerol-3-phosphate acyl transferase (as an oxidative injury of critical enzyme sulfhydryls) (53). Pulmonary endothelial and epithelial cells may respond differently to the lethal effects of oxidants. Whereas alveolar type II cells are sensitive to both NO- and peroxynitrite-mediated oxidative injury, endothelial cells appear to be sensitive to peroxynitrite, and much less sensitive to NO. In this setting exogenous N-acetyl-cysteine appears to be protective (54).

The classic pulmonary oxidative stress directly related to ALI involves oxidative inactivation of alpha-1-antiprotease by myleoperoxidase (55). Oxidative fragmentation of this key antiprotease at a critical methionine residue produces an altered protein now incapable of binding to the active site of neutrophil elastase. This fosters a local environment characterized by not only oxidant/anti-oxidant imbalance, but also protease and antiprotease imbalance.

Lipid peroxidation has been known to influence β-adrenergic receptor density (56) and superoxide and hydrogen peroxide concentration are known to affect intracellular redox balance and alterations in cell proliferation (57). Similarly, gene expression in general is sensitive to intracellular redox alteration and in particular intracellular thiols (glutathione, thioredoxin) (58,59). Transmembrane alterations in cellular redox potential are known to activate NF-κB in macrophages. In turn NF-κB regulates the transcription of a variety of inflammatory, immune, acute phase response, and apototic genes (60). It is known that hydrogen peroxide generated from phagocytic respiratory burst can also activate this system. Similarly 4-hydroxy-2-nonenal can activate the transcription factor AP-1. Excessive NO generated from enhanced expression of inducible NO synthase can provide a negative feedback servo mechanism for inhibition of NF-κB via nitrosylation of a critical cysteine (61). Another key action related to alterations in intracellular redox status relates to peroxynitrite-mediated nitration of tyrosine residues, for example, in proteins with activity affected by phosphorylation (e.g. kinases). Peroxynitrite may facilitate nitration of the carbon adjacent to the tyrosyl hydroxyl group, with the bulky nitro group now preventing phosphorylation, hence directly interfering with this mechanism of signal transduction (62).

Although many of the metabolic affects of enhanced reactive oxygen/nitrogen species flux can alter cellular metabolism without affecting cell viability, massive oxyradical insult may result in cellular necrosis, while insults of lesser magnitude are capable of inducing apoptosis. For example, oxidants such as NO may enhance expression of the proapoptotic, tumor suppressor gene, P_{53} (63). It is also well established that various antioxidants can suppress apoptosis (64). Following an apoptotic signal, there ensues a progressive accumulation of intracellular lipid peroxide. It is known that if Bcl-2 is overexpressed, a resultant decrease in intracellular lipid peroxides occurs and apoptosis is averted. An alternate effect of excessive reactive oxygen/nitrogen species stress relates to activation of nuclear proteosome (by poly-ADP ribose polymerase) to degrade oxidatively damaged histones. Excessive ADP ribosylation may inadvertently lead to necrotic cell death secondary to ATP depletion (65). On the other hand, this same oxidant stress may result in the induction of heat shock proteins, which exert protective effects for mitochondria (66).

Hyperoxic Lung Injury

As oxidative stress in simple terms is probably best expressed as the product of magnitude of oxidant insult and duration of insult, increases in fractional inspired oxygen commonly used in the hospital setting would be expected to cause or exacerbate ALI. In a model of hyperoxia in adult baboons, 100% oxygen for 5-7 days caused ALI, including bilateral alveolar infiltrates, severe hypoxemia, diffuse alveolar damage, decreased total lung capacitance, decreased static compliance, and increased bronchoalveolar lavage (BAL) protein, neutrophils, and elastase (67). An additional aspect of hyperoxic lung injury involves reduced clearance/recycling of surfactant. For example, in hyperoxia-exposed animals, radiolabeled phosphatidylcholine clearance was reduced in excess of four-fold following 24 hours of 100% oxygen. Type II alveolar pneumocytes isolated from such oxygen-exposed animals exhibited decreased phosphatidylcholine synthesis (68).

Hyperoxia can induce fibroblasts to synthesize and release transforming growth factor-β, which has a role in both epithelial and mesenchymal hyperplasia commonly seen in the latter phases of ALI (69,70). Leukotoxin, an epoxide metabolite of linoleic acid has been found in increased concentrations in rats exposed to 100% oxygen. As noted earlier this metabolite mediates smooth muscle contraction as well as uncoupling of mitochondrial respiration (71).

Neonatal Acute Lung Injury

Although bronchopulmonary dysplasia (BPD) is generally considered a chronic lung disease, it has its origins during the first week of life. Previously BPD was attributed to alveolar-type II cell immaturity, and hence pulmonary surfactant deficiency. However, since the introduction of exogenous pulmonary surfactant, the incidence of BPD has not decreased significantly. Rather, infants of increasing prematurity are now routinely rescued even at the cannalicular stage of lung development. At the clinical level, premature infants are subject to the "triple threat" of pulmonary immaturity (both structural and functional): iatrogenic injury from ventilator-associated events, oxygen toxicity, and inflammatory host autoinjury not unlike that seen in ALI in older patients. Moreover, with specific reference to redox aspects of the inflammatory armamentarium, the premature infant is subject to increased flux of reactive oxygen/nitrogen species stress in a setting where the infant's antioxidant repertoire is undeveloped and the ability to up regulate antioxidant defenses is also impaired (72-74).

Infants with classic respiratory distress syndrome or extreme prematurity can be identified for their propensity to develop BPD by enhanced concentrations of alveolar lining fluid xanthine oxidase, myleoperoxidase, active elastase, and a predominance of neutrophils (75). Protein carbonyls assayed from endotracheal tube aspirates are early indicators of tendency to develop chronic lung disease, and length of mechanical ventilation appears to correlate with extent of protein carbonylation (76,77). Other indicators of the involvement of oxidant stress in neonatal ALI predisposing to chronic lung disease include, elevated 8-epi-isoprostane $F_2\alpha$ (78), increased exhalation of the lipid peroxidation breakdown products ethane and pentane (79), and increased concentration of aldehyde lipid peroxidation by-products, malondialdyde, nonenal, heptanal and 4-hydroxy nonenal, (80,81).

Acute Respiratory Distress Syndrome/Acute Lung Injury

Multiple aspects of ALI pathophysiology including pulmonary vasoreactivity, bronchoconstriction, increased mucus plugging, pulmonary capillary leak, diaphragmatic dysfunction, decreased airway ciliary activity, and pulmonary surfactant inactivation all appear to have some origins related to enhanced reactive oxygen/nitrogen species flux. A variety of compelling data demonstrate the importance of reactive oxygen and nitrogen species in the pathophysiology of ALI including: 1) decreased plasma vitamin E and polyunsaturated fatty acids associated with increased lipid peroxidation products (82); 2) decreased plasma antioxidants ascorbate and ubiquinol 10 levels associated with nM levels of lipid peroxides in BAL fluid (83); 3) widespread staining for nitrotyrosine in the lung autopsy specimens from patients with ALI (84); 4) enhanced hydroxylation, nitration, and chlorination of BAL fluid proteins (85, 86, and Table 1); 5) decreased plasma linoleic acid associated with increased plasma 4-hydroxy-2-nonenal (87); 6) increased expired hydrogen peroxide and isoprostanes (88, 89); and 7) altered alveolar lining fluid glutathione with decreases in total glutathione and increases in oxidized glutathione (90, 91). In addition, antioxidant activity of BAL fluid for certain components (transferrin, ceruloplasmin) is actually increased in patients with ALI as a result of alveolar capillary membrane leak (92). Finally, patients with ALI have decreased exhaled NO levels compared to control patients, leading to speculation that this reflects enhanced formation of nitrotyrosine (93).

Table 1. Tyrosine metabolites (nmol/mg) in BAL protein from controls and patients with ALI (Ref. 92)

Metabolite	Control Patients	ALI Patients
Ortho-tyrosine	0.67 ± 0.67	7.98 ± 3.78
Chloro-tyrosine	1.55 ± 1.34	4.82 ± 1.07
Nitro-tyrosine	0.29 ± 0.29	2.21 ± 0.65

Correlation analysis of Chloro-tyr vs myeloperoxidase, $r = 0.67$, $p<0.01$

Antioxidant Therapy in ALI

Numerous supportive measures commonly utilized in the setting of ALI would be expected to have a beneficial therapeutic effect with respect to reactive oxygen/nitrogen species injury (94). For example, minimizing oxygen debt by avoiding low flow and ischemia-reperfusion events and by insuring adequate oxygen delivery would be expected to reduce nitro/oxyradical flux from all sources. Whereas transfusion of erythrocytes enhances oxygen delivery linearly with hemoglobin levels, such transfusions also augment enzymatic oxidant-scavenging systems in terms of superoxide dismutase, catalase, and glutathione peroxidase. Similarly, infusions of colloid solutions not only aid in expanding intravascular volume, but also provide a variety of free radical scavengers commonly found in plasma. β-adrenergic agonists and phosphodiesterase inhibitors have a number of direct anti-inflammatory effects. In addition, epinephrine has been demonstrated to enhance transcription of superoxide dismutase (95). While optimal nutritional support promotes healing and growth and provides a variety of antioxidants including β-carotene, vitamin E, vitamin C, and selenium, it should be realized that parenteral nutrition also represents a source of oxyradical stress in terms of lipid peroxides that are contained in lipid emulsions, as well as various free radical species that can be generated from the interaction of riboflavin and light (96).

Utilization of lung protective strategies including avoiding over-distension and repeated alveolar collapse and re-expansion are mechanical ventilation maneuvers that will reduce lung inflammation and associated nitro/oxyradical flux directly. Minimizing inhaled oxygen concentration will reduce univalent electron bleed from the mitochondrial respiratory transport chain and formation of superoxide anion (97). Similarly, tolerating some degree of respiratory acidosis will not only reduce mechanical lung injury, but there is now increasing evidence that therapeutic hypercapnia may actually mediate an inflammatory effect (98).

As glutathione represents a key non-enzymatic antioxidant found in high concentrations in alveolar lining fluid, and since clinical investigations have ascertained a decrease in total alveolar glutathione and increase oxidized glutathione in the setting of ALI, interventional trials to replete lung sulfhydryls seem appropriate. A randomized controlled trial examining the efficacy of N-acetylcysteine in ALI demonstrated a trend towards improved lung compliance that was not statistically significant, but no changes in chest radiograph or survival (99). Moreover, an anticoagulant effect and depressed cardiac performance were noted in patients receiving N-acetylcysteine. This clinical study was disappointing given the fact that N-acetylcysteine has documented benefit in animal models of ALI. More recently, combination therapy with N-acetylcysteine and a flavenoid iron chelator, rutin, was associated with a decreased length of stay in the intensive care unit but no changes in mortality. Relevant surrogate markers for this study included diminished expired ethane and malondialdehyde as well as increased alveolar lining fluid glutathione in treated patients (100).

Animal models have indicated a potential beneficial effect of inhaled NO in hyperoxia-induced ALI. For example, survival of adult rats at 120 hours in 95-100% oxygen was only 8.3% compared to a survival of 70% in rats similarly exposed to 95-100% oxygen plus 100 ppm NO (101). In this setting the beneficial effects of NO may include: reaction with the coordination site of iron to limit Fenton chemistry, stimulation of guanalate cyclase resulting in inhibition of neutrophils and platelets, termination of radical chain reactions, diversion of superoxide metabolism, and alteration of transcription (102). As noted previously, NO may reduce availability of NF-κB for binding to regulatory regions of proinflammatory cytokine genes (103), and NO may inhibit neutrophil migration and oxidative activity by interfering with F-actin assembly and C18 up-regulation (104). Both pediatric (105,106) and adult (107,108) clinical investigations have indicated at least short-term improvements in oxygenation in ALI patients treated with inhaled NO. No long-term beneficial effects have been demonstrated. One investigation ascertained decreased spontaneous neutrophil superoxide production and decreased β-integrin expression associated with decreased levels of IL-8 and IL-6 (109). Another investigation indicated no change in BAL levels of lipid peroxides, elastase, myeloperoxidase, IL-8, or leukotriene B_4 in patients treated with NO up to 40 ppm (110). Similarly, arterial levels of malondialdehyde, hexanal, and pentanal were shown to exceed mixed venous levels by some 2 to 10-fold in ALI patients and this transpulmonary gradient was not affected by NO up to 40 ppm (111).

Provision of antioxidant enzymes, for example Cu/Zn superoxide dismutase encapsulated in pulmonary surfactant, has been demonstrated to mitigate hyperoxic lung injury in premature rabbits (112). An initial clinical trial in 45 premature infants examined subcutaneous dosing of bovine superoxide dismutase. For the 14 surviving patients treated with superoxide dismutase the incidence of BPD was 21%, while for the 17 surviving placebo-treated infants the incidence of BPD was 70% ($p = 0.008$) (113). However, a subsequent large study (301 patients) involving intratracheal recombinant human Cu/Zn superoxide dismutase showed no benefit (114).

Finally, at the threshold of the gene therapy era, exogenous administration of heme oxygenase-1 by gene transfer was shown to provide protection against hyperoxia-induced lung injury. The heme oxygenase gene was delivered by adenoviral vector and resulted in decreased pulmonary neutrophil leukosequestration and reduced lung epithelial apoptosis in this animal investigation (115). Similarly, adenovirus-mediated transfer of cDNAs for superoxide dismutase and catalase was shown to protect animals against hyperoxia-mediated lung injury, but not against ischemia-reperfusion lung injury (116). Other potentially beneficial pulmonary-targeted gene therapy interventions related to reactive oxygen/nitrogen species associated ALI might include enhanced expression of alpha-1-antiprotease, IL-10, heat shock proteins, lipocortin and glutathione peroxidase.

REFERENCES

1. Fridovich, I. (1978) The biology of oxygen radicals. *Science* 201, 875-880
2. Kowaltowski, A.J., and Vercesi, A.E. (1999) Mitochondrial damage induced by conditions of oxidative stress. *Free Radic Biol Med* 26, 463-471
3. Freeman, B.A., and Crapo, J.D. (1981) Hyperoxia increases oxygen radical production in rat lungs and rat mitochondria. *J Biol Chem* 256, 10986-10992
4. Kukreja, R.C., Kontos, H.A., Hess, M.L., and Ellis, E.F. (1986) PGH synthase and lipoxygenase generate superoxide in the presence of NADH or NADPH. *Circ Res* 59, 612-619
5. Chanock, S.J., Benna, J.E., Smith, R.M., and Babior, B.B. (1994) The respiratory burst oxidase. *J Biol Chem* 269, 24519-24522
6. Miller, R.A., Britigan, B.E. (1995) The formation and biologic significance of phagocyte-derived oxidants. *J Invest Med* 43, 39-49
7. Jones, R.D., Hancock, J.T., and Morice, A.H. (2000) NADPH oxidase: A universal oxygen sensor? *Free Radic Biol Med* 29, 416-424
8. Harris, C.M., and Massey, V. (1997) The reaction of reduced xanthine dehydrogenase with molecular oxygen. *J Biol Chem* 272, 8370-8379
9. Phan, S.H., Gannon, D.E., Ward, P.A., and Karmiol, S. (1992) Mechanism of neutrophil-induced xanthine dehydrogenase to xanthine oxidase in endothelial cells. Evidence of a role for elastase. *Am J Respir Cell Mol Biol* 6, 270-278.
10. Linder, N., Rapola, J., and Raivio, K.O. (1999) Cellular expression of xanthine oxidoreductase protein in normal human tissues. *Lab Invest* 79, 967-974
11. Tan, S., Yokoyama, Y., Dickens E., Cash, T.G., Freeman, B.A., and Parks, D.A. (1993) Xanthine oxidase activity in the circulation of rats following hemorrhagic shock. *Free Radic Biol Med* 15, 407-414
12. Nielsen, V.G., Tan, S., Weinbroum, A., McCammon, A.T., Samuelson, P.N., Gelman, S., and Parks, D.A.. (1996) Lung injury after hepatoenteric ischemia-reperfusion: Role of xanthine oxidase. *Am J Respir Crit Care Med* 154, 1364-1369
13. Tan, S., Gelman, S., Wheat, J.K., and Parks, D.A. (1995) Circulating xanthine oxidase in human ischemia reperfusion. *South Med J* 88, 479-482
14. Anderson, B.O., Moore, E.E., Moore, F.A., Left, J.A., Terada, L.S., Harken, A.H., and Repine, J.E. (1991) Hypovolemic shock promotes neutrophil sequestration in lungs by a xanthine oxidase-related mechanism. *J Appl Physiol* 71, 1862-1865
15. Poss, W.B., Hueckstеadt, T.P., Panus, P.C., Freeman, B.A., and Hoidal, J.R. (1996) Regulation of xanthine oxidase by hypoxia. *Am J Physiol* 270, L941-L946
16. Cross, A.R., and Jones, O.T.G. (1991) Enzymatic mechanisms of superoxide production. *Biochem Biophys Acta* 1057, 281-298
17. Moncada, S., and Higgs, A. (1993) The L-arginine-nitric oxide pathway. *N Engl J Med* 329, 2002-2012
18. Gutierrez, H.H., Pitt, B.R., Schwarz, M., Watkins, S.C., Lowenstein, C., Caniggia, I., Chumley, P., and Freeman, B.A. (1995) Pulmonary alveolar epithelial inducible NO synthase gene expression: regulation by inflammatory mediators. *Am J Physiol* 268, L501-L508
19. Tracey, W.R., Xue, C., Klinghofer, V., Barlow, J., Pollock, J.S., Fostermann, U., and Johns, R.A. (1994) Immunochemical detection of inducible NO synthase in human lung. *Am J Physiol* 266, L722-L727
20. Taylor, B.S., and Geller, D.A. (2000) Molecular regulation of the human inducible nitric oxide synthase (iNOS) gene. *Shock* 13, 413-424
21. Pryor, W.A., and Squadrito, G.L. (1995) The chemistry of peroxynitrite: a product from the reaction of nitric oxide with superoxide. *Am J Physiol* 268, L699-L722
22. Beckman, J.S., and Koppenol, W.H. (1996) Nitric oxide, superoxide and peroxynitrite: the good, the bad and the ugly. *Am J Physiol* 271, C1424-C1437

23. Dweik, R.A., Laskowski, D., Abu-Soud, H.M., Kaneko, F.T., Hutte, R., Stuehr, D.J., and Erzurum, S.C. (1998) Nitric oxide synthesis in the lung. Regulation by oxygen through a kinetic mechanism. *J Clin Invest* 101, 660-666
24. Wink, D.A., and Mitchell, J.B. (1998) Chemical biology of nitric oxide: Insights into regulatory, cytotoxic, and cytoprotective mechanisms of nitric oxide. *Free Radic Biol Med* 25, 434-456
25. Mohsenin, V. (1991) Lipid peroxidation and anti-elastase activity in the lung under oxidant stress: role of antioxidant defenses. *J Appl Physiol* 70, 1456-1462
26. Rubbo, H., Radi, R., Trujillo, M., Telleri, R., Kalyanaraman, B., Barnes, S., Kirk, M., and Freeman, B.A. (1994) Nitric oxide regulation of superoxide and peroxynitrite-dependent lipid peroxidation. *J Biol Chem* 269, 26066-26075
27. Kettle, A.J., and Winterbourn, C.C. (1997) Myeloperoxidase: a key regulator of neutrophil oxidant production. *Redox Rep* 3, 3-15
28. Winterbourn, C.C., and Kettle, A.J. (2000) Biomarkers of myeloperoxidase-derived hypochlorous acid. *Free Radic Biol Med* 29, 403-409
29. Schraufstatter, I.U., Browne, K., Harris, A., Hyslop, P.A., Jackson, J.H., Quehenberger, O., and Cochrane, C.G. (1990) Mechanisms of hypochlorite injury of target cells. *J Clin Invest* 85, 554-562
30. Thomas, C.E., Morehouse, Z.A., and Aust, S.D. (1985) Ferratin and superoxide-dependent lipid peroxidation. *J Biol Chem* 260, 3275-3280
31. Gardner, H.W. (1989) Oxygen radical chemistry of polyunsaturated fatty acids. *Free Radic Biol Med* 7, 65-86
32. Esterbauer, H., Schaur, R.J., and Zollner, H. (1991) Chemistry and biochemistry of 4-hydroxynonenal, malonaldehyde and related aldehydes. *Free Radic Biol Med* 11, 81-128
33. Itabe, H., Yamamoto, H., Suzuki, M., Kawai, Y., Nakagawa, Y., Suzuki, A., Imanaka, T., and Takano, T. (1996) Oxidized phosphatidylcholines that modify proteins. *J Biol Chem* 271, 33208-33217
34. Smiley, P.L., Stremler, K.E., Prescott, S.M., Zimmerman, G.A., and McIntyre, T.M. (1999) Oxidatively fragmented phosphatidylcholines activate human neutrophils through the receptor for platelet-activating factor. *J Biol Chem* 266, 11104-11110
35. Roberts, L.J., and Morrow, J.D. (2000) Measurement of F_2-isoprostanes as an index of oxidative stress *in vivo*. *Free Radic Biol Med* 28, 505-513
36. Kaduce, T.L., Figard, P.H., Leifur, R., and Spector, A.A. (1989) Formation of 9-hydroxy-octadecadienoic acid from linoleic acid in endothelial cells. *J Biol Chem* 264, 6823-6830
37. Ozawa, T., Sugiyama, S., Hayakawa, M., Satake, T., Taki, F., Iwata, M., and Taki, K. (1988) Existence of leukotoxin 9, 10-epoxy-12-octadecanoate in lung lavages from rats breathing pure oxygen and from patients with the adult respiratory distress syndrome. *Am Rev Respir Dis* 137, 535-540
38. Hazen, S.L., d'Avignon, A., Anderson, M.M., Hsu, F.F., and Heinecke, J.W. (1998) Human neutrophils employ the myeloperoxidase-hydrogen peroxide-chloride system to oxidize amino acids to a family of reactive aldehydes. *J Biol Chem* 273, 4997-5005
39. Jacob, J.S., Cistola, D.P., Hsu, F.F., Muzaffar, S., Mueller, D.M., Hazen, S.L., and Heinecke, J.W. (1996) Human phagocytes employ the myeloperoxidase-hydrogen peroxide system to synthesize dityrosine, trityrosine, pulcherosine, and isodityrosine by a tyrosyl radical-dependent pathway. *J Biol Chem* 271, 19950-19956
40. Savenkova, M.I., Mueller, D.M., and Heineke, J.W. (1994) Tyrosyl radical generated by myeloperoxidase is a physiological catalyst for the initiation of lipid peroxidation in low-density lipoprotein. *J Biol Chem* 269, 20394-20400
41. Hassoun, P.M., Yu, F.S., Zulueta, J.J., White, A.C., and Lanzillo, J.J. (1995) Effect of nitric oxide and cell redox status on the regulation of endothelial cell xanthine oxidase. *Am J Physiol* 268, L809-L817
42. Haddad, I.Y., Ischiropoulos, H., Holm, B.A., Beckman, J.S., Baker, J.R., and Matalon, S. (1993) Mechanisms of peroxynitrite-induced injury to pulmonary surfactants. *Am J Physiol* 265, L555-L564

43. Winterbourn, C.C., Buss, H., Chan, T.P., Plank, L.D., Clark, M.A., and Windsor, J.A. (2000) Protein carbonyl measurements show evidence of early oxidative stress in critically ill patients. *Crit Care Med* 28, 143-149
44. Beckman, K.B., and Ames, B.N. (1997) Oxidative decay of DNA. *J Biol Chem* 272, 19633-19636
45. Zingarelli, B., O'Connor, M., Wong, H., Salzman, A.L., and Szabo, C. (1996) Peroxynitrite-mediated DNA strand breakage activates poly-adenosine diphosphate ribosyl synthetase and causes cellular energy depletion in macrophages stimulated with bacterial lipopolysaccharide. *J Immunol* 156, 350-358
46. Jackson, R.M., Parish, G., and Helton, E.S. (1998) Peroxynitrite modulates MnSOD gene expression in lung epithelial cells. *Free Radic Biol Med* 25, 463-478
47. Breen, A.P., and Murphy, J.A. (1995) Reactions of oxyl radicals with DNA. *Free Radic Biol Med* 18, 1033-1077
48. Freeman, B.A., Rosen, G.M., and Barber, M.J. (1985) Superoxide perturbation of the organization of vascular endothelial cell membranes. *J Biol Chem* 261, 6590-6593
49. Ochoa, L., Waypa, G., Mahoney, J.R., and Rodriguez, L., (1997) Minnear, F.L. Contrasting effects of hypochlorous acid and hydrogen peroxide on endothelial permeability. *Am J Respir Crit Care Med* 156, 1247-1255
50. Elliott, S.J., and Koliwad, S.K. (1995) Oxidant stress and endothelial membrane transport. *Free Radic Biol Med* 19, 649-658
51. Radi, R., Rodriguez, M., Castro, L., and Telleri, R. (1994) Inhibition of mitochondrial electron transport by peroxynitrite. *Arch Biochem Biophys* 308, 89-95
52. Baker, R.R., Panus, P.C., Holm, B.A., Engstrom, P.C., Freeman, B.A., and Matalon, S. (1990) Endogenous xanthine oxidase-derived O_2 metabolites inhibit surfactant metabolism. *Am J Physiol* 259, L328-L334
53. Holm, B.A., Hudak, B.B., Keicher, L., Cavanaugh, C., Baker, R.R., Hu, P., and Matalon, S. (1991) Mechanisms of H_2O_2-mediated injury to type II cell surfactant metabolism and protection with PEG-catalase. *Am J Physiol* 261, C751-C757
54. Gow, A.J., Thom, S.R., and Ischiropoulos, H. (1998) Nitric oxide and peroxynitrite-mediated pulmonary cell death. *Am J Physiol* 274, L112-L118
55. Ossanna, P.J., Test, S.T., Matheson, N.R., Regiani, S., and Weiss, S.J. (1986) Oxidative regulation of neutrophil elastase – alpha-1-proteinase inhibitor interactions. *J Clin Invest* 77, 1939-1951
56. Kramer, K., Rademaker, B., and Rozendal, W. (1986) Influence of lipid peroxidation on β-adrenoreceptors. *FEBS Lett* 198, 80-84
57. Burdon, R.H. (1995) Superoxide and hydrogen peroxide in relation to mammalian cell proliferation. *Free Radic Biol Med* 18, 775-794
58. Arrigo, A-P. (1999) Gene expression and the thiol redox state. *Free Radic Biol Med* 27, 936-944
59. Sen, C.K., and Packer, L. (1996) Antioxidant and redox regulation of gene transcription. *FASEB* 10, 709-720
60. Kaul, N., Choi, J., and Forman, H.J. (1998) Transmembrane redox signaling activates NF-κB in macrophages. *Free Radic Biol Med* 24, 202-207
61. Janssen-Heininger, Y.M.W., Poynter, M.E., and Baeuerle, P.A. (2000) Recent advances towards understanding redox mechanisms in the activation of nuclear factor κB. *Free Radic Biol Med* 28, 1317-1327
62. Berlett, B.S., Friguet, B., Yim, M.B., Chock, P.B., and Stadtman, E.R. (1996) Peroxynitrite-mediated nitration of tyrosine residues in *Escherichia coli* glutamine synthetase mimics adenylation relevance to signal transduction. *Proc Natl Acad Sci USA* 93, 1776-1780
63. Messmer, U.K., Ankarcrona, M., Nicotera, P., and Brune, B. (1994) P_{53} expression in nitric oxide-induced apoptosis. *FEBS Letters* 355, 23-26
64. Hockenberry, D.M., Oltvai, Z.N., Yin, X-M., Milliman, C.L., and Korsmeyer, S.J. (1993) Bcl-2 functions in an antioxidant pathway to prevent apoptosis. *Cell* 75, 241-245

65. Ullrich, O., Reinheckel, T., Sitte, N., Hass, R., Grune, T., and Davies, K.J.A. (1999) Poly-ADP ribose polymerase activates nuclear proteasome to degrade oxidatively damaged histones. *Proc Natl Acad Sci USA* 96, 6223-6228
66. Polla, B.S., Kantengwa, S., Francious, D., Salvioli, S., Franceschi, C., Marsac, C., and Cossarizza, A. (1996) Mitochondria are selective targets for the protective effects of heat shock against oxidative injury. *Proc Natl Acad Sci USA* 93, 6458-6463
67. DeLosSantos, R., Seidenfeld, J.J., Anzueto, A., Collins, J.F., Coalson, J.J., Johanson, W.G., and Peters JI. (1987) One hundred percent oxygen lung injury in adult baboons. *Am Rev Respir Dis* 136, 657-661
68. Novotny, W.E., Hudak, B.B., Matalon, S., and Holm, B.A. (1995) Hyperoxic lung injury reduces exogenous surfactant clearance *in vivo*. *Am J Respir Crit Care Med* 151, 1843-1847
69. Vivekananda, J., Lin, A., Coalson, J., and King, R.J. (1994) Acute inflammatory injury in the lung precipitated by oxidant stress induces fibroblasts to synthesize and release transforming growth factor-β. *J Biol Chem* 269, 25057-25061
70. Leonarduzzi, G., Scavazza, A., Biasi, F., Chiarpetto, E., Camandola, S., Vogl, S., Dargel, R., and Poli, G. (1997) The lipid peroxidation end product 4-hydroxy-2, 3-nonenal up-regulates transforming growth factor β1 expression in the macrophage lineage: a link between oxidative injury and fibrosclerosis. *FASEB J* 11, 851-857
71. Ozawa, T., Hayakawa, M., Takamura, T., Sugiyama, S., Suzuki, K., Iwata, M., Taki, F., and Tomita, T. (1986) Biosynthesis of leukotoxin, 9, 10-epoxy-12 octadecanoate, by leukocytes in lung lavages of rat after exposure to hyperoxia. *Biochem Biophys Res Commun* 134, 1071-1078
72. Zimmerman, J.J. (1995) Bronchoalveolar inflammatory pathophysiology of bronchopulmonary dysplasia. *Clin Perinatol* 22, 429-456
73. Saugstad, O.D. (1996) Mechanisms of tissue injury by oxygen radicals: implications for neonatal disease. *Acta Pediatr* 85, 1-4
74. Speer, C.P., and Groneck, P. (1998) Oxygen radicals, cytokines, adhesion molecules and lung injury in neonates. *Semin Neonatol* 3, 219-228
75. Contreras, M., Hariharan, N., Lewandoski, J.R., Ciesielski, W., Koscik, R., and Zimmerman, J.J. (1996) Bronchoalveolar oxyradical inflammatory elements herald bronchopulmonary dysplasia. *Crit Care Med* 24, 29-37
76. Varsila, E., Personen, E., and Andersson, S. (1995) Early lung protein oxidation in the neonatal lung is related to development of chronic lung disease. *Acta Paediatr* 84, 1296-1299
77. Gladstone, I.M., and Levine, R.L. (1994) Oxidation of proteins in the neonatal lung. *Pediatrics* 93, 764-768
78. Goil, S., Troug, W.E., Barnes, C., Norberg, M., Rezaiekbaligh, M., and Thibeault, D. (1998) Eight-epi $PGF_2\alpha$: A possible marker of lipid peroxidation in term infants with severe pulmonary disease. *J Pediatr* 132, 349-351
79. Pitkanen, O.M., and Hallman, M. (1998) Evidence for increased oxidative stress in preterm infants eventually developing chronic lung disease. *Semin Neonatol* 3, 199-205
80. Inder, T.E., Darlow, B.A., Sluis, K.B., Winterbourn, C.C., Graham, P., Sanderson, K.J., and Taylor, B.J. (1996) The correlation of elevated levels of an index of lipid peroxidation (MDA-TBA) with adverse outcomes in the very low birthweight infant. *Acta Pediatr* 85, 1116-1122
81. Ogihara, T., Hirano, K., Morinobu, T., Kim, H-S., Hiroi, M., Ogihara, H., and Tamai, H. (1999) Raised concentrations of aldehyde lipid peroxidation products in premature infants with chronic lung disease. *Arch Dis Child Fetal Neonatal Ed* 80, F21-F25
82. Richard, C., Lemonnier, F., Thibault, M., Couturier, M., and Auzepy, P. (1990) Vitamin E deficiency and lipoperoxidation during adult respiratory distress syndrome. *Crit Care Med* 18, 4-9
83. Cross, C.E., Forte, T., Stocker, R., Louie, S., Yamamato, Y., Ames, B., and Frei, B.(1990) Oxidative Stress and abnormal cholesterol metabolism in patients with adult respiratory distress syndrome. *J Lab Clin Med* 115, 396-404

84. Kooy, N.W., Royall, J.A., Ye, Y.Z., Kelly, D.R., and Beckman, J.S. (1995) Evidence for *in vivo* peroxy-nitrite production in human acute lung injury. *Am J Respir Crit Care Med* 151, 1250-1254
85. Lamb, N.J., Quinlan, G.J., Westerman, S.T., Gutteridge, J.M.C., and Evans, T.W. (1999) Nitration of proteins in bronchoalveolar lavage fluid from patients with acute respiratory distress syndrome receiving inhaled nitric oxide. *Am J Respir Crit Care Med* 160, 1031-1034
86. Lamb, N.J., Gutteridge, J.M.C., Baker, C., Evans, T.W., and Quinlan, G.J. (1999) Oxidative damage to proteins of bronchoalveolar lavage fluid in patients with acute respiratory distress syndrome: Evidence for neutrophil-mediated hydroxylation, nitration and chlorination. *Crit Care Med* 27, 1738-1744
87. Quinlan, G.J., Lamb, N.J., Evans, T.W., and Gutteridge, J.M.C. (1996) Plasma fatty acid changes and increased lipid peroxidation in patients with adult respiratory distress syndrome. *Crit Care Med* 24, 241-246
88. Sznajder, J.I., Fraiman, A., Hall, J.E., Sanders, W., Schmidt, G., Crawford, G., Nahum, A., Factor, P., and Wood, L.D.H. (1989) Increased hydrogen peroxide in the expired breath of patients with acute hypoxemic respiratory failure. *Chest* 96, 606-612
89. Carpenter, C.T., Price, P.V., and Christman, B.W. (1998) Exhaled breath condensate isoprostanes are elevated in patients with acute lung injury or ARDS. *Chest* 114, 1653-1659
90. Pacht, E.R., Timerman, A.P., Lykens, M.G., and Merola, M.G. (1991) Deficiency of alveolar fluid glutathione in patients with sepsis and the adult respiratory distress syndrome. *Chest* 100, 1397-1403
91. Bunnell, E., and Pocht, E.R. (1993) Oxidized glutathione is increased in the alveolar fluid of patients with the adult respiratory distress syndrome. *Am Rev Respir Dis* 148, 1174-1178
92. Lykens, M.G., Davis, W.B., and Pacht, E.R. (1992) Antioxidant activity of broncho-alveolar lavage fluid in the adult respiratory distress syndrome. *Am J Physiol* 262, L169-L175
93. Brett, S.J., and Evans, T.W. (1993) Measurement of endogenous nitric oxide in the lungs of patients with the acute respiratory distress syndrome. *Am Rev Respir Dis* 148, 955-60
94. Youn, Y.K., LaLonde, C., and Demling, R. (1991) Use of antioxidant therapy in shock and trauma. *Circ Shock* 35, 245-249
95. Mehta, J.L., and Dayuan, L. (2001) Epinephrine up-regulates superoxide dismutase in human coronary artery endothelial cells. *Free Radic Biol Med* 30, 148-153
96. Laborie, S., Lavoie, J-C., and Chessex, P. (2000) Increased urinary peroxides in newborn infants receiving parenteral nutrition exposed to light. *J Pediatr* 16, 628-32
97. Lee, W.L., Detsky, A.S., and Stewart, T.E. (2000) Lung-protective strategies in ARDS. *Intens Care Med* 26, 1151-1155
98. Laffey, J.G., Tanaka, M., Engelberts, D., Luo, X., Yuan, S., Tanswell, A.K., Post, M., Lindsay, T., and Kavanagh, B.P. (2000) Therapeutic hypercapnia reduces pulmonary and systemic injury following *in vivo* lung reperfusion. *Am J Respir Crit Care Med* 162, 2287-2294
99. Jepsen, S., Herlevsen, P., Knudsen, P., Bud, M.I., and Klausen, N-O. (1992) Antioxidant treatment with N-acetylcysteine during adult respiratory distress syndrome: A prospective, randomized, placebo-controlled study. *Crit Care Med* 20, 918-923
100. Ortolani, O., Conti, A., DeGaudio, A.R., Masoni, M., and Novelli, G. (2000) Protective effects of N-acetyl cysteine and rutin on the lipid peroxidation of the lung epithelium during the adult respiratory distress syndrome. *Shock* 13, 14-18
101. Nelin, L., Wetty, S.E., Morrisey, J.F., Gotuaco, C., and Dawson, C.A. (1998) Nitric oxide increases the survival of rats with a high oxygen exposure. *Pediatr Res* 43, 727-732

102. Gutierrez, H.H., Nieves, B., Chumley, P., Rivera, A., and Freeman, B.A. (1996) Nitric oxide regulation of superoxide dependent lung injury: oxidant-protective actions of endogenously produced and exogenously administered nitric oxide. *Free Radic Biol Med* 21, 43-52
103. Walley, K.R., McDonald, T. E., Higashimoto, Y., and Hayashi, S. (1999) Modulation of proinflammatory cytokines by nitric oxide in murine acute lung injury. *Am J Respir Crit Care Med* 160, 698-704
104. Bloomfield, G.L., Holloway, S., Ridings, P.C., Fisher, B.J., Blocher, C.R., Sholley, M., Bunch, T., Sugerman, H.J., and Fowler, A.A. (1997) Pretreatment with inhaled nitric oxide inhibits neutrophil migration and oxidative activity resulting in attenuated sepsis-induced lung injury. *Crit Care Med* 25, 584-593
105. Dobyns, E.L., Cornfield, D.N., Anas, N.G., Fortenberry, J.D., Tasker, R.C., Lynch, A., Liu, P., Eells, P., Riebel, J., Baier, M., Kinsella, J.P., and Abman, S.H. (1999) Multicenter randomized controlled trial of the effects of inhaled nitric oxide therapy on gas exchange in children with hypoxemic respiratory failure. *J Pediatr* 134, 406-412
106. Ream, R.S., Hauver, J.F., Lynch, R.E., Kountzman, B., and Gale, G.B. (1999) Low-dose inhaled nitric oxide improves the oxygenation and ventilation of infants and children with acute, hypoxemic respiratory failure. *Crit Care Med* 27, 989-996
107. Dellinger, R.P., Zimmerman, J.L., Taylor, R.W., Straube, R.C., Hauser, D.L., Criner, G.J., Davis, K., Hyers, T.M., and Papadakas, P. (1998) Inhaled nitric oxide in ARDS study group. Effects of inhaled nitric oxides in patients with acute respiratory distress syndrome: results of a randomized phase II trial. *Crit Care Med* 26, 15-23
108. Michael, J.R., Barton, R.G., Saffle, J.R., Mone, M., Markewitz, B.A., Hillier, K., Elstad, M.R., Campbell, E.J., Troyer, B.E., Whatley, R.E., Liou, T.G., Samuelson, W.M., Carveth, H.J., Hinson, D.M., Morris, S.E., Davis, B.L., and Day, R.W. (1998) Inhaled nitric oxide versus conventional therapy. *Am J Respir Crit Care Med* 157, 1372-1380
109. Chollet-Martin, S., Gatecel, C., Kermarrec, N., Gougerot-Pocidalo, M-A., and Payen, D. Alveolar neutrophil functions and cytokine levels in patients with the adult respiratory distress syndrome during nitric oxide inhalation. *Am J Respir Crit Care Med* 53, 985-990
110. Cuthbertson, B.H., Galley, H.F., and Webster, N.R. (2000) Effect of inhaled nitric oxide on key mediators of the inflammatory response in patients with acute lung injury. *Crit Care Med* 28, 1736-1741
111. Weigand, M.A., Snyder-Ramos, S.A., Möllers, A.G., Bauer, J., Hansen, D., Kochen, W., Martin, E., and Motsch, J. (2000) Inhaled nitric oxide does not enhance lipid peroxidation in patients with acute respiratory distress syndrome. *Crit Care Med* 28, 3429-3435
112. Walther, F.J., David-Cu, R., and Lopez, S.L. (1995) Antioxidant-surfactant liposomes mitigate hyperoxic lung injury in premature rabbits. *Am J Physiol* 269, L613-L617
113. Rosenfeld, W.N., Evans, H., Concepcion, L., Jhaveri, R., Schaeffer, H., and Friedman, A. (1984) Prevention of bronchopulmonary dysplasia by administration of bovine super-oxide dismutase in preterm infants with respiratory distress syndrome. *J Pediatr* 105, 781-786
114. Davis, J.M., Rosenfeld, W.N. et al, North American rh SOD Study Group. (1999) The effects of multiple doses of recombinant human CuZn superoxide dismutase (rh SOD) in premature infants with respiratory distress syndrome (RDS). *Pediatr Res* 45, 193A (Abstract 1129)
115. Otterbein, L.E., Kolls, J.K., Mantell, L.L., Cook, J.L., Alam, J., and Choi, A.M.K. (1999) Exogenous administration of heme oxygenase-1 by gene transfer provides protection against hyperoxia-induced lung injury. *J Clin Invest* 103, 1047-1055
116. Danel, C., Erzurum, E., Prayssac, P., Eissa, N.T., Crystal, R.G., Herve, P., Baudet, B., Mazmanian, M., and Lemarchand, P. (1998) Gene therapy for oxidant injury-related diseases: Adenovirus-mediated transfer of superoxide dismutase and catalase cDNAs protects against hyperoxia but not against ischemia-reperfusion lung injury. *Hum Gene Ther* 9, 1487-1496

Chapter 7

ENDOGENOUS NITRIC OXIDE IN ACUTE LUNG INJURY

Neil W. Kooy
Division of Critical Care Medicine, Children's Hospital Medical Center and Children's Hospital Research Foundation, Cincinnati, OH

INTRODUCTION

Nitric oxide (NO) is a ubiquitous signal transduction molecule formed from the oxidative deamination of one of the equivalent guanidine nitrogens of the amino acid L-arginine via the activity of at least four known isoforms of NO synthase (1). Within the lung, NO is synthesized by endothelial NO synthase (NOS-3) localized in airway epithelial cells and pulmonary vascular endothelial cells, and by neuronal NO synthase (NOS-1) localized in noncholinergic, nonadrenergic nerve fibers (2). During states of inflammation leading to the development of acute lung injury (ALI), cellular stimuli such as bacterial endotoxin, or the inflammatory cytokines TNF-α, IL-1β, or INF-γ, induce alveolar macrophages, vascular smooth muscle cells, airway epithelial cells, and invading inflammatory cells, to transcribe the inducible isoform of NO synthase (NOS-2) via activation of the transcription factor NF-κB (3,4). Similar to NOS-1 and NOS-3, NOS-2 requires calmodulin for activity. Unlike NOS-1 or NOS-3, however, NOS-2 binds calmodulin tightly and does not require addition of exogenous calmodulin to exert its full biological activity (5). Therefore, NOS-2 is not regulated by intracellular calcium levels and the resultant production of NO in high concentration has been demonstrated to play both cytotoxic and cytoprotective roles in the pathogenesis of ALI.

Despite intense investigation into the role of endogenous NO in the pathophysiology of ALI and other inflammatory disease states, considerable controversy continues to exist regarding NO-mediated cytoprotection versus cytotoxicity. This chapter will emphasize the basic biochemistry of NO and its reaction product peroxynitrite in an attempt to establish those reactivities that are most likely to occur within the cellular milieu during states of inflammation.

Biochemistry of NO

NO is an uncharged, paramagnetic, diatomic gas composed of seven electrons from nitrogen and eight electrons from oxygen. Since electron orbitals contain a maximum of two electrons, the electron in the highest occupied orbital of NO must be unpaired, making NO a free radical species. Although most nitrogen oxide compounds are strong oxidizing species, NO is a relatively mild oxidant (ε'_o (pH 7.0) = 0.39 V) and at the low concentrations encountered *in vivo*, has limited reactivity with most biological molecules, resulting in a biological half-life on the order of seconds. Prolonged biological half-life, charge neutrality, small molecular radius, and hydrophilicity, allowing for facile transmembrane diffusion, are important chemical characteristics dictating the function of NO as a signal transduction molecule.

In the gaseous phase, high concentrations of NO react rapidly with oxygen to form the potent oxidant nitrogen dioxide ($\bullet NO_2$), an important toxic component of air pollution. Extrapolation of this observation to the biological milieu, however, is somewhat presumptive and has led to the erroneous description of NO as a highly reactive molecule. As will be discussed, within the biological milieu, where NO is present at low concentrations in the aqueous phase, the reaction of NO with molecular oxygen has relatively little significance.

The rates of reaction for NO with many biological molecules are known, permitting focus on those reactions that ensue expeditiously enough to predominate under biologically relevant conditions. Through the use of chemical rates of reaction, the biological chemistry of NO can be simplified within a reasonable approximation to three predominant reactions: 1) direct, reversible reaction with transition metals, in which the unpaired electron of NO is partially transferred to the metal forming a metal-nitrosyl complex (6),

$$M^n + \bullet NO \leftrightarrow M^n\text{–}NO$$

2) reaction with dioxygen-metal complexes, resulting in metal oxidation and chemical conversion of NO to the inert nitrate anion (NO_3^-),

$$M^n\text{–}O_2 + \bullet NO \rightarrow M^{n+1} + NO_3^-$$

and 3) reaction with oxygen-derived free radical species, which may result in increased oxidant activity via the formation of peroxynitrite (7,8), or decreased oxidant activity via the termination of lipid peroxidation reactions (9,10):

$$\bullet NO + O2^{\bullet -} \rightarrow ONOO^- \qquad k = 1.9 \times 10^{10}\ M^{-1}\ s^{-1}$$

$$LOO\bullet + \bullet NO \rightarrow LOONO \qquad k = 10^9\ M^{-1}\ s^{-1}$$

While undoubtedly this is an oversimplification of the number and types of biochemical reactions occurring within the complexity of the biological milieu, we will limit our discussion to the above reactions and focus on them in more detail in the following sections. The reactions of NO and its reaction product, peroxynitrite, are summarized in Figure 1

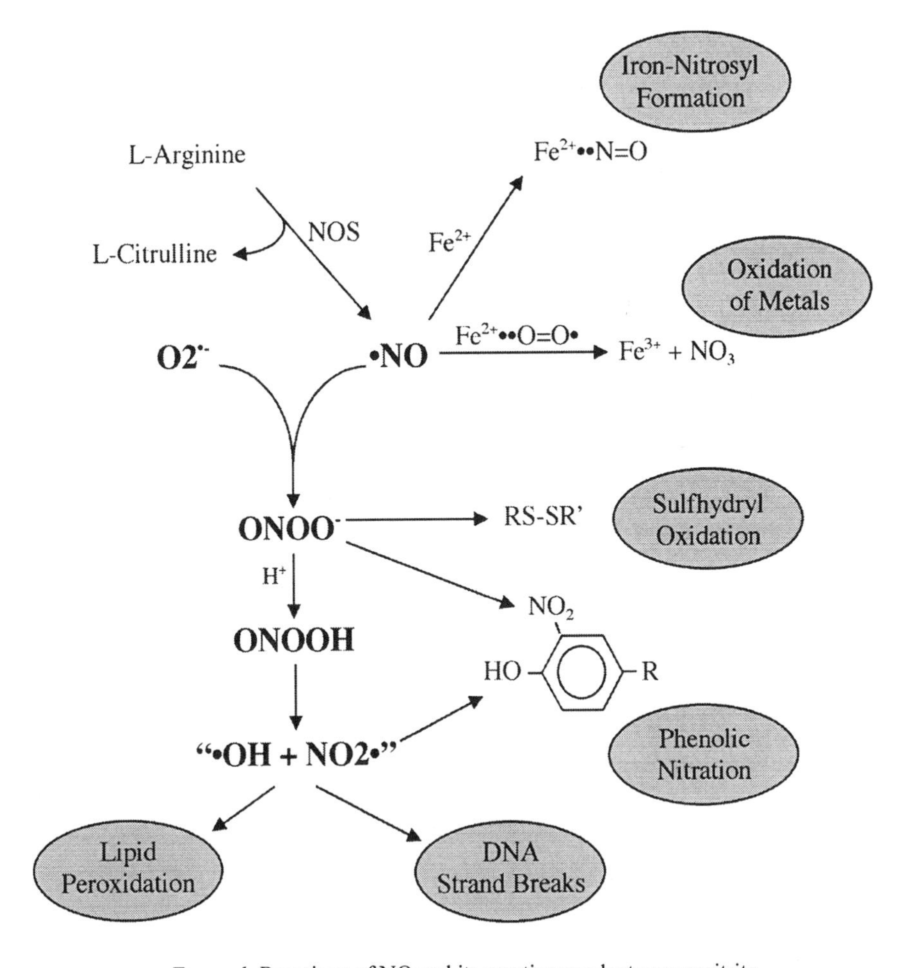

Figure 1. Reactions of NO and its reaction product peroxynitrite.

Reaction of NO with Oxygen

Molecular oxygen has two unpaired electrons, one on each oxygen atom, with parallel spins, making oxygen a biradical species allowing rapid reaction with NO at high concentrations, resulting in the formation of nitrogen dioxide. The overall reaction for formation of nitrogen dioxide from NO and oxygen is:

$$2 \bullet NO + O_2 \rightarrow 2 \bullet NO_2$$

The first step in the overall reaction is the reversible reaction of NO with molecular oxygen to form the nitrosodioxyl radical:

$$\bullet NO + O_2 \leftrightarrow ONOO\bullet$$

In an aqueous environment, this intermediate may be stabilized by hydrogen bonding with water. The subsequent rate limiting reaction involves the reaction of the nitrosodioxyl radical with a second molecule of NO:

$$ONOO\bullet + \bullet NO \rightarrow ONOONO$$

Homolytic cleavage of the resulting species results in the formation of 2 molecules of nitrogen dioxide:

$$ONOONO \rightarrow 2 \bullet NO_2$$

Therefore, the overall rate for the formation of nitrogen dioxide from NO and oxygen, at any given concentration of oxygen, is determined by the square of the NO concentration, $k_3[\bullet NO]^2[O_2]$. Based upon this equation for the rate of nitrogen dioxide formation, the half-life of NO, $1/(k_3[\bullet NO][O_2])$, is not constant, but is inversely proportional to the concentration of NO. As a consequence, as the NO concentration decreases, NO half-life increases and nitrogen dioxide formation decreases. As the concentration of NO in biological systems is on the order of 10^{-9} to 10^{-6} M, the reaction of NO with oxygen to form nitrogen dioxide is extremely slow, and therefore may be of little biological significance.

Direct Reaction with Transition Metals

NO avidly forms metal bond complexes with transition metals. In the presence of ferrous iron (Fe^{2+}), NO forms a ferrous-nitrosyl complex, while reaction with ferric iron (Fe^{3+}) does not occur. The unpaired electron of NO is partially transferred to the ferrous iron, reversibly forming an iron-nitrosyl complex with the reactivity of nitrosonium cation (NO^+).

$$Fe^{2+} + \bullet NO \leftrightarrow Fe^{2+}\text{–}NO$$

Formation of a ferrous-nitrosyl complex with heme proteins, including guanylate cyclase (11) and cytochrome P450 (12), is one of the most facile and relevant reactions of NO in biological systems.

NO acting in either an autocrine or paracrine fashion binds to the heme moiety contained within soluble guanylate cyclase resulting in the removal of the distal histidine and results in a five coordinate iron-nitrosyl complex that activates the enzyme (13, 14). Activation of guanylate cyclase results in the intracellular formation of cyclic guanosine monophosphate (cGMP) and the secondary activation of cGMP-dependent protein kinases (15). Most of the physiological actions of NO, modulation of vascular tone, neurotransmission, inhibition of platelet activation and adherence, and the killing of pathogens, are mediated by the NO-induced activation of soluble guanylate cyclase.

The cytochrome P450 family of enzymes are involved in the synthesis and catabolism of numerous biomolecules, including fatty acids, steroids, prostaglandins, and leukotrienes (16). In contrast to NO-induced iron-nitrosyl formation leading to the activation of guanylate cyclase, NO-induced heme iron-nitrosyl formation inhibits the activity of cytochrome P450 (12, 17, 18), demonstrating different biological results from analogous chemical reactions depending upon the target of NO interaction.

Cytochrome P450 activity may be reversibly or irreversibly inhibited by NO. Reversible inhibition results when NO forms an iron-nitrosyl complex with the ferrous heme, preventing oxygen binding (12). Mechanistically, irreversible inhibition may result from binding of NO to the ferrous heme forming a five-coordinate adduct in which the axial cysteine ligand is removed from the coordination sphere of the iron. Upon sufficient opening of the active site, reactive oxygen species may oxidize the cysteine residue resulting in heme removal from the protein (19).

Reaction of NO with Metal-Oxygen Complexes

The reaction of NO with metal-oxygen species results in oxidation of the metal center and the resultant formation of nitrate through peroxynitrite as an intermediate (20). An important direct effect of NO is the reaction between NO and oxyferrohemoglobin to form methemoglobin, which is a principal modulatory event regulating NO signal transduction (21).

$$Hb(Fe—O_2) + \bullet NO \rightarrow Hb(Fe^{3+}) + NO_3^-$$

Aided by the high concentration of red blood cell hemoglobin in the vascular compartment, the rapid reaction of hemoglobin with NO provides one of the primary metabolic fates and detoxification mechanisms for NO (22). Because hemoglobin reacts with alkyl hydroperoxides to yield oxidant species and free radical intermediates, NO reactions with oxyheme may provide a protective antioxidant mechanism via reducing oxidant-induced oxoferryl species formed *in vivo* (21).

Reaction of NO with Superoxide Anion and Formation of Peroxynitrite

During the inflammatory response, vascular endothelial and smooth muscle cells, macrophages, neutrophils, platelets, and pulmonary epithelial cells are critical sources and targets of reactive oxygen species, including superoxide anion, hydrogen peroxide, and NO (23). Live *Escherichia coli* administration to primates leads to the systemic production of reactive oxygen species, which is augmented following the administration of the inflammatory cytokine TNF-α (24). In dogs and rats, experimentally induced sepsis results in lipid peroxidation, which is attenuated by the prior administration of superoxide dismutase, a scavenger of superoxide anion radical (25, 26, 27). Moreover, the administration of superoxide dismutase prior to inducement of sepsis results in improved survival (28, 29). The ability of superoxide dismutase to reduce endothelial injury (30), and decrease mortality in animal models of sepsis (27, 28, 29), indirectly implicates the participation of superoxide anion radical with the pathological processes of the systemic inflammatory response syndrome. Superoxide anion radical can be directly toxic (31), however, it has limited reactivity with most biological molecules *in vivo* (32). Specifically, superoxide anion radical is incapable of abstracting a hydrogen atom from polyunsaturated fatty acids to initiate lipid peroxidation. To account for the apparent toxicity of superoxide anion radical in inflammatory states, secondary reactions leading to more toxic reactive oxygen species must be considered.

Because NO contains an unpaired electron and is paramagnetic, it reacts at a near diffusion-limited rate with superoxide anion to form peroxynitrite anion (7,8):

$$\bullet NO + O2^{\bullet -} \longrightarrow ONOO^- \quad k = 1.9 \pm 0.2 \times 10^{10}\ M^{-1}\ s^{-1}$$

This rate of reaction is approximately ten times greater than the rate of reaction for superoxide anion with superoxide dismutase (33), making peroxynitrite formation a favored reaction under conditions of systemic inflammation where cellular production of NO and superoxide are increased.

Biochemistry of Peroxynitrite

Peroxynitrite is a complex molecule which exists in both *cis* and *trans* stereoisomeric conformations (Figure 2). In the *cis* conformation, the negative charge is partially delocalized over all four atoms, with a significant interaction between the two terminal peroxide oxygens. The terminal oxygen of peroxynitrite cannot directly approach the nitrogen to form nitrate without substantial rearrangement of the other oxygen atoms making peroxynitrite an unusually stable molecule (34). The terminal peroxide oxygen on the *trans* anion can directly approach the nitrogen by a stretching of the O–O bond and bending of the N–O–O bond. Although the N–O bond in the middle of *cis* peroxynitrite is represented as a single bond, the bond order is closer to 1.5. Consequently, there is a significant energy barrier for isomerization and *cis* and *trans* peroxynitrite cannot directly interconvert (35). Peroxynitrite has a

pK_a of 6.8 and is approximately 20% protonated under physiological conditions (34). The addition of a hydrogen ion neutralizes the negative charge which is partially delocalized over the anion, reducing the barrier for isomerization to the *trans* conformation. The resulting trans-peroxynitrous acid decomposes to yield a vibrationally excited intermediate with the reactivity of hydroxyl radical and nitrogen dioxide (7, 36).

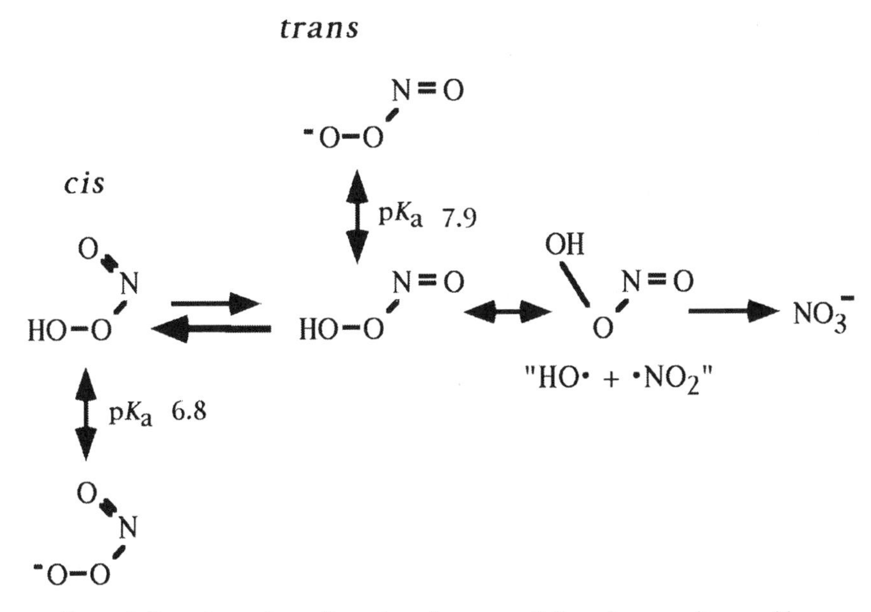

Figure 2. Stereoisomeric configurations for peroxynitrite and peroxynitrous acid.

Peroxynitrite anion directly reacts with protein and non-protein sulfhydryls to form disulfides (37), an important mechanism for peroxynitrite-mediated inactivation of enzymes, including tryptophan hydroxylase (38), sarcoplasmic reticulum calcium ATPase (39), glyceraldehyde-3-phosphate dehydrogenase (40), amiloride-sensitive sodium channels (41), and creatine kinase (42). Once protonated to peroxynitrous acid, this unstable species will initiate one- and two-electron oxidation reactions with target molecules. The ground state form of peroxynitrous acid reacts with methionine (43), cytochrome c^{2+} (44), ascorbate (45, 46), and tryptophan (47). The vibrationally excited intermediate formed upon decomposition of *trans*-peroxynitrous acid is capable of oxidizing lipids (48), deoxyribonucleic acid (49), and a number of other organic molecules (50). Therefore, the conformation of peroxynitrite and the pH of the local environment can dramatically alter the reactivity of peroxynitrite and affect its toxicity.

In addition to its role in oxidative reactions, peroxynitrite nitrates free or protein-associated tyrosines and other phenols via the low molecular mass metal-, superoxide dismutase-, or carbon dioxide-catalyzed formation of a nitronium ion-like species (34, 51-54). Tyrosine nitration decreases surfactant protein A function (55), inhibits cytochrome P-450 (56), manganese superoxide dismutase (57), prostacyclin synthase (58), protein tyrosine phosphorylation (59), and inactivates the complement subcomponent C1q binding of human immunoglobulin (60).

Reaction of NO with Lipid Radicals

NO has been observed to play a critical role in regulating lipid oxidation induced by reactive oxygen and nitrogen species (9, 61 and Figure 3). Under conditions where the rate of NO formation is less than, or equivalent to, the rate of superoxide formation, NO stimulates superoxide-induced lipid and lipoprotein oxidation (61). Under conditions of higher, yet biologically relevant rates of NO formation, which supersede superoxide formation, NO mediates protective reactions within lipid bilayers by inhibiting superoxide and peroxynitrite-induced oxidation (9, 61, 62). Therefore, the prooxidant versus antioxidant outcome of lipid oxidation reactions is critically dependent on the relative concentrations of NO and superoxide.

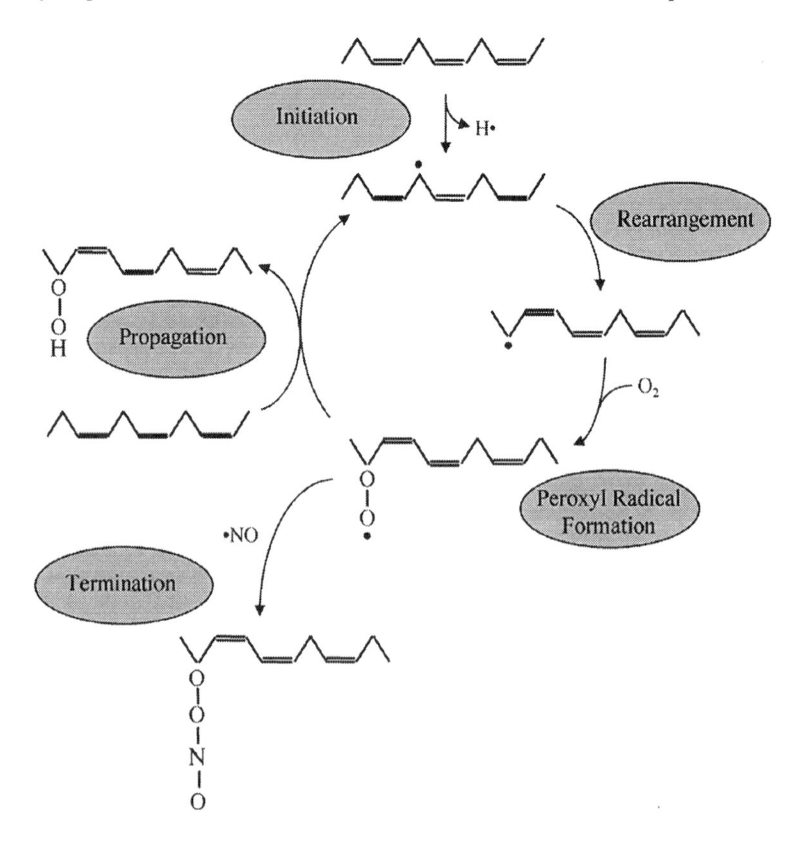

Figure 3. Lipid peroxidation and mechanisms of NO-mediated termination reactions.

Lipid peroxidation results in the formation of alkoxyl or peroxyl radicals that react with NO at near diffusion-limited rates (10):

$$LOO\bullet + \bullet NO \rightarrow LOONO \qquad k = 1.3 \times 10^{9}\ M^{-1}\ s^{-1}.$$

Through this reaction NO is capable of inhibiting lipid peroxidation via the termination of lipid radical chain propagation reactions with the resultant formation of novel organic peroxynitrites. Because of their instability and potential to decompose through radical or oxidative pathways, however, these nitrogen-containing lipid oxidation products may serve as inflammatory mediators or intermediates in as yet unidentified oxidant-mediated pathological processes.

Therefore, a dynamic competition exists between superoxide and lipid radicals for reaction with NO. When available for reaction with lipid radicals, NO may act as an inhibitor of chain propagation reactions via radical-radical annihilation reactions occurring at near diffusion-limited rates (10). NO significantly concentrates in lipophilic cell compartments, a property that may further enhance the ability of NO to regulate oxidant-induced lipid oxidation (63).

NO-Mediated Inhibition of Mitochondrial Respiration

A principal mechanism of NO-induced cellular and tissue injury is inhibition of mitochondrial respiration (3, 64, 65, and Figure 4). In early studies utilizing cells in culture, the susceptible targets for NO-dependent inhibition were thought to contain iron-sulfur centers as integral components of electron transfer, including complex I of the respiratory chain and aconitase, a key enzyme in the tricarboxylic acid cycle (66, 67). More recently, however, the direct effects of NO on mitochondrial function have been questioned. In isolated mitochondria, NO has little or no direct effect on components of the respiratory chain containing iron-sulfur centers but does inhibit cytochrome c oxidase via NO binding to the heme moiety of cytochrome a_3, the terminal member of the respiratory chain (68, 69, 70). Moreover, NO-mediated inhibition of electron transport chain components requires the reaction of NO with superoxide to form peroxynitrite (70). Peroxynitrite irreversibly inactivates complex I and complex II of the electron transport chain, preferentially succinate dehydrogenase and ATPase (70, 71). Indeed, inhibition of isolated mitochondrial respiration induced by peroxynitrite closely resembles the inhibition pattern observed during NO treatment of intact cells (67, 70, 71). Moreover, it has also been demonstrated that mitochondrial aconitase is readily inactivated by peroxynitrite but not significantly by NO (72, 73).

During states of inflammation, mitochondria are a key cellular source of superoxide anion, with oxidant-mediated mitochondrial injury resulting in a subsequent increase in substrate-supported superoxide formation (71). Moreover, oxidant-mediated mitochondrial injury also leads to increased intramitochondrial calcium (74), stimulating the recently discovered constitutive, calcium-dependent NO synthase within the inner mitochondrial membrane to increase NO formation (75, 76). When coupled with the increase in superoxide production, the increased NO formation will lead to the site specific formation of peroxynitrite, further inhibition of electron transport, and enhancement of superoxide, thereby continuing a self-propagation of peroxynitrite-dependent mitochondrial and cellular injury.

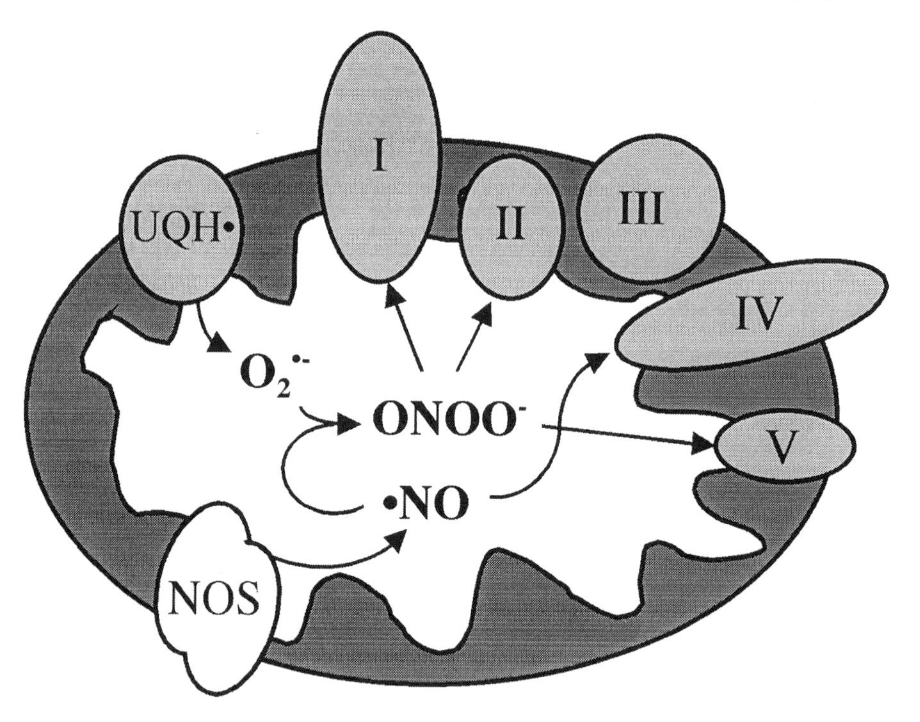

Figure 4. Sites of inhibition of NO and peroxynitrite within the mitochondrion.

NO AND PEROXYNITRITE IN ALI

The administration of bacterial endotoxin in animal models results in the development of ALI. Increases in NO production within the lungs of can be demonstrated within 60 minutes through the measurement of exhaled NO (77). The early detection of NO suggests increased production from the constitutive isoforms, as the transcription and translation of NOS-2 does not occur within this time period. Moreover, following the administration of bacterial endotoxin, the presence of NOS-2 is demonstrable within invading neutrophils, alveolar macrophages, and alveolar epithelial cells, suggesting further increase in the local production of NO (78).

Whether augmented pulmonary NO production serves positive or deleterious roles during the development of ALI remains a debatable issue. A potential beneficial role for enhanced NO production is based upon the observations that NO inhibits neutrophil migration (79), cytokine production (80), and NF-κB activation (81). Moreover, NO synthase inhibition has been demonstrated to increase lung edema and worsen indices of ALI (82). Physiologic formation of NO by NOS-3 is believed to protect against endotoxin-induced damage by counteracting aggregation of leukocytes via decreased expression of adhesion molecules (83), and by inhibiting platelet activation (84) and aggregation (3, 85). Furthermore, stimulation of cGMP production is suggested to enhance the barrier function of endothelial cells (86).

There is ample evidence, however, that excessive NO production by iNOS may be deleterious and contribute to the induction of ALI. First, iNOS expression in many pulmonary cell types coincides with the development of lung injury in endotoxemic humans and animals (87, 88). Second, infusion of NOS inhibitors attenuates the microvascular leakage and edema formation induced by endotoxemia (89, 90) improves gas exchange (91, 92), and improves survival by counteracting circulatory failure (93). Moreover, mice lacking NOS-II are more resistant to endotoxin-induced ALI when compared to wild-type animals (94). Third, extensive protein tyrosine nitration, a footprint for peroxynitrite formation, has been detected in lung sections of humans and animals with ALI and coincides with iNOS expression (95, 96)

Other animal models of ALI suggest that NO plays a prominent deleterious role. For example, studies of immune complex-mediated lung injury suggest that NO plays a prominent role in the inflammatory cell-mediated component of ALI. IgG immune complex-mediated injury is neutrophil-dependent (97, 98), whereas IgA immune complex-mediated injury is macrophage-dependent (99). In both models, lung injury is reduced when endogenous NO production is inhibited by N-monomethyl-L-arginine. Moreover, studies utilizing smoke inhalation-induced ALI in the rat (100) and paraquat-induced ALI in the guinea pig (101) have also identified endogenous NO production as an important mediator of lung injury. Furthermore, ischemia-reperfusion of isolated rat lung is associated with increases in nitrite/nitrate, nitrotyrosine, thiobarbituric acid reactive substances, conjugated dienes, and dinitrophenylhydrazine-reactive protein carbonyls, indicating increased NO and oxidant-mediated injury (102). The ischemia-reperfusion-induced changes were inhibited in the presence of the NO synthase inhibitor N-nitro-L-arginine methyl ester (102), demonstrating not only NO-mediated injury, but NO-mediated oxidant injury via the formation of peroxynitrite.

Based upon the specificity of tyrosine nitration for the activity of peroxynitrite (103), specific antibodies to nitrotyrosine have been utilized to demonstrate the presence of nitrotyrosine in animal models of endotoxemia and ALI (104, 105, 106), as well as in human ALI (95, 96). In human sepsis-induced ALI, nitrotyrosine residues were apparent throughout the lung specimen including the proteinaceous alveolar exudate, intra-alveolar macrophages, vascular endothelium and vascular smooth muscle (95). The pattern and intensity of nitrotyrosine detected seemed to correlate with the severity and mechanism of the pulmonary disease. For example, the patient with the most severe lung injury clinically had the most extensive nitrotyrosine staining (95). The presence of nitrotyrosine residues within the vasculature of the lungs, as well as within the lung interstitium and alveolar space, is consistent with the notion that sepsis represents diffuse intravascular inflammation and ALI is the pulmonary expression of this process. These results demonstrate unequivocally the presence and activity of peroxynitrite during the process of ALI. The precise mechanisms underlying peroxynitrite-induced cellular injury as a process in the development of ALI remain to be elucidated.

REFERENCES

1. Bredt, D.S., and Snyder, S.H. (1990) Isolation of nitric oxide synthetase, a calmodulin-requiring enzyme. *Proc Natl Acad Sci USA 87*, 682-685
2. Gaston, B.J., Drazen, J.M., Loscalzo, J., and Stamler, J.S. (1994) The biology of nitrogen oxides in the airways. *Am J Respir Crit Care Med* 149, 538-551
3. Moncada, S., Palmer, R.M., and Higgs, E.A. (1991) Nitric oxide: physiology, pathophysiology, and pharmacology. *Pharmacol Rev* 43, 109-142
4. Xie, Q.W., Kashiwabara, Y., and Nathan, C. (1994) Role of transcription factor NF-κB/Rel in induction of nitric oxide synthase. *J Biol Chem* 269, 4705-4708
5. Cho, H.J., Xie, Q., Calaycay, J., Mumford, R.A., Swidereck, K.M., Lee, T.D., and Nathan, C. (1992). Calmodulin as a tightly bound subunit of calcium-, calmodulin-independent nitric oxide synthase. *J Exp Med* 176, 599-604
6. Wink, D.A., and Ford, P.C. (1995) Nitric oxide reactions important to biological systems: a survey of some kinetics investigations. *Methods: A Companion to Methods Enzymol* 7, 14-20
7. Beckman, J.S., Beckman, T.W., Chen, J., Marshall, P.A., and Freeman, B.A. (1990) Apparent hydroxyl radical production by peroxynitrite: implications for endothelial injury from nitric oxide and superoxide. *Proc Natl Acad Sci USA* 87, 1620-1624
8. Kissner, R., Nauser, T., Bugnon, P., Lye, P.G., and Koppenol, W.H. (1997) Formation and properties of peroxynitrite as studied by laser flash photolysis, high-pressure stopped-flow technique, and pulse radiolysis. *Chem Res Toxicol* 10, 1285-1292
9. Rubbo, H., Parthasarathy, S., Kalyanaraman, B., Barnes, S., Kirk, M., and Freeman, B. A. (1996) Nitric oxide inhibition of lipoxygenase-dependent liposome and low density lipoprotein oxidation: termination of radical chain propagation reaction and formation of nitrogen-containing oxidized lipid derivatives. *Arch Biochem Biophys* 324, 15-25
10. Padmaja, S., and Huie, R.E. (1993) The reaction of nitric oxide with organic peroxyl radicals. *Biochem Biophys Res Commun* 195, 539-544
11. Ignarro, L.J., Degnan, J., Baricos, W., Kadowitz, P., and Wolin, M.S. (1982) Activation of purified guanylate cyclase by nitric oxide requires heme: comparison of the heme-deficient, heme-reconstituted, and heme-containing forms of soluble enzyme from bovine lung. *Biochim Biophys Acta* 718, 49-59

12. Wink, D.A., Osawa, Y., Darbyshire, J.F., Jones, C.R., Eshenaur, S.C., and Nims, R.W. (1993) Inhibition of cytochromes P450 by nitric oxide and a nitric oxide-releasing agent. *Arch Biochem Biophys* 300, 115-123
13. Yu, A.E., Hu, S., Spiro, T.G., and Burstyn, J.N. (1994) Resonance raman spectroscopy of soluble guanylyl cyclase reveals displacement of distal and proximal heme ligand by NO. *J Am Chem Soc* 116, 4117-4118
14. Stone, J.R., and Marletta, M.A. (1994) Soluble guanylate cyclase from bovine lung: activation with nitric oxide and carbon monoxide and spectral characterization of the ferrous and ferric state. *Biochemistry* 33, 5636-5640
15. Lincoln, T.M., and Cornwell, T.L. (1993) Intracellular cyclic GMP receptor proteins. *FASEB J* 7, 328-338
16. Wink, D.A., and Mitchell, J.B. (1998) Chemical biology of nitric oxide: insights into regulatory, cytotoxic, and cytoprotective mechanisms of nitric oxide. *Free Rad Biol Med* 25, 434-456
17. Khatsenko, O.G., Gross, S.S., Rifkind, A.B., and Vane, J.R. (1993) Nitric oxide is a mediator of the decrease in cytochrome P450-dependent metabolism caused by immunostimulants. *Proc Natl Acad Sci USA* 90, 11147-11151
18. Stadler, J., Trockfeld, J., Shmalix, W.A., Brill, T., Siewert, J.R., Greim, H., and Doehmer, J. (1994) Inhibition of cytochromes P450 by nitric oxide. *Proc Natl Acad Sci USA* 91, 3559-3563
19. Kim, Y.M., Begonia, H.A., Muller, C., Pitt, B.R., Watkins, W.D., and Lancaster, J.R. (1995) Loss and degradation of enzyme-bound heme induced by cellular nitroxide synthesis. *J Biol Chem* 270, 5710-5713
20. Doyle, M.P., and Hoekstra, J.W. (1981) Oxidation of nitrogen oxides by bound dioxygen in hemeproteins. *J Inorg Biochem* 14, 351-358
21. Gorbunov, N., Osipov, A., Day, B., Zayas, B., Kagan, V., and Elsayed, N. (1995) Reduction of ferrylmyoglobin and ferrylhemoglobin by nitric oxide: a protective mechanism against ferryl hemoprotein-induced oxidations. *Biochemistry* 34, 6689-6699
22. Lancaster, J. (1994) Simulation of the diffusion and reaction of endogenously produced nitric oxide. *Proc Natl Acad Sci USA* 91, 8137-8141
23. Radi, R., Panus, P.C., Royall, J.A., Paler-Martinez, A., and Freeman, B.A. (1992) Generation of reactive species by vascular endothelium. *Cellular and Molecular Mechanisms of Inflammation* 4, 83-118
24. Lloyd, S.S., Chang, A.K., Taylor, Jr., F.B., Janzen, E.G., and McCay, P.B. (1993) Free radicals and septic shock in primates: the role of tumor necrosis factor. *Free Radic Biol Med* 14, 223-242
25. Morgan, R.A., Manning, P.B., Coran, A.G., Drongwoski, R.A., Tillm, G.O., Ward, P.D., and Oldham, K.T. (1988) Oxygen free radical activity during live *E. coli* septic shock in the dog. *Circ Shock* 25, 319-323
26. Takeda, K., Shimada, Y., Okada, T., Amano, T., Sakai, T., and Yoshiya, I. (1986) Lipid peroxidation in experimental septic rats. *Crit Care Med* 14, 719-723
27. Kunimoto, F., Morita, T., and Fujita, T. (1987) Inhibition of lipid peroxidation improves survival rate of endotoxemic rats. *Circ Shock* 21, 15-22
28. Broner, C.W., Shenep, J.L., Stidham, G.L., Stokes, D.C., and Hildner, W.K. (1988) Effect of scavengers of oxygen-derived free radicals on mortality in endotoxin-challenged mice. *Crit Care Med* 16, 848-851
29. McKechnie, K., Furman, B.L., and Parratt, J.R. (1986) Modification by oxygen free radical scavengers of the metabolic and cardiovascular effects of endotoxin infusion in conscious rats. *Circ Shock* 19, 429-439
30. Freeman, B.A., Jackson, R.M., Matalon, S., and Harding, S.M. (1988) Biochemical and Functional Aspects of Oxygen-mediated Injury to Vascular Endothelium in Endothelial Cells, ed. U. S. Ryan (CRC, Boca Raton, FL), pp. 13-32
31. Fridovich, I. (1986) Biological effects of superoxide radical. *Arch Biochem Biophys* 247, 1-11

32. Sawyer, D.T., and Valentine, J.S. (1981) How super is superoxide? *Acct Chem Res* 14, 393-400
33. Cudd, A., and Fridovich, I. (1982) Electrostatic interactions in the reaction mechanism of bovine erythrocyte superoxide dismutase. *J Biol Chem* 257, 11443-11447
34. Beckman, J.S., Ischiropoulos, H., Zhu, L., van der Woerd, M., Smith, C.D., Chen, J., Harrision, J., Martin, J.C., and Tsai, J.H.M. (1992) Kinetics of superoxide dismutase and iron catalyzed nitration of phenolics by peroxynitrite. *Arch Biochem Biophys* 298, 438-445
35. Beckman, J.S. (1996) Oxidative damage and tyrosine nitration from peroxynitrite. *Chem Res Toxicol* 9, 836-844
36. Koppenol, W.H., Moreno, J.J., Pryor, W.A., Ischiropoulos, H., and Beckman, J.S. (1992) Peroxynitrite, a cloaked oxidant formed by nitric oxide and superoxide. *Chem Res Toxicol* 5, 834-842
37. Radi, R., Beckman, J.S., Bush, K.M., and Freeman, B.A. (1991) Peroxynitrite oxidation of sulhydryls. The cytotoxic potential of superoxide and nitric oxide. *J Biol Chem* 266, 4244-4250
38. Kuhn, D.M., and Geddes, T.J. (1999) Peroxynitrite inactivates tryptophan hydroxylase via sulfhydryl oxidation. Coincident nitration of enzyme tyrosyl residues has minimal impact on catalytic activity. *J Biol Chem* 274, 29726-29732
39. Viner, R.I., Williams, T.D., Schoneich, C. (1999) Peroxynitrite modification of protein thiols: oxidation, nitrosylation, and S-glutathiolation of functionally important cysteine residue(s) in the sarcoplasmic reticulum Ca-ATPase. *Biochemistry* 21, 12408-12415
40. Souza, J.M., and Radi, R. (1998) Glyceraldehyde-3-phosphate dehydrogenase inactivation by peroxynitrite. *Arch Biochem Biophys* 15, 187-194
41. Bauer, M.L., Beckman, J.S., Bridges, R.J., Fuller, C.M., and Matalon, S. (1992) Peroxynitrite inhibits sodium uptake in rat colonic membrane vesicles. *Biochim Biophys Acta* 1104, 87-94
42. Konorev, E.A., Hogg, N., and Kalyanaraman, B. (1998) Rapid and irreversible inhibition of creatine kinase by peroxynitrite. *FEBS Lett* 427, 171-174
43. Pryor, W.A., Jin, X., and Squadrito, G.L. (1995) One- and two-electron oxidations of methionine by peroxynitrite. *Proc Natl Acad Sci USA* 91, 11173-11177
44. Thomson, L., Trujillo, M., Telleri, R., and Radi, R. (1995) Kinetics of cytochrome c^{2+} oxidation by peroxynitrite: implications for superoxide measurements in nitric oxide-producing biological systems. *Arch Biochem Biophys* 319, 491-507
45. Bartlett, D., Church, D.F., Bounds, P.L., and Koppenol, W.H. (1994) The kinetics of the oxidation of L-ascorbic acid by peroxynitrite. *Free Rad Biol Med* 18, 85-92
46. Squadrito, G.L., Jin, X., Pryor, W.A. (1995) Stopped-flow kinetic study of the reaction of ascorbic acid with peroxynitrite. *Arch Biochem Biophys* 322, 53-59
47. Alverez, B., Rubbo, H., Kirk, M., Barnes, S., and Freeman, B.A. (1996) Peroxynitrite-dependent tryptophan nitration. *Chem Res Toxicol* 9, 390-396
48. Radi, R., Beckman, J.S., Bush, K.M., and Freeman, B.A. (1991) Peroxynitrite-induced membrane lipid peroxidation: the cytotoxic potential of superoxide and nitric oxide. *Arch Biochem Biophys* 288, 481-487
49. King, P.A., Anderson, V.E., Edwards, J.O., Gustafson, G., Plumb, R.C., and Suggs, J.W. (1992) A stable solid that generated hydroxyl radical upon dissolution in aqueous solutions: reaction with proteins and nucleic acid. *J Am Chem Soc* 114, 5430-5432
50. Crow, J.P., Spruell, C., Chen, J., Gunn, C., Ischiropoulos, H., Zhu, L., Tsai, J.H.M., Smith, C.D., Radi, R., Koppenol, W.H., Freeman, B.A., Matalon, S., and Beckman, J.S. (1994) On the pH-dependent yield of hydroxyl radical products from peroxynitrite. *Free Rad Biol Med* 16, 331-338
51. Ischiropoulos, H., Zhu, L., Chen, J., Tsai, J.H.M., Martin, J.C., Smith, C.D., and Beckman, J.S. (1992) Peroxynitrite-mediated tyrosine nitration catalyzed by superoxide dismutase. *Arch Biochem Biophys* 298, 431-437

52. Gow, A., Duran, D., Thom, S.R., and Ischiropoulos, H. (1996) Carbon dioxide enhancement of peroxynitrite-mediated protein tyrosine nitration. *Arch Biochem Biophys* 333, 42-48
53. Uppu, R.M., Squadrito, G.L., and Pryor, W.A. (1996) Acceleration of peroxynitrite oxidations by carbon dioxide. *Arch Biochem Biophys* 327, 335-343
54. Denicola, A., Freeman, B.A., Trujillo, M., and Radi, R. (1996) Peroxynitrite reaction with carbon dioxide/bicarbonate: kinetics and influence on peroxynitrite-mediated oxidations. *Arch Biochem Biophys* 333, 49-58
55. Haddad, I.Y., Ischiropoulos, H., Holm, B.A., Beckman, J.S., Baker, J.R., and Matalon, S. (1993) Mechanisms of peroxynitrite-induced injury to pulmonary surfactants. *Am J Physiol* 265, L555-L564
56. Janig, G.R., Kraft, R., Blanck, J., Ristau, O., Rabe, H., and Ruckpaul, K. (1987) Chemical modification of cytochrome P-450 LM4. Identification of functionally linked tyrosine residues. *Biochim Biophys Acta* 91, 512-523
57. MacMillan-Crow, L.A., Crow, J.P., and Thompson, J.A. (1998) Peroxynitrite-mediated inactivation of manganese superoxide dismutase involves nitration and oxidation of critical tyrosine residues. *Biochemistry* 37, 1613-1622
58. Zou, M., Martin, C., and Ullrich, V. (1997) Tyrosine nitration as a mechanism of selective inactivation of prostacyclin synthase by peroxynitrite. *J Biol Chem* 378, 707-713
59. Gow, A.J., Duran, D., Malcolm S., and Ischiropoulos, H. (1996) Effects of peroxynitrite-induced protein modifications on tyrosine phosphorylation and degradation. *FEBS Lett* 385, 63-66
60. McCall, M.N., and Easterbrook-Smith, S.B. (1989) Comparison of the role of tyrosine residues in human IgG and rabbit IgG in binding of complement subcomponent C1q. *Biochem J* 257, 845-851
61. Rubbo, H., Radi, R., Trujillo, M., Telleri, R., Kalyanaraman, B., Barnes, S., Kirk, M., and Freeman, B.A. (1994) Nitric oxide regulation of superoxide and peroxynitrite-dependent lipid peroxidation: formation of novel nitrogen-containing oxidized lipid derivatives. *J Biol Chem* 269, 26066-26075
62. Hogg, N., Darley-Usmar, V.M., Wilson, M.T., and Moncada, S. (1993) The oxidation of a-tocopherol in human low density lipoprotein by the simultaneous generation of superoxide and nitric oxide. *FEBS Lett* 326, 199-203
63. Rubbo, H., Darley-Usmar, V., and Freeman, B.A. (1996) Nitric oxide regulation of tissue free radical injury. *Chem Res Toxicol* 9, 809-820
64. Hibbs, J.B., Vavrin, Z., and Taintor, R.R., (1987) L-arginine is required for the expression of the activated macrophage effector mechanism causing selective metabolic inhibition in target cells. *J Immunol* 138, 550-565
65. Lancaster, J.R., and Hibbs, J.B. (1990) EPR demonstration of iron-nitrosyl complex formation by cytotoxic activated macrophages. *Proc Natl Acad Sci USA* 87, 1223-1227
66. Hibbs, J.B., Taintor, R.R., Vavrin, Z., and Rachlin, E.M. (1988) Nitric oxide: a cytotoxic activated macrophage effector molecule. *Biochem Biophys Res Commun* 57, 87-94
67. Stadler, J., Billiar, T.R., Curran, R.D., Stuehr, D.J., Ochoa, J.B., and Simmons, R.L. (1991) Effect of exogenous and endogenous nitric oxide on mitochondrial respiration of rat hepatocytes. *Am J Physiol* 260, C910-C916
68. Cleeter, M.W.J., Cooper, J.M., Darley-Usmar, V.M., Moncada, S., and Schapira, A.H.V. (1994) Reversible inhibition of cytochrome c oxidase, the terminal enzyme of the mitochondrial respiratory chain, by nitric oxide. *FEBS Lett* 345, 50-54
69. Brown, G.C., and Cooper, C.E. (1994) Nanomolar concentrations of nitric oxide reversibly inhibit synaptosomal respiration by competing with oxygen at cytochrome c oxidase. *FEBS Lett* 356, 295-298
70. Cassina, A., and Radi, R. (1996) Differential inhibitory action of nitric oxide and peroxynitrite on mitochondrial electron transport. *Arch Biochem Biophys* 328, 309-316

71. Radi, R., Rodriguez, M., Castro, L., and Telleri, R. (1994) Inhibition of mitochondrial electron transport by peroxynitrite. *Arch Biochem Biophys* 308, 89-95
72. Hausladen, A., and Fridovich, I. (1994) Superoxide and peroxynitrite inactivate aconitases, but nitric oxide does not. *J Biol Chem* 269, 29405-29408
73. Castro, L., Rodriguez, M., and Radi, R. (1994) Aconitase is readily inactivated by peroxynitrite, but not by its precursor, nitric oxide. *J Biol Chem* 269, 29409-29415
74. Duchen, M.R. (2000). Mitochondria and calcium: from cell signaling to cell death. *J Physiol* 15, 57-68
75. Bates, T.E., Loesch, A., Burnstocl, G., and Clark, J.B. (1995) Immunocytochemical evidence for a mitochondrially located nitric oxide synthase in brain and liver. *Biochem Biophys Res Commun* 213, 896-900
76. Ghafourifar, P., and Richter, C. (1997) Nitric oxide synthase activity in mitochondria. *FEBS Lett* 418, 291-296
77. Stitt, J.T., Dubois, A.B., Douglas, J.S., and Shimada, S.G. (1997) Exhalation of gaseous nitric oxide by rats in response to endotoxin and its absorption by the lungs. *J Appl Physiol* 82, 305-16
78. Numata, M., Suzuki, S., Miyazawa, N., Miyashita, A., Nagashima, Y., Inoue, S., Kaneko, T., and Okubo T. (1998) Inhibition of inducible nitric oxide synthase prevents LPS-induced acute lung injury in dogs. *J Immunol* 160, 3031-3037
79. Bloomfield, G.L., Holloway, S., Ridings, P.C., Fisher, B.J., Blocher, C.R., Sholley, M., Bunch, T., Sugarman, H.J., and Fowler, A.A. (1997) Pretreatment with inhaled nitric oxide inhibits neutrophil migration and oxidative activity resulting in attenuated sepsis-induced acute lung injury. *Crit Care Med* 25, 584-593
80. Iuvone, T.F., D'Aquisto, F.D., Carnuccio, R., and Di Rosa, M. (1996) Nitric oxide inhibits LPS-induced tumor necrosis factor synthesis *in vitro* and *in vivo*. *Life Sci* 59, 207-211
81. Peng, H.B., Libby, P., and Liao, J.K. (1995) Induction and stabilization of IκBα by nitric oxide mediated inhibition of NF-κB. *J Biol Chem* 270, 14214-14219
82. Pheng, L.H., Francoeur, C., and Denis, M. (1995) The involvement of nitric oxide in a mouse model of adult respiratory distress syndrome. *Inflammation* 19, 599-610
83. De Caterina, R., Libby, P., Peng, H.B., Thannickal, V.J., Rajavashisth, T.B., Gimbrone, M.A. Jr, Shin, W.S., Liao, J.K. (1995) Nitric oxide selectively reduces endothelial expression of adhesion molecules and proinflammatory cytokines. *J Clin Invest* 96, 60-68
84. May, G.R., Crook, P., Moore, P.K., Page, C.P. (1991) The role of nitric oxide as an endogenous regulator of platelet and neutrophil activation within the pulmonary circulation of the rabbit. *Br J Pharmacol* 102, 759-63
85. McQuaid, K.E., and Keenan, A.K. (1997) Endothelial barrier dysfunction and oxidative stress: roles for nitric oxide? *Exp Physiol* 82, 369-376
86. Westendorp, R.G., Draijer, R., Meinders, A.E., and Van Hinsbergh, V.W. (1994) Cyclic-GMP-mediated decrease in permeability of human umbilical and pulmonary artery endothelial cell monolayers. *J Vasc Res* 31, 42-51
87. Kobzik, L., Bredt, D.S., Lowenstein, C.J., Drazen, J., Gaston, B., Sugarbaker, D., and Stamler, J.S. (1993) Nitric oxide synthase in human and rat lung: immunocytochemical and histochemical localization. *Am J Respir Cell Mol Biol* 9, 371-377
88. Goode, H.F., Howdle, P.D., Walker, B.E., and Webster, N.R. (1995) Nitric oxide synthase activity is increased in patients with septic syndrome. *Clin Sci* 88, 131-133
89. Arkovitz, M.S., Wispe, J.R., Garcia, V.F., and Szabo, C. (1996) Selective inhibition of the inducible isoform of nitric oxide synthase prevents pulmonary transvascular flux during acute endotoxemia. *J Pediatr Surg* 31, 1009-1015
90. Akaike, T., Noguchi, Y., Ijiri, S., Setoguchi, K., Suga, M., Zheng, Y.M., Dietzschold, B., and Maeda, H. (1996) Pathogenesis of influenza virus-induced pneumonia: involvement of both nitric oxide and oxygen radicals. *Proc Natl Acad Sci USA* 93, 2448-2453

91. Mikawa, K., Nishina, K., Tamada, M., Takao, Y., Maekawa, N., and Obara, H. (1998) Aminoguanidine attenuates endotoxin-induced acute lung injury in rabbits. *Crit Care Med* 26, 905-911
92. Evgenov, O.V., Hevroy, O., Bremnes, K.E., and Bjertnaes, L.J. (2000) Effect of aminoguanidine on lung fluid filtration after endotoxin in awake sheep. *Am J Respir Crit Care Med* 162, 465-470
93. Wu, C.C., Chen, S.J., Szabo, C., Thiemermann, C., and Vane, J.R. (1995) Aminoguanidine attenuates the delayed circulatory failure and improves survival in rodent models of endotoxic shock. *Br J Pharmacol* 114, 1666-72
94. Kritof, A.S., Goldberg, P., Laubach V., and Hussain, S.N.A. (1998) Role of inducible nitric oxide synthase in endotoxin-induced acute lung injury. *Am J Repir Crit Care Med* 158, 1883-1889
95. Kooy, N.W., Royall, J.A., Ye, Y.Z., Kelly, D.R., and Beckman, J.S. (1994) Evidence for in vitro peroxynitrite production in human acute lung injury. *Am J Respir Crit Care Med* 151, 1250-1254
96. Haddad, I.Y., Pataki, G., Galliani C., Beckman, J.S., and Matalon, S. (1994) Quantitation of nitrotyrosine levels in lung sections of patients and animals with acute lung injury. *J Clin Invest* 94, 2407-2413
97. Mulligan, M.S., Hevel, J.M., Marletta, M.A., and Ward, P.A. (1991) Tissue injury caused by deposition of immune complexes is L-arginine dependent. *Proc Natl Acad Sci USA* 88, 6338-6342
98. Mulligan, M.S., Moncada, S., and Ward, P.A. (1992) Protective effect of inhibitors of nitric oxide synthase in immune complex-induced vasculitis. *Br J Pharmacol* 107, 1159-1162
99. Mulligan, M.S., Warren, J.S., Smith, C.W., Anderson, D.C., Yeh, C.G., Rudolph, A.R., and Ward, P.A. (1992) Lung injury after decomposition of IgA immune complexes. *J Immunol* 148, 3086-3092
100. Ischiropoulos, H., Mendiguren, I., Fisher, D., Fisher, A.B., and Thom, S.R. (1994) Role of neutrophils and nitric oxide in lung alveolar injury from smoke inhalation. *Am J Respir Crit Care Med* 150, 337-341
101. Berisha, H.I., Pakbaz, H., Absod, A., and Said, S. (1994) Nitric oxide as a mediator of oxidant lung injury due to paraquat. *Proc Natl Acad Sci USA* 91, 7445-7449
102. Ischiropoulos, H., Al-Mehdi, A., and Fisher, A.B. (1995) Reactive species in lung ischemia-reperfusion injury: contribution of peroxynitrite. *Am J Physiol* 269, L158-L164
103. Beckman, J.S., Ye, Y.Z., Anderson, P., Chen, J., Accavetti, M.A., Tarpey, M.M., and White C.R. (1994) Extensive nitration of protein tyrosines observed in human atherosclerosis detected by immunohistochemistry. *Biol Chem Hoppe-Seyler* 375, 81-88
104. Szabo, C., Salzman, A.L., and Ischiropoulos, H. (1992) Endotoxin triggers the expression of an inducible isoform of nitric oxide synthase and the formation of peroxynitrite in the rat aorta *in vivo*. *FEBS Lett* 363, 235-238
105. Wizeman, T.M., Gardiner, C.R., Laskin, J.D., Quinones, S., Durham, S.K., Goller, N.L., Ohnishi, S.T., and Laskin, D.L. (1994) Production of nitric oxide and peroxynitrite in the lung during endotoxemia. *J Leukoc Biol* 56, 759-768
106. Ischiropoulos, H., Al-Mehdi, A.B., and Fisher, A.B. (1995) Reactive species in rat lung injury: contribution of peroxynitrite. *Am J Physiol* 269, L158-L164

Chapter 8

HEME OXYGENASE-1 IN ACUTE LUNG INJURY

Jigme M. Sethi and Augustine M.K. Choi
Division of Pulmonary, Allergy and Critical Care Medicine, University of Pittsburgh Medical Center, Pittsburgh, PA

INTRODUCTION

Heme oxygenase (HO) was first described, in 1968, as the enzyme responsible for the rate-limiting step in the catalytic breakdown of the heme moiety of hemoglobin, and understandably, initial work focused on the regulation and function of this enzyme in heme metabolism (1). That HO could be important for much more than the color changes of a bruise as heme was broken down through biliverdin to bilirubin, was not suspected until the work of Tyrrell and his colleagues. These researchers showed that HO-1 could be vigorously induced not just by heme, its natural substrate, but by a variety of agents that had as their common feature the ability to generate reactive oxidant species (ROS) (2). Examples of the broad spectrum of agents capable of inducing HO include lipopolysaccharide (LPS), phorbol esters, sodium arsenite, hydrogen peroxide, ultraviolet radiation, hyperthermia, hyperoxia, sulfhydryl reagents, heat shock, and heavy metals. This extraordinary variety of inducers is surprising for an enzyme subserving a seemingly housekeeping function of heme turnover, and speculation arose that the enzyme was also vital to cellular homeostasis.

Much subsequent research on the regulation and function of HO-1 in *in vitro* and *in vivo* models of oxidant-mediated tissue and cellular injury has firmly established this ubiquitous enzyme as a key cytoprotective, stress-response protein with a major role in protection against oxidative stress (3). However, there are even more intriguing aspects of this remarkable enzyme that reflect its crucial importance in nature. First, the enzyme is highly conserved, with approximately 90% identity between the rat and human genes for HO-1 and HO-2, for example. Second, HO is present ubiquitously through nature, not only in mammals but also in algae (4), plants (5), and bacteria (6), and despite this ubiquitous distribution, no known mutant forms of the enzyme exist. Indeed, targeted deletions of the HO-1 gene in mice result in HO-1 null (-/-) mice that usually do not survive to term (7). The few mice that are born alive are abnormal, with growth retardation, anemia, and leukocytosis, and die within a year of birth. Finally, the enzyme system has been retained evolutionarily in all animals that use hemoglobin for oxygen transport, despite the fact that carbon monoxide (CO), a potentially lethal gas capable of poisoning cytochrome a_3, is a by-product of this enzymatic reaction.

This chapter will focus on the current understanding of the physiological significance of HO-1 induction in response to inflammatory and oxidative stress, particularly with respect to acute lung injury (ALI).

Heme Oxygenase and its Isoforms

HO is a microsomal enzyme that catalyzes the initial and rate-limiting step in the degradation of heme to bilirubin. Utilizing NADPH and molecular oxygen, the enzyme cleaves the α-methenyl bridges between pyrroles I and II of the heme porphyrin ring, producing biliverdin, and liberating free iron and CO in equi-molar amounts (Figure 1). The biliverdin is subsequently reduced by biliverdin reductase to bilirubin, while free iron is rapidly sequestered in ferritin, and CO is simply exhaled from the body. Since the degradation of heme is mostly carried out in the reticulo-endothelial system, under physiological conditions the activity of this enzyme is highest in the spleen where senescent erythrocytes are trapped and destroyed.

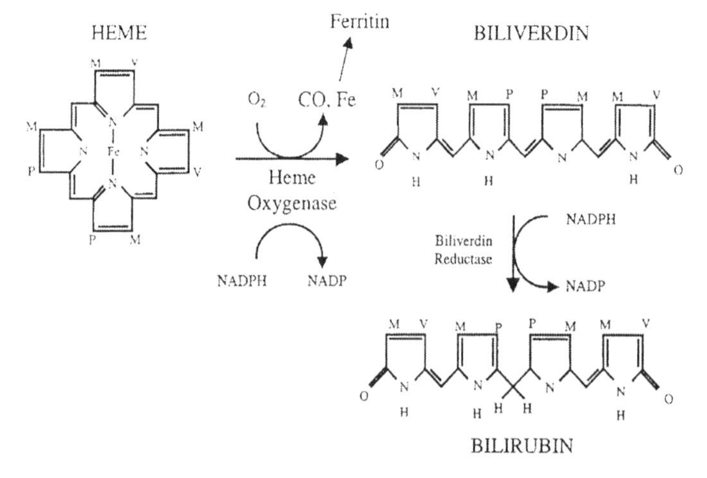

Figure 1. Catalytic reaction of heme oxygenase

Distinct genes code for each of the three isoforms of HO that have been discovered to date (8, 9). While HO-2 and HO-3 are constitutively expressed, HO-1 is highly inducible by both heme and non-heme inducers, as described above. In contrast to the other HO isoforms, HO-3 is a poor heme catalyst but has two potential heme binding sites and therefore may regulate heme-dependent processes (9). The distribution and difference of each isoform are summarized in Table 1.

Table 1. Summary of heme oxygenase isozymes

	HO-1	HO-2	HO-3
Localization	Ubiquitous	Brain and testes	Ubiquitous*
Inducibility	High	Constitutive	Constitutive
Molecular mass (kD)	~32-33	~36	~32-33
Number of transcripts	1	2	1
Known inducers	Heme/Metalloporphyrins Transition metals Ultraviolet light Lipopolysaccharide Hyperoxia and hypoxia Electrophiles; Thiol scavengers Chemotherapy; Prostaglandins Cytokines; Phorbol esters Hydrogen peroxide Heat shock; Sodium arsenite	Corticosterone	?

*Highest in liver, prostate, and kidneys

Regulation of Heme Oxygenase

The genes for rat (10) and human HO-1 (11) have been cloned, but the mouse HO-1 is the best characterized among the HO-1 genes of different species (12). There are five exons in the HO-1 gene, and the relative positions of these exons are maintained in all the three species. The deduced amino acid sequence of the mouse HO-1 gene exhibits 93.4 % and 82.3% identity with the rat and human proteins, respectively, and retains the histidine residues at positions 25 and 132 which are critical to heme binding at the functional site of the enzyme.

In 1978, Shibahara et al demonstrated that heme-dependent stimulation of HO activity in porcine alveolar macrophages was inhibited by prior treatment with the RNA synthesis inhibitor, actinomycin D, suggesting that induction of enzyme activity was regulated at the level of gene transcription (13). It appears now that regulation of HO-1 induction by most agents, not just heme, occurs at the level of gene transcription, and research has focused on identifying the *cis*-acting DNA elements and their cognate proteins that mediate the induction response. The sheer number and diversity of inducers would argue against a single, common transcription activation pathway, but since many of these agents generate oxidative stress it may well be that in some cases a promiscuous activation pathway or element does exist.

Various regulatory elements exist in the 5' flanking region of the mouse rat and human HO-1 gene. Most of these exhibit motifs that are variations of recognition sites for known DNA binding proteins, such as the *fos/jun* (AP-1) and NFκB/*Rel* family of oxidative stress-response transcription factors. Only a few of these putative regulatory elements, however, actually mediate inducer-dependent HO-1 gene activation. In the case of the mouse HO-1 gene, the best characterized of the different species, three inducer responsive regions exist: the proximal enhancer (PE), directly upstream of the promoter and two distal enhancers (DE), DE 1 and 2, located 4 and 10 kb upstream of the transcription initiation site, respectively. The dominant sequence is a 10-bp sequence, called the stress response element (StRE). This sequence is present 5 times in the two DE's, contains an AP-1 binding site, and is necessary for induction of the mouse HO-1 by all agents tested except hypoxia. Mutation of the AP-1 binding heptad abolishes HO-1 activation by heavy metals, hydrogen peroxide, arsenite, and LPS (14, 15).

The first three residues of the StRE conform to the anti-oxidant/electrophile response elements, or ARE's, and are necessary for induction by electrophiles (16). Similarly, induction by heme is abrogated by disruption of the StRE within or outside of the AP-1 heptad, suggesting that the heme inducer recognizes the entire 10 bp of the StRE (17). The v-Maf oncoprotein and the NF-E2 transcription activating factors, or the closely related and ubiquitously expressed Nrf1 and Nrf2 transcription activating factors, are likely proteins involved in this response.

The StRE's are unresponsive to hypoxic stress, however. In vascular smooth muscle cells, hypoxic induction of HO-1 is mediated via two motifs in DE2 (the HypoREs) that bind the hypoxia inducible factor (HIF-1), and this response to hypoxia can be abrogated by mutating the hypoRE site (18). HIF-1 is the nuclear factor that activates transcription of a variety of hypoxia-inducible genes including erythropoetin and vascular endothelial growth factor.

IL-6 induces HO-1 transcription by binding to a surface receptor and activating a JAK factor that, in turn, phosphorylates and dimerizes a STAT factor which translocates to the nucleus and binds to an IL6 response element in the PE region of the mouse gene (19). This IL-6 response element is necessary but not sufficient for IL6 induction of HO-1 in RAW 264.7 (mouse peritoneal macrophages) cells (20). Instead, co-operativity is needed with a protein(s) bound to the StRE's, located about 4 kb away. This unique situation is not clearly understood, but may also pertain to HO-1 activation by interferon-γ, which also uses the JAK-STAT pathways.

Notably, in RAW 264.7 cells, the hyperoxia response element maps to the IL-6 response element. It could be that IL-6 is induced in a paracrine or autocrine manner by hyperoxia (peak accumulation of mRNA 4 to 8 hours), leading to subsequent HO-1 induction (peak induction 24 to 48 hours). On the other hand, hyperoxia does induce c-*jun* and c-*fos* (component proteins of the AP-1 complex), both of which induce STAT-DNA binding.

HO-1 is also known as heat shock protein-32, reflecting the transcriptional activation of the rat gene by heat shock, mediated by two synergistic heat shock elements in the PE region. Human HO-1 induction by heat shock has only been consistently seen in the Hep3B hepatoma cell line, and to a minimal extent in HeLa cells, and skin fibroblasts. A single HSE does exist in the human HO-1 gene, but recent evidence suggests it is constitutively repressed *in vivo* under the action of a sequence downstream to the heat shock element (21).

Induction of Heme Oxygenase in ALI

ALI is characterized by diffuse inflammatory damage to the alveoli of the lung in response to a variety of insults, including bacteria, viruses and other infectious pathogens, environmental toxins, mineral and metal dusts, noxious gases, and drugs. The pathological hallmark of this process is diffuse alveolar damage characterized by destruction of alveolar epithelium, exudation of eosinophilic proteinaceous material into the alveolar space, and deposition of coagulated fibrin along the alveolar walls, forming hyaline membranes. A fundamental principle underlying the pathogenesis of this disorder is that overwhelming release of ROS contributes directly to the cellular and tissue damage seen in this condition (22 and Chapters 6, 7, and 9). For example, many injurious agents, such as LPS derived from Gram-negative bacteria, host-derived TNFα, and cytotoxic drugs like bleomycin, contribute to lung injury through ROS production. Much of the ROS produced comes from the oxidative burst of infiltrating inflammatory cells in the lung, but the supplemental oxygen that patients require to support organ function also contributes to the oxidant burden of the injured lung.

These ROS, such as singlet oxygen, hydrogen peroxide, hydroxyl radical and superoxide, are extremely reactive, rapidly oxidizing proteins, membrane lipids, and nucleic acids. Since ROS production is the price that has to be paid for having evolved aerobic metabolic machinery, organisms have also developed complex enzymatic and non-enzymatic anti-oxidant mechanisms to scavenge these ROS. Non-enzymatic defenses include vitamins C and E, sulfhydryl-containing glutathione, bilirubin, uric acid, the transferrin-lactoferrin system, and ceruloplasmin, to name a few. Although manganese-superoxide dismutase (MnSOD), copper-zinc superoxide dismutase (Cu-ZnSOD), catalase and glutathione peroxidase are well-established enzymes responsible for the scavenging of ROS, it is increasingly evident that many other stress response genes are involved in antioxidant defense, among them heat shock proteins, metallothioneins, transcription factors, and in particular, HO-1.

Two of the most instructive and established *in vivo* models of oxidant-induced ALI involve the administration of hyperoxia (>95% O_2) and endotoxin, both of which produce changes in rats that are remarkably similar to those seen in the human disease. In the rat hyperoxia model minimal lung edema or pleural effusion is observed in the first 48 hours of exposure to >95% oxygen, but increases significantly between 48 and 60 hours. Rats uniformly die between 60 and 72 hours of continuous hyperoxia (23). In the rat endotoxin model, rats given sub-lethal intravenous injections of LPS develop marked neutrophilic alveolitis and lung edema and hemorrhage. Not surprisingly, marked up-regulation of lung HO-1 mRNA, protein, and enzymatic activity occurs in both the hyperoxic- and LPS-mediated lung injury models (15, 24). In both models, immunohistochemical staining for HO-1 protein increases diffusely in the alveolar and respiratory bronchiolar epithelium, and in infiltrating inflammatory cells. Of note, similar induction of MnSOD mRNA after hyperoxia does not correlate with increased enzymatic activity, while mRNA levels of Cu-ZnSOD, catalase, and glutathione peroxidase do not increase, despite variable changes in enzymatic activity (23). Thus, HO-1 appears to be uniquely positioned to defend against ROS-mediated lung injury, in these models. *In vitro* experiments have confirmed the induction by hyperoxia of HO-1 mRNA (and protein) in cultured lung fibroblasts, pulmonary epithelial cells, alveolar macrophages, and rat aortic vascular smooth muscle cells, and have also shown that this mRNA induction is due to increased gene transcription and not increased mRNA stability (24).

Functional Significance of HO-1 Induction in ALI

Proof that induction of HO-1 confers protection against hyperoxia comes from several studies. Pulmonary alveolar epithelial cells stably transfected with HO-1 cDNA express increased basal levels of HO-1 mRNA and enzymatic activity and are markedly resistant to hyperoxic injury (25,

26). Inhibiting HO activity with tin protoporphyrin (SnPP), a selective HO inhibitor, reverses this protection. Instillation of an adenoviral vector containing the coding region of the HO-1 gene into rat lungs resulted in increased HO-1 protein expression in the bronchiolar epithelium. The rats with lung overexpression of HO-1 exhibited a > 90% reduction in pleural effusions, lung edema, and inflammation and survived longer in hyperoxic environments than control rats that received the adenoviral vector without the HO-1 gene insert (27).

Similarly, pretreatment of rats with hemoglobin (an inducer of HO-1) conferred considerable protection against LPS-mediated neutrophilic alveolitis and lung edema (28). It is probable that the 20-fold induction of HO-1 in the lung by lethal doses of LPS occurs too late (16-24 hours) to protect the animal from the rapid development of shock and death. However, treatment with hemoglobin to induce HO in advance of the lethal insult does confer a statistically significant survival benefit after a lethal dose of intravenous LPS. Most importantly, pretreatment with SnPP abrogated the induction of lung HO-1 by LPS, increased susceptibility to a lethal dose of LPS, and significantly ablated the survival benefit conferred by prior treatment with hemoglobin (28).

The beneficial effects of HO-1 in ALI may extend much beyond its role as an anti-oxidant. There is accumulating evidence that HO-1 exerts anti-apoptotic effects, and can modulate cell growth and the inflammatory cascade. Each of these mechanisms has relevance for ALI. Apoptosis, or programmed non-inflammatory cell death, for example, plays an important role in both human ARDS and hyperoxic lung injury in rodents (see Chapter 15). For example, the pro-apoptotic peptide, soluble Fas Ligand (sFasL), is present in the bronchoalveolar lavage fluid from patients in early ARDS but not in those at risk for ARDS, and mediates apoptotic death of distal airspace epithelium that expresses the receptor for this “death” ligand (29). Indeed, higher levels of sFasL correlate with a fatal outcome. Proliferating fibroblasts and endothelial cells from the lungs of ARDS patients have also been shown to undergo apoptosis (30). In this context, it is exciting that HO-1 upregulation can protect both cultured fibroblasts from apoptosis induced by TNFα (31), and bovine aortic endothelial cells from peroxynitrite-induced apoptosis (32). Exogenous gene transfer of HO-1 into mouse lungs, referred to earlier, attenuates the degree of hyperoxia induced apoptosis *in vivo*, and these effects can be replicated by CO alone, a by-product of the reaction catalyzed by HO (see below) (27). These anti-apoptotic properties of HO-1 may not be surprising in light of the fact that ROS are believed to mediate apoptosis, and another key anti-apoptotic protein, bcl-2, is also an anti-oxidant (33, 34).

Induction of HO-1 may also serve to modulate the immune system in the body. The phenotype of the HO-1 (-/-) null mice, and the single reported HO-1 deficient patient, are both remarkable for increased inflammatory profiles including leukocytosis and thrombocytosis (35, 36). Further, HO-1, acting via CO, attenuates eosinophilic infiltration in a model of allergic airway disease (37). Even more impressively, expression of HO-1 has been shown to abrogate transplant rejection in an *in vivo* heart xenotransplantation model, which is strong evidence for a potent anti-inflammatory, or immune-modulating, role for HO-1 (38). Indeed, increased levels of CO are found in the exhaled breath of critically ill patients admitted to the intensive care unit (39), probably produced by the lung itself (40), reflecting the induction of HO-1 in clinical states of systemic or pulmonary inflammation.

HO-1 can also modulate the growth of cells *in vitro*. In the stably transfected HO-1 over expressing pulmonary epithelial cell line that was resistant to hyperoxic stress, decreased cell proliferation was associated with an increase in the number of cells in the G_0/G_1 phase and decreased entry into the S phase (25). The HO-1 overexpressing cells accumulated at the G_2/M phase and failed to progress through the cell cycle with serum stimulation, and importantly, SnPP reversed both the growth arrest and the survival benefit, confirming that these effects were the result of HO-1 activation. It may well be that growth arrest confers a protection against hyperoxic stress. Again, transfer of HO-1 via an adenoviral vector to vascular smooth muscle cells also slows their growth (41). CO derived from the HO-1 in vascular smooth muscle cells exerts a paracrine effect on co-cultured endothelial cells, decreasing the expression of mitogens like endothelin-1 and PGDF on these endothelial cells (42). Since smooth muscle hypertrophy contributes to the pulmonary hypertension of hypoxemia, HO-1 may serve to ameliorate this process.

Potential mechanisms for protection

The issue of just how HO-1 exerts this multitude of beneficial effects is a currently unresolved, but avidly researched. The catabolic activity of the enzyme generates three biologically versatile by-products, any or all of which have the potential to execute some of the cytoprotective effects of heme oxygenase.

Bilirubin. Traditionally thought of as a waste product since it is excreted in the bile and accumulates in cholestatic diseases, bilirubin is now known to be a potent free radical scavenger (43). It is demonstrably active in nanomolar concentrations against hydrogen peroxide-induced injury in neuronal cells (44), and may mediate the protective effects of HO-1 in a rat liver ischemia-reperfusion injury model (45), and in an ischemic heart disease model (46). Furthermore, biliverdin, the precursor of bilirubin, can modulate the expression of adhesion molecules (P- and E-selectin) on vascular endothelial beds, illustrating the considerable anti-inflammatory capability of these molecules (47).

Ferritin and free iron. Ferrous iron is able to donate one electron to hydrogen peroxide, ultimately leading to the generation of the highly potent hydroxyl radical via the Fenton reaction, as well as acting as a catalyst for lipid peroxidation. This apparent conundrum, whereby an anti-oxidant enzyme like HO-1 catalyzes the production of a potent oxidant, can be resolved by noting the natural fate of free iron. Free iron is rapidly sequestered in ferritin, thereby reducing the pro-oxidant state of the cell (48). Both methemoglobin-mediated and heme-mediated induction of HO-1 result in simultaneous increases in ferritin (49, 50), suggesting that ferritin acts as a sink to prevent iron's involvement in oxidative injury. Human skin fibroblasts are protected against the oxidative stress of ultraviolet irradiation, by a mechanism involving the heme oxygenase-dependent induction of ferritin, thus providing clear evidence for the biological importance of this sequestration pathway (51). Finally, a unique Fe^{2+} ATPase iron transporter responsible for vigorous iron uptake into cells, has been shown to co-localize with HO-1 to the microsomal fraction, and is possibly repressed by HO-1. Mice with genomic deletion of HO-1 have augmented iron transport and accumulation (52), suggesting another mechanism for the anti-oxidant effect of HO-1.

While this evidence does implicate ferritin in mediating the protective effects of HO-1, an elegant set of experiments by Otterbein et al provides evidence that this is not the only mechanism. Otterbein et al pretreated rats with heme and desferrioxamine to induce HO-1 but not ferritin, and were able show full protection against LPS-mediated, lethal endotoxic shock in these animals (53). Pretreatment with inorganic iron alone strongly induced ferritin in the tissues (without induction of HO-1), but failed to protect against lethal dose LPS, and exogenous apoferritin also did not confer protection in the model.

Carbon Monoxide. This "poisonous gas", so called because of its ability to poison cytochrome a_3 and impair the oxygen-carrying capacity of hemoglobin, was first suspected to be endogenously produced in 1927, well before the description of the heme oxygenase reaction as its major source. Claude Bernard noted the affinity of CO for hemoglobin in 1857, and the one and a half centuries worth of investigation since has incessantly emphasized the more readily apparent toxic aspects of this gaseous molecule. That nature and evolution did not entirely subscribe to this viewpoint became apparent only in the last few years, as researchers began working with CO in the minute quantities more relevant to endogenous production in the HO-1 reaction. Indeed, animals exposed to doses of CO as low as 250 ppm, are protected against hyperoxic (54) and LPS-induced lung injury, in a comparable manner to overexpression of HO-1. *In vitro*, exposure of cultured macrophage cells to 250 ppm CO blunts the LPS-induced production of the pro-inflammatory cytokines TNFα, IL-1β, MIP-1α, and IL-6, while augmenting the production of the anti-inflammatory cytokine IL-10 (55). Likewise, TNFα and IL-6 are inhibited, and IL-10 production is augmented, in *in vivo* models of CO treatment. In the cardiac xenotransplantation model referred to above, inhibition of HO-1 activity with SnPP inhibits the long term survival conferred by overexpression of HO-1, but survival can be rescued by exposure to 250ppm CO (56). In other words, CO alone can provide the protection against rejection that is attributed to HO-1 overexpression.

The mechanisms whereby CO exerts these effects are being defined. CO can activate guanyl cyclase in the lung (57), but the anti-inflammatory effects mediated by CO do not use this pathway (55). In an *in vitro* system, exogenous CO by itself replicates the ability of HO-1 to modulate the MAP kinase pathways (58), which may be the mechanism by which cytokine production is affected (59 and Chapter 1).

Could CO be the major mediator of the cytoprotective effects of HO-1? Although the wealth of data would support such a contention, conclusions can only be reached if CO could be shown to replicate all the effects of HO-1 in a variety of cell types and *in vivo*. More likely, the three byproducts (bilirubin, free iron/ferritin, and CO) interact in complex ways and contribute collectively to the observed effects of HO-1 as an anti-inflammatory and cytoprotective agent. Elucidation of the roles of these molecules is crucial to the design of effective therapeutic strategies and agents based upon the mechanism of action of HO-1.

Conclusion

Obligate aerobes have to face the potential for damage caused by free oxygen radicals. Perhaps it is no accident that the enzyme that degrades the oxygen transporter hemoglobin also functions as an anti-oxidant, and acts via an intermediary gas that itself reduces oxygen transport. This simple, but exquisite, natural logic might have been obscure at first, but recent years have witnessed the remarkable rehabilitation of heme oxygenase from lowly catabolic enzyme to a powerful anti-inflammatory and cytoprotective gene product. Bilirubin and ferritin are now established anti-oxidants in cellular metabolism, and as more research focuses on CO perhaps this gas will also be elevated from poison to pharmacological agent, explaining why its physiological production in the body has withstood the test of time and evolution. It is certain, however, that heme oxygenase and its products are vitally important to the cell and the organism, and with the benefit of more research may change current paradigms of protection against oxidative and inflammatory injury.

REFERENCES

1. Tenhunen, R., Marver, H.S., and Schmid, R. (1968) The enzymatic conversion of heme to bilirubin by microsomal heme oxygenase. *Proc Natl Acad Sci USA* 61, 748-755
2. Applegate, L.A., Luscher, P., and Tyrrell, R.M. (1991) Induction of heme oxygenase: a general response to oxidant stress in cultured mammalian cells. *Cancer Res* 51, 974-978
3. Otterbein, L.E., and Choi, A.M. (2000) Heme oxygenase: colors of defense against cellular stress. *Am J Physiol Lung Cell Mol Physiol* 279, L1029-L1037
4. Richaud, C., and Zabulon, G. (1997) The heme oxygenase gene (pbsA) in the red alga Rhodella violacea is discontinuous and transcriptionally activated during iron limitation. *Proc Natl Acad Sci USA* 94, 11736-11741
5. Cornejo, J., Willows, R.D., and Beale, S.I. (1998) Phytobilin biosynthesis: cloning and expression of a gene encoding soluble ferredoxin-dependent heme oxygenase from Synechocystis sp. PCC 6803. *Plant J* 15, 99-107
6. Wilks, A., and Schmitt, M.P. (1998) Expression and characterization of a heme oxygenase (Hmu O) from Corynebacterium diphtheriae. Iron acquisition requires oxidative cleavage of the heme macrocycle. *J Biol Chem* 273, 837-841
7. Poss, K.D., and Tonegawa, S. (1997) Heme oxygenase 1 is required for mammalian iron reutilization. *Proc Natl Acad Sci USA* 94, 10919-10924
8. Cruse, I., and Maines, M.D. (1988) Evidence suggesting that the two forms of heme oxygenase are products of different genes. *J Biol Chem* 263, 3348-3353
9. McCoubrey, W.K., Huang, T.J., and Maines, M.D. (1997) Isolation and characterization of a cDNA from the rat brain that encodes hemoprotein heme oxygenase-3. *Eur J Biochem* 247, 725-732
10. Shibahara, S., Muller, R., Taguchi, H., and Yoshida, T. (1985) Cloning and expression of cDNA for rat heme oxygenase. *Proc Natl Acad Sci USA* 82, 7865-7869
11. Shibahara, S., Sato, M., Muller, R.M., and Yoshida, T. (1989) Structural organization of the human heme oxygenase gene and the function of its promoter. *Eur J Biochem* 179, 557-563
12. Alam, J., Cai, J., and Smith, A. (1994) Isolation and characterization of the mouse heme oxygenase-1 gene. Distal 5' sequences are required for induction by heme or heavy metals. *J Biol Chem* 269, 1001-1009
13. Shibahara, S., Yoshida, T., and Kikuchi, G. (1978) Induction of heme oxygenase by hemin in cultured pig alveolar macrophages. *Arch Biochem Biophys* 188, 243-250
14. Alam, J., Camhi, S. and Choi, A.M. (1995) Identification of a second region upstream of the mouse heme oxygenase-1 gene that functions as a basal level and inducer-dependent transcription enhancer. *J Biol Chem* 270, 11977-11984
15. Camhi, S.L., Alam, J., Otterbein, L., Sylvester, S.L., and Choi , A.M. (1995) Induction of heme oxygenase-1 gene expression by lipopolysaccharide is mediated by AP-1 activation. *Am J Respir Cell Mol Biol* 13, 387-338
16. Prestera, T., Talalay, P., Alam, J., Ahn, Y.I., Lee, P.J., and Choi, A.M. (1995) Parallel induction of heme oxygenase-1 and chemoprotective phase 2 enzymes by electrophiles

and antioxidants: regulation by upstream antioxidant-responsive elements (ARE). *Mol Med* 1, 827-837
17. Inamdar, N.M., Ahn, Y.I., and Alam, J. (1996) The heme-responsive element of the mouse heme oxygenase-1 gene is an extended AP-1 binding site that resembles the recognition sequences for MAF and NF-E2 transcription factors. *Biochem Biophys Res Commun* 221, 570-576
18. Lee, P., et al., Regulation of heme oxygenase-1 (HO-1) after hypoxia. (1996) *Am J Respir Crit Care Med* 153, A29 (Abstract)
19. Mitani, K., Fujita, H., Kappas, A., and Sassa, S. (1992) Heme oxygenase is a positive acute-phase reactant in human Hep3B hepatoma cells. *Blood* 79, 1255-1259
20. Camhi, S., Alam, J., and Choi, A. (1996) Transcriptional activation of the mouse heme oxygenase-1 gene (HO-1) by interleukin-6 (IL-6) requires cooperation between the 5' distal enhancer and the proximal promoter. *Am J Respir Crit Care Med* 153, A29
21. Okinaga, S., Takahashi, K., Takeda, K., Yoshizawa, M., Fujita, H., Sasaki, H., Shibahara, S. (1996) Regulation of human heme oxygenase-1 gene expression under thermal stress. *Blood* 87, 5074-5084
22. Kinnula, V.L., Crapo, J.D., and Raivio, K.O. (1995) Generation and disposal of reactive oxygen metabolites in the lung. *Lab Invest* 73, 3-19
23. Clerch, L.B. and Massaro, D. (1993) Tolerance of rats to hyperoxia. Lung antioxidant enzyme gene expression. *J Clin Invest* 91, 499-508
24. Lee, P.J., Alam, J., Sylvester, S.L., Inamdar, N., Otterbein, L., and Choi, A.M. (1996) Regulation of heme oxygenase-1 expression in vivo and in vitro in hyperoxic lung injury. *Am J Respir Cell Mol Biol* 14, 556-568
25. Lee, P.J., Alam, J., Wiegand, G.W., and Choi, A.M. (1996) Overexpression of heme oxygenase-1 in human pulmonary epithelial cells results in cell growth arrest and increased resistance to hyperoxia. *Proc Natl Acad Sci USA* 93, 10393-10398
26. Dennery, P.A., Wong, H.E., Sridhar, K.J., Rodgers, P.A., Sim, J.E., and Spitz, D.R. (1996) Differences in basal and hyperoxia-associated HO expression in oxidant-resistant hamster fibroblasts. *Am J Physiol* 271, L672-679
27. Otterbein, L.E., Kolls, J.K., Mantell, L.L., Cook, J.L., Alam, J., and Choi, A.M. (1999) Exogenous administration of heme oxygenase-1 by gene transfer provides protection against hyperoxia-induced lung injury. *J Clin Invest* 103, 1047-1054
28. Otterbein, L., Sylvester, S.L., and Choi, A.M. (1995) Hemoglobin provides protection against lethal endotoxemia in rats: the role of heme oxygenase-1. *Am J Respir Cell Mol Biol* 13, 595-601
29. Matute-Bello, G., Liles, W.C., Steinberg, K.P., Kiener, P.A., Mongovin, S., Chi, E.Y., Jonas, M., and Martin, T.R. (1999) Soluble Fas ligand induces epithelial cell apoptosis in humans with acute lung injury (ARDS). *J Immunol* 163, 2217-2225
30. Polunovsky, V.A., Chen, B., Henke, C., Snover, D., Wendt, C., Ingbar, D.H., and Bitterman, P.B. (1993) Role of mesenchymal cell death in lung remodeling after injury. *J Clin Invest* 1993 92, 388-397
31. Petrache, I., Otterbein, L.E., Alam, J., Wiegand, G.W., and Choi, A.M. (2000) Heme oxygenase-1 inhibits TNF-alpha-induced apoptosis in cultured fibroblasts. *Am J Physiol Lung Cell Mol Physiol* 278, L312-319
32. Foresti, R., Sarathchandra, P.. Clark, J.E., Green, C.J., and Motterlini, R. (1999) Peroxynitrite induces haem oxygenase-1 in vascular endothelial cells: a link to apoptosis. *Biochem J* 339, 729-736
33. Kluck, R.M., Bossy-Wetzel, E., Green, D.R., and Newmeyer, D.D. (1997) The release of cytochrome c from mitochondria: a primary site for Bcl-2 regulation of apoptosis. *Science* 275, 1132-1136
34. Yang, J., Liu, X., Bhalla, K., Kim, C.N., Ibrado, A.M., Cai, J., Peng, T.I., Jones, D.P., and Wang, X. (1997) Prevention of apoptosis by Bcl-2: release of cytochrome c from mitochondria blocked. *Science* 275, 1129-1132

35. Poss, K.D. and Tonegawa, S. (1997) Reduced stress defense in heme oxygenase 1-deficient cells. *Proc Natl Acad Sci USA* 94, 10925- 10930
36. Yachie, A., Niida, Y., Wada, T., Igarashi, N., Kaneda, H., Toma, T., Ohta, K., Kasahara, and Y., Koizumi, S. (1999) Oxidative stress causes enhanced endothelial cell injury in human heme oxygenase-1 deficiency. *J Clin Invest* 103, 129-135
37. Chapman, J., et al. (1999) Exogenous carbon monoxide attenuates aeroallergen-induced eosinophilic inflammation in mice. *Am J Respir Crit Care Med* 159, A218 (Abstract)
38. Soares, M.P., Lin, Y., Anrather, J., Csizmadia, E., Takigami, K., Sato, K., Grey, S.T., Colvin, R.B., Choi, A.M., Poss, K.D., and Bach, F.H. (1998) Expression of heme oxygenase-1 can determine cardiac xenograft survival. *Nat Med* 4, 1073-1077
39. Scharte, M., Bone, H.G., Van Aken, H., and Meyer, J. (2000) Increased carbon monoxide in exhaled air of critically ill patients. *Biochem Biophys Res Commun* 267, 423-426
40. Meyer, J., Prien, T., Van Aken, H., Bone, H.G., Waurick, R., Theilmeier, G., and Brooke, M. (1998) Arterio-venous carboxyhemoglobin difference suggests carbon monoxide production by human lungs. *Biochem Biophys Res Commun* 244, 230-232
41. Thau, S., Sasidhar, M., and Choi, A. (1999) Heme oxyenase-1 modulates cellular proliferation of rat pulmonary artery smooth muscle cells. *Am J Respir Crit Care Med* 159, A345
42. Morita, T., and Kourembanas, S. (1995) Endothelial cell expression of vasoconstrictors and growth factors is regulated by smooth muscle cell-derived carbon monoxide. *J Clin Invest* 96, 2676-2682
43. Stocker, R., Yamamoto, Y., McDonagh, A.F., Glazer, A.N., and Ames, B.N. (1987) Bilirubin is an antioxidant of possible physiological importance. *Science* 235, 1043-1046
44. Dore, S., Takahashi, M., Ferris, C.D., Zakhary, R., Hester, L.D., Guastella, D., and Snyder, S.H. (1999) Bilirubin, formed by activation of heme oxygenase-2, protects neurons against oxidative stress injury. *Proc Natl Acad Sci USA* 96, 2445-2450
45. Yamaguchi, T., Terakado, M., Horio, F., Aoki, K., Tanaka, M., and Nakajima, H. (1996) Role of bilirubin as an antioxidant in an ischemia-reperfusion of rat liver and induction of heme oxygenase. *Biochem Biophys Res Commun* 223, 129-135
46. Clark, J.E., Foresti, R., Sarathchandra, P., Kaur, H., Green, C.J., and Motterlini, R. (2000) Heme oxygenase-1-derived bilirubin ameliorates postischemic myocardial dysfunction. *Am J Physiol Heart Circ Physiol* 278, H643-651
47. Vachharajani, T.J., Work, J., Issekutz, A.C., and Granger, D.N. (2000) Heme oxygenase modulates selectin expression in different regional vascular beds. *Am J Physiol Heart Circ Physiol* 278, H1613-1617
48. Balla, G., Jacob, H.S., Balla, J., Rosenberg, M., Nath, K., Apple, F., Eaton, J.W., and Vercellotti, G.M. (1992) Ferritin: a cytoprotective antioxidant stratagem of endothelium. *J Biol Chem* 267, 18148-18153
49. Balla, J., Jacob, H.S., Balla, G., Nath, K., Eaton, J.W., and Vercellotti, G.M. (1993) Endothelial-cell heme uptake from heme proteins: induction of sensitization and desensitization to oxidant damage. *Proc Natl Acad Sci USA* 90, 9285-9289.
50. Eisenstein, R.S., Garcia-Mayol, D., Pettingell, W., and Munro, H.N. (1991) Regulation of ferritin and heme oxygenase synthesis in rat fibroblasts by different forms of iron. *Proc Natl Acad Sci USA* 88, 688-692
51. Vile, G.F. and Tyrrell, R.M. (1993) Oxidative stress resulting from ultraviolet A irradiation of human skin fibroblasts leads to a heme oxygenase-dependent increase in ferritin. *J Biol Chem* 268, 14678-14681
52. Baranano, D.E., Wolosker, H., Bac, B.I., Barrow, R.K., Snyder, S.H., and Ferris, C.D. (2000) A mammalian iron ATPase induced by iron. *J Biol Chem* 275, 15166-15173
53. Otterbein, L., Chin, B.Y., Otterbein, S.L., Lowe, V.C., Fessler, H.E., and Choi, A.M. (1997) Mechanism of hemoglobin-induced protection against endotoxemia in rats: a ferritin-independent pathway. *Am J Physiol* 272, L268-275

54. Otterbein, L.E., Mantell, L.L., and Choi, A.M. (1999) Carbon monoxide provides protection against hyperoxic lung injury. *Am J Physiol* 276, L688-694
55. Otterbein, L.E., Bach, F.H., Alam, J., Soares, M., Tao Lu, H., Wysk, M., Davis, R.J., Flavell, R.A., and Choi, A.M. (2000) Carbon monoxide has anti-inflammatory effects involving the mitogen-activated protein kinase pathway. *Nat Med* 6, 422-428
56. Sato, K., Balla, J., Otterbein, L., Smith, R.N., Brouard, S., Lin, Y., Csizmadia, E., Sevigny, J., Robson, S.C., Vercellotti, G., Choi, A.M., Bach, F.H., and Soares, M.P. (2001) Carbon Monoxide generated by Heme oxygenase-1 suppresses the rejection of mouse to rat cardiac transplants: A mechanism for graft accommodation. *J Immunol* 166, 4185-4194
57. Cardell, L.O., Lou, Y.P., Takeymama, K., Ueki, I.F., Lausier, J., and Nadel, J.A. (1998) Carbon monoxide, a cyclic GMP-related messenger, involved in hypoxic bronchodilation in vivo. *Pulm Pharmacol Ther* 11, 309-315
58. Sethi, J., Otterbein, L., and Choi, A. (1999) Differential effects of exogenous carbon monoxide on TNF-alpha induced Mitogen Activated Protein Kinase signaling in rat pulmonary artery endothelial cells. *Am J Respir Crit Care Med* 159, A350
59. Dyck, E., et al. (2000) Differential function effect of the C-Jun N-terminal kinase (JNK) Mitogen Activated Protein Kinase (MAPK) in hyperoxic lung injury: Discordant relationship between lung injury and neutrophil alveolitis. *Am J Respir Crit Care Med* 161, A 514

Chapter 9

PULMONARY ENDOTHELIAL SURFACE REDOX ACTIVITY: ROLES IN PROPAGATION OF AND PROTECTION FROM INJURY

Marilyn P. Merker, Robert D. Bongard and Christopher A. Dawson
Medical College of Wisconsin, Marquette University, and Veteran's Administration Medical Center, Milwaukee, WI

INTRODUCTION

Endothelial cells, like other cells, have endogenous enzymatic sources of reactive oxygen (ROS) species (e.g., superoxide, hydrogen peroxide, hydroxyl radical) and nitrogen species (e.g., nitric oxide, peroxynitrite) generated from the mitochondrial electron transport chain, nitric oxide synthases, xanthine dehydrogenase/xanthine oxidase, NAD(P)H oxidases, cytochrome P450 enzymes, and the enzymes of arachadonic acid metabolism, lipoxygenase and cyclooxygenase (1). The ROS generated in these reactions can be important in host defense and in signal transduction, but their generation can also be self-destructive (2, 3). Their respective roles in propagation of pulmonary endothelial and lung injury have been evaluated extensively (4-7). Similarly, roles for antioxidant enzyme systems such as superoxide dismutase, catalase, glutathione peroxidase, and heme oxygenase, and for low molecular weight oxidant scavengers, such as glutathione, urate, and ascorbate, in protecting the endothelium and lung from injury are well established (8-11).

Research into these redox activities has mainly focused on intracellular processes, but interest in redox reactions associated with the endothelial cell plasma membrane has been growing. Because of their location at the blood-tissue interface, redox reactions carried out at the endothelial plasma membrane surface are in a unique position to influence blood and blood-tissue functions as well as the endothelial cells themselves. The lung endothelium is unique in that it makes up a much larger fraction of the whole organ than does the endothelium of other organs, and its surface area is nearly as large as the combined endothelial surfaces of all the rest of the body. Since virtually all of the venous blood crosses this surface before entering the systemic circulation, the pulmonary endothelium is situated to modify the redox status of blood-borne compounds and generate redox active substances that can have access to other other organs as well. This chapter will focus on some aspects of these reactions occuring at or near the endothelial plasmalemmal surface as they relate to pulmonary and systemic endothelial cell physiology and pathophysiology.

NAD(P)H Oxidases

Probably the best-characterized plasma membrane redox system is the NADPH oxidase of macrophages and neutrophils, which transfers electrons from intracellular NADPH to extracellular or phagosome oxygen resulting in superoxide ($O_2^{\bullet -}$) formation as a part of the inflammatory response (12). This is one contributor to pulmonary endothelial injury mediated by activated neutrophils and macrophages within the lung (13). Activation of the phagocyte enzyme involves binding of the cytosolic subunits $p47^{phox}$, $p67^{phox}$, $p40^{phox}$, and the G-protein rac-2 to the $gp91^{phox}$ and $p22^{phox}$ membrane subunits. Endothelial cells also contain several of these cytosolic subunits as well as the key membrane flavohemeprotein components $gp91^{phox}$ and $p22^{phox}$, and the b_{558} cytochrome (14-17). While the physiological roles and mechanisms of activation of endothelial NADPH oxidase activity may not be completely analogous to the phagocyte enzyme (18), these components, or analogous ones, apparently contribute to observed endothelial NAD(P)H oxidase activities (19-22). A number of stimuli result in NAD(P)H oxidase mediated $O_2^{\bullet -}$ release from vascular endothelium, including angiotensin II, endothelin, ischemia, and mechanical forces (23-28).

While $O_2^{\bullet -}$ release is one functional consequence of NAD(P)H oxidase activity, the redox active components of NADPH oxidase have the potential for a diversity of electron acceptor specificities, as demonstrated by the behavior of these components in different experimental environments. For example, when the one electron reduction of oxygen to $O_2^{\bullet -}$ carried out by NADPH oxidase is inhibited by the flavoprotein inhibitor, diphenylene iodonium (DPI), the activity of an NAD(P)H diaphorase domain capable of catalyzing a two electron reduction of various redox active substances can be observed (29). The diaphorase activity may play an important role in the bioactivity of redox active blood-borne substances, with the potential for both pro- and antioxidant effects (30-32). Molecular homologies between the yeast ferric reductase redox component FRE1 and $gp91^{phox}$ have led to the suggestion that NADPH oxidase in vascular cells may also be involved in reduction of extracellular ferric to ferrous iron chelates required for transferrin independent cellular iron uptake (1).

A role for pulmonary endothelial NADPH oxidase in lung pathophysiology was demonstrated by Al Mehdi et al (33) by comparing lung ROS production in a mouse strain lacking the $gp91^{phox}$ gene with the gp91^{phox} +/+ strain. Ischemia increased ROS production in gp91^{phox} +/+ mice but not gp91^{phox} −/− mice (Figure 1). The increase in ROS production in gp91^{phox} +/+ lungs was attenuated by DPI and a specific NADPH oxidase inhibitor, PR-39. Colocalization of the ROS induced fluorescent signal and a fluorescent endothelial cell marker revealed the endothelium as the dominant site of ROS production following these treatments. NADPH oxidase involvement in ROS production during ischemia was also demonstrated in pulmonary endothelial cells in culture (27, 34).

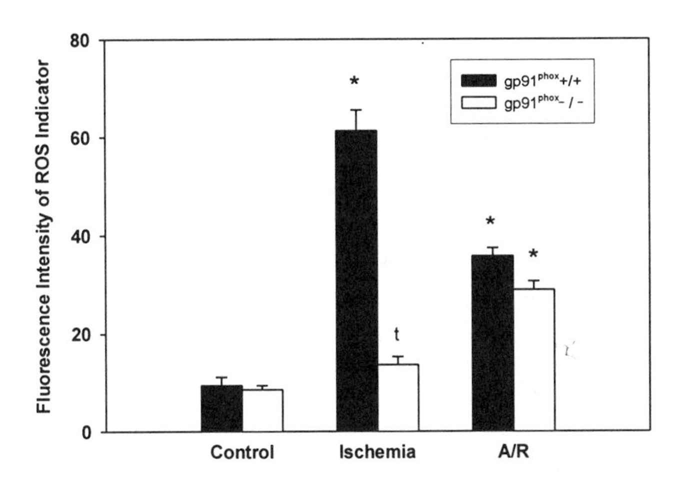

Figure 1. ROS generation (mean ± SE) in isolated lungs from gp91^{phox} +/+ and gp91^{phox} −/− mice. Mouse lungs were subjected to 1 hour of ischemia (no flow and ventilation with 95%O_2/5%CO_2) or to 1 hour of anoxia (flow and ventilation with 95%N_2/5% CO_2) followed by 1 hour of reoxygenation (A/R). *$P<0.05$ vs corresponding control; $^{t}P<0.05$ vs corresponding wild type. *Reprinted with permission from Al Mehdi et al (33) with modifications.*

In contrast to other organs, lung ischemia is not generally associated with hypoxia or anoxia. Rather, it has been hypothesized that the drop in shear stress is the stimulus for activation of the endothelial cell NADPH oxidase (27). The distinction between pathways of ROS production in the ischemic and anoxia/reoxygenation lungs is further emphasized by the results shown in Figure 1, in which increased ROS resulting from anoxia/reoxygenation, which may be more analogous to ischemia/ reperfusion in other organs, was apparently not of NADPH oxidase origin. Instead, allopurinol inhibitable xanthine oxidase activity was implicated (35). The apparent source of ROS in anoxia/reoxygenation is not as consistent between intact lung studies and studies of endothelial cells in culture as it seems to be for models of ischemia (33). While ROS production in lung anoxia/reoxygenation is apparently dominated by xanthine oxidase, in cultured cells both xanthine oxidase and NADPH oxidase have been implicated (22,36).

In addition to these NAD(P)H oxidases that utilize intracellular pyridine nucleotides, vascular cells can oxidize extracellular NAD(P)H to reduce oxygen to $O_2^{\bullet -}$ (37, 38). Oxidases for which both electron donor and electron acceptor are extracellular have been characterized more extensively in other cell types (39-41). Since NAD(P)H is not normally found in the extracellular fluid and is virtually cell membrane impermeable, the natural electron donors for these oxidases are not known.

Xanthine Oxidase

ROS generation from xanthine oxidase has been implicated in various experimental models of endothelial dysfunction and lung injury (4, 35, 36, 42). Endothelial xanthine oxidase is derived from at least two sources. One is the proteolytic conversion of intracellular xanthine dehydrogenase to xanthine oxidase that occurs under hypoxic or anoxic conditions (43, 44). The other is the deposition of xanthine oxidase on the luminal endothelial plasma membrane as a result of circulating enzyme binding to the cell surface (45-48). Plasma xanthine oxidase levels are increased in response to ischemic or inflammatory injury to systemic organs (46, 49-51), and ROS production catalyzed by the resulting increase of endothelial plasma membrane-bound xanthine oxidase is one contributor to lung injury following systemic organ injury (52, 53).

Because lung ischemia is generally associated with an increase rather than a decrease in lung tissue PO_2, it is not expected that lung ischemia would promote xanthine dehydrogenase to xanthine oxidase conversion. This is borne out by the results of the study described above in which xanthine oxidase was not the effector of ROS production in the ischemic lungs (33). On the other hand, under anoxic conditions, the conversion from xanthine dehydrogenase to xanthine oxidase would be expected to occur in the lung as in other organs (43). Anoxia also favors catabolism of ATP to hypoxanthine, which serves as a xanthine oxidase substrate, and upon reoxygenation, when xanthine oxidase is supplied with oxygen, ROS production ensues (43). Thus, this sequence provides an explanation for the apparent involvement of xanthine oxidase in the anoxia/reoxygenation model of lung injury (33, 35).

Like NAD(P)H oxidases, xanthine oxidase has multiple redox centers and the potential to utilize alternative electron acceptors and donors (54, 55). Thus, an allopurinol-insensitive, DPI-sensitive xanthine oxidase activity that utilizes NADH to generate $O_2^{\bullet -}$ has been implicated as a mechanism contributing to reperfusion injury (56). Xanthine oxidase can also catalyze glycosidic cleavage of anthracyclines such as doxorubicin and redox cycling of the pulmonary toxin paraquat and quinone type drugs (57-59).

Nitric Oxide Synthase

Endothelial nitric oxide (NO) synthase is present in specialized regions of the endothelial plasma membrane known as caveoli (60), producing NO at or near the cell surface. NO synthase has several redox active centers that work in concert to produce NO from arginine and oxygen, but the sequence of events leading to NO production can be altered depending on available substrates and cofactors such that, for example, a diaphorase activity or a $O_2^{\bullet -}$ generating activity is accentuated (61-63). It is not clear that any of these potential reactions can involve extracellular substrates, but the proximity of the enzyme to the cell surface ensures that its highly diffusible reaction products are involved in redox reactions occurring at the cell surface.

In addition to its well known role as a signalling molecule in regulating vascular tone and endothelial function (64), NO can also undergo reactions with ROS (see Chapter 7). These reactions can amplify the pro-oxidant effects of ROS, inactivate ROS, or lead to inactivation of NO itself. For example, NO reacts with $O_2^{\bullet -}$ to form the highly reactive oxidant peroxynitrite ($ONOO^-$) (65, 66). In addition to producing $ONOO^-$, the reaction consumes NO. Reduced levels of NO are associated with endothelial dysfunction, in part due to the loss of direct NO effects, but also due to the loss of secondary effects such as NO-dependent scavenging of peroxyl radicals (67). A role for the reaction of NO and $O_2^{\bullet -}$ and the subsequent generation of $ONOO^-$ in lung pathophysiology, is strongly suggested by the demonstration of nitrotyrosine formation in the lungs of patients with chronic obstructive pulmonary disease (68), pulmonary fibrosis (69), and acute lung injury (70).

Other Transplasma Membrane Electron Transport Systems

In addition to the phox-type NADPH oxidase, there are a number of other transplasma membrane electron transport (TPMET) systems in various cell types that transfer intracellular reducing equivalents to a broad range of extracellular electron acceptors (71, 72). These redox systems have been implicated in control of cell proliferation (40, 73) and other transmembrane signalling processes (74), protection from oxidant stress (75-77), iron metabolism (78, 79), antioxidant function of ascorbate and coenzyme Q_{10} (77, 80), release of NO from NO donors (81), and other functions (72, 82). In comparison with some other cell types, in which TPMET roles consistent with specific functions of those cells have been identified, the physiological role and mechanisms of pulmonary endothelial TPMET systems are less well understood. A variety of observations, however, suggest unique pro- and antioxidant functions of endothelial TPMET-mediated reduction of blood-borne redox active substances.

We were first introduced to endothelial TPMET by our observation that when certain electron acceptors are infused into the pulmonary artery, they exit the pulmonary veins in the reduced form (83). We found that the rate of reduction by pulmonary arterial endothelial cells in culture was sufficient to account for the rate observed in the intact lung, indicating that this cell type effects reduction within the pulmonary circulation (84). Several approaches were used to demonstrate that the reduction occurs on the endothelial cell surface (84, 85). The simplest involved the use of a cell impermeant electron acceptor, a polyacrylamide polymer (average molecular weight of 35,000 Daltons) containing covalently attached redox active toluidine blue (TBO) residues (84). This compound, referred to as TBOP, is colored when oxidized and colorless when reduced and because of its size, cannot enter cells in either the oxidized or reduced form.

When the oxidized form of TBOP, $TBOP^+$, is added to the medium surrounding cultured bovine pulmonary arterial endothelial cells, it begins to disappear from the medium until a steady state is reached (84) (Figure 2). The steady state represents the opposing rates of $TBOP^+$ reduction by the cells and autooxidation of the colorless reduced form in the medium. When the cells are removed from medium (at the time represented by the vertical dashed line in Figure 2), the $TBOP^+$ reappears as a result of unopposed autooxidation. The complete recovery of $TBOP^+$ in the medium confirms that neither the oxidized or reduced forms entered the cells.

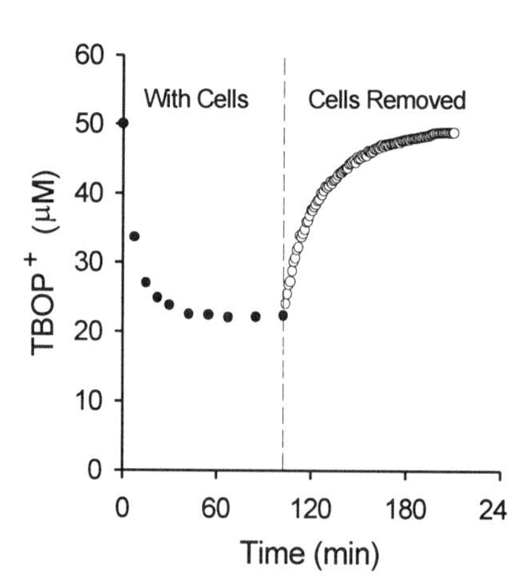

Figure 2. $TBOP^+$ reduction in the presence of endothelial cells (left of the vertical dashed line) and autooxidation of the reduced form when the cells are removed from the medium (right of dashed line).

Phenolic or polyphenolic compounds, including natural products such as coenzyme Q_{10}, tocopherols, vitamin K, other ingested plant products, xenobiotic drugs, and toxins, along with their quinonoid redox couples, can have anti- and/or pro-oxidant properties that depend, to a large extent, on the stability of their semiquinone or hydroquinone forms under aerobic conditions. The fact that thiazine reductases are commonly also quinone reductases (86) led us to investigate the impact of endothelial cells on the extracellular redox status of quinones introduced into the cell medium (87, 88). As an initial step, we evaluated the possibility that when endothelial cells are incubated with certain quinones, the reduced hydroquinone appears in the medium. Figure 3 shows the results using coenzyme Q_1 quinone, a water soluble analog of coenzyme Q_{10} quinone. The pattern of the graph is reminiscent of the $TBOP^+$ graph (Figure 2), with a rapid reduction phase that finally reaches equilibrium. Again, the reappearance of the quinone in the cell-free medium demonstrates its reduction by the cells. The key difference between TBOP (Figure 2) and coenzyme Q_1 (Figure 3) is that when the coenzyme Q_1 hydroquinone containing medium was removed from the cells, the reooxidation did not result in complete recovery of the initial coenzyme Q_1 quinone concentration. This implies that some of the quinone entered the cells in either the oxidized or reduced forms, or both, confounding evaluation of a role for TPMET. Thus, direct evidence for TPMET involvement is more difficult to acquire for low molecular weight amphipathic or lipophilic compounds because without a clearly cell impermeant form, it is difficult to rule out cell entry and intracellular reduction of the oxidized forms.

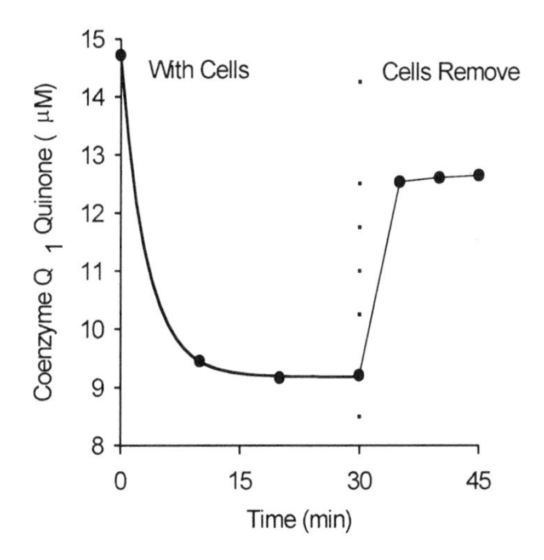

Figure 3. Coenzyme Q_1 is reduced by the endothelial cells. Before 30 mins, coenzyme Q_1 quinone was in the presence of the cells. At 30 mins, the cells were removed and an oxidizing agent added to the medium. The difference in the amount of quinone detected before and after addition of the oxidizing agent represents the amount of the reduced form (hydroquinone) in the medium when it was in contact with the cells.

To evaluate the possibility that TPMET might have contributed to the appearance of the hydroquinone in the medium, we studied quinone reduction by an endothelial cell plasma membrane preparation. The method takes advantage of a cell membrane impermeable amine reactive reagent, sulfosuccinimidyl 6-biotinamido hexanoate, which results in biotin-labelling of proteins exposed on the outer cell surface (89). Solubilized extracts of cells with labelled external plasma membrane proteins were incubated with beads that have avidin bound to their surface. Because of the high affinity of biotin for avidin, the biotin-labelled cell surface proteins bind to the beads, and can then be separated from intracellular proteins by centrifugation. Coenzyme Q_1 reductase activity of the biotin-labelled endothelial plasma membrane protein coated beads was detected by the oxidation of NADH in the presence of the quinone form as shown in Figure 4.

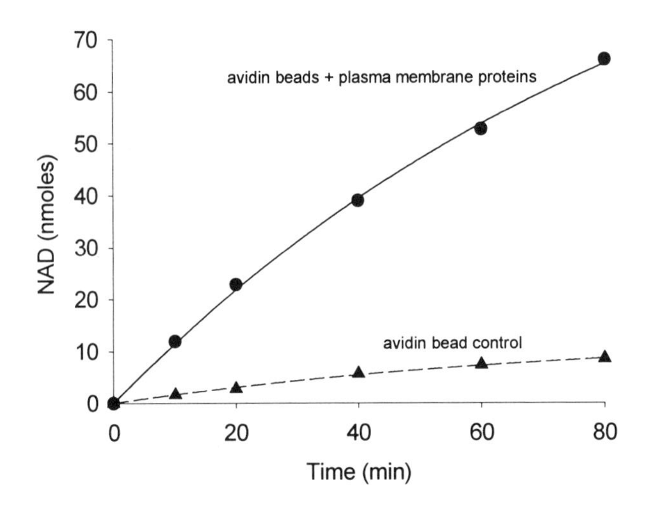

Figure 4. Coenzyme Q_1 reductase activity of biotin-labelled endothelial plasma membrane proteins on avidin beads. Both reaction mixtures contained NADH, coenzyme Q_1 quinone, and either avidin beads incubated with cell extracts from cells that contained biotin-labelled endothelial plasma membrane proteins (•), or avidin beads incubated with cell extracts from cells that were not exposed to the biotin-labelling reagent (▲). The latter is a control for non-specific binding of cell proteins and for NADH stability in the assay mixture.

There are at least two separate TPMET systems on pulmonary endothelial cells, the thiazine reductase system, which may also be the quinone reducing system described above, and another that utilizes ascorbate as the intracellular electron donor and can use ferricyanide, but not thiazines, as an electron acceptor (83-85, 88, 90, 91). By analogy with ferricyanide reductases found on other cell types, the endothelial ferricyanide reductase may have a number of functions (74, 79, 92), including catalysis of NO release from the NO donor sodium nitroprusside (81). Thus, like the other redox systems discussed in this chapter, pulmonary endothelial TPMET systems have the potential to reduce a range of extracellular electron acceptors. Although these endothelial TPMET systems are probably not unique to pulmonary endothelial cells, the size and location of the pulmonary endothelium makes it a unique reactor system for affecting the redox status of natural and xenobiotic blood borne substances in a manner analogous to its more well known role in controlling blood levels of other vasoactive agents (93-95).

Involvement of pulmonary TPMET in systemic vascular function may be exemplified by considering the fact that plasma lipoprotein oxidation has several detrimental effects on blood vessels, causing endothelial dysfunction (96-98), participating in atherosclerotic plaque formation (99, 100) and other pathological changes (101). There has been considerable interest in the concept that endothelial cells, as well as macrophages, which are the other major cell type involved in atherosclerotic plaque formation, can promote oxidation of low density lipoproteins (102-104). This prooxidant activity only occurs when free transition metals, such as copper or iron, are present in the medium (100), and it can be explained by TPMET catalyzed reduction of Cu^{++} at the cell surface (104, 105).

The *in vivo* relevance of this prooxidant effect of endothelial cells may be questionable since free transition metal concentrations are normally very low in plasma. In fact, in the absence of these metal ions, both endothelial cells and macrophages can protect plasma lipoproteins from oxidation (100, 104). One explanation for this protective effect is implied by the demonstration of the endothelial plasma membrane-associated quinone reductase activity described above. Low density lipoproteins contain small quantities of antioxidants such as coenzyme Q_{10} hydroquinone and tocopherols that protect against lipid oxidation. Plasma antioxidants such as ascorbate (80), reduced thioctic acid (106), various quinols (107), and others (108, 109) also play a protective role as co-antioxidants utilized to regenerate the lipoprotein antioxidants. Reduction of oxidized or partially oxidized forms of these hydrophilic and amphipathic co-antioxidants on the endothelial surface would provide, in turn, a means of restoring them to their reduced, antioxidant forms, thus maintaining their capacity to prevent depletion of lipoprotein antioxidants. The dual-edged sword of such redox reactions is again emphasized by considering the consequences of reduction of redox active toxins, such as paraquat. Redox cycling of paraquat at or near the endothelial surface is responsible for the presence of $O_2^{\bullet -}$ found in the medium surrounding paraquat-exposed pulmonary endothelial cells (110). This, along with the fact that the reduced monocation form of paraquat is more cell permeant than the parent paraquat di-cation (111), suggests that reduction at the endothelial cell surface probably contributes to the mechanism of action of this pulmonary toxin.

Summary

We have discussed various scenarios in which plasma membrane redox systems have been implicated in pulmonary endothelial function. Any of these functions may in fact be secondary to more basic cellular processes involving these systems. For example, it has been suggested that the phlogistic phagocyte form of NADPH oxidase evolved from a more ubiquitous ancestral plasma membrane redox system whose primary function was in cell signaling (12). In the case of endothelial cell function, plasma membrane redox systems might provide a mechanism for endothelial sensing and adaptation to changes in blood composition (74). Alternatively, or in addition, it has also been suggested that plasma membrane redox systems provide a means of maintaining intracellular energy flow under conditions such as aging or hypoxia, wherein the mitochondrial respiratory chain may be impaired (82). These examples, along with the others presented in the chapter, are meant to provide an indication of the potential scope of plasma membrane redox system function in pulmonary endothelium, and to motivate research directed at the physiological, pathophysiological and pharmacological roles of these systems in the lung.

ACKNOLEGEMENT

The authors' research is supported by NHLBI HL-24349 and HL65537, the American Heart Association and the Department of Veterans' Affairs.

REFERENCES

1. Sanders, K.A., Hueckstеadt, T., Xu, P., Sturrock, A.B., and Hoidal, J.R. (1999) Regulation of oxidant production in acute lung injury. *Chest* 116, 56S-61S
2. Babior, B.M. (1997) Superoxide: a two-edged sword. *Braz J Med & Biol Res* 30, 141-155
3. Irani, K. (2000) Oxidant signaling in vascular cell growth, death, and survival : a review of the roles of reactive oxygen species in smooth muscle and endothelial cell mitogenic and apoptotic signaling. *Circ Res* 87, 179-183
4. Hassoun, P.M., Yu, F.S., Cote, C.G., Zulueta, J.J., Sawhney, R., Skinner, K.A., Skinner, H.B., Parks, D.A., and Lanzillo, J.J. (1998) Upregulation of xanthine oxidase by lipopolysaccharide, interleukin-1, and hypoxia. Role in acute lung injury. *Am J Respir Crit Care Med* 158, 299-305
5. Rubbo, H., Darley-Usmar, V., and Freeman, B.A. (1996) Nitric oxide regulation of tissue free radical injury. *Chem Res Toxicol* 9, 809-820
6. Lum, H. and Roebuck, K. (2001) Oxidant stress and endothelial dysfunction. *Am J Physiol* 280, C219-C741
7. Zhu, S., Manuel, M., Tanaka, S., Choe, N., Kagan, E., and Matalon, S. (1998) Contribution of reactive oxygen and nitrogen species to particulate-induced lung injury. *Environ Health Perspect* 106 Suppl 5, 1157-1163

8. Heffner, J.E., and Repine, J.E. (1989) Pulmonary strategies of antioxidant defense. *Am Rev Resp Dis* 140, 531-554
9. Maritz, G.S. (2000) (1996) Ascorbic acid. Protection of lung tissue against damage. *SubCell Biochem* 25, 265-291
10. Otterbein, L.E., Kolls, J.K., Mantell, L.L., Cook, J.L., Alam, J., and Choi, A.M. (1999) Exogenous administration of heme oxygenase-1 by gene transfer provides protection against hyperoxia-induced lung injury. *J Clin Invest* 103, 1047-1054
11. Quinlan, T., Spivack, S., and Mossman, B.T. (1994) Regulation of antioxidant enzymes in lung after oxidant injury. *Environ Health Persp* 102, Suppl 2, 79-87
12. Babior, B.M. (1999) NADPH oxidase: an update. *Blood* 93, 1464-1476
13. Ward, P.A. (1997) Phagocytes and the lung. *Ann NY Acad Sci* 832, 304-310
14. Bayraktutan, U., Blayney, L., and Ajay, M. (2000) Molecular characterization and localization of the NAD(P)H oxidase components pp91-phox and p22-phox in endothelial cells. *Arterioscler Thromb Vasc Biol*:20, 1903-1911
15. Gorlach, A., Brandes, R.P., Nguyen, K., Amidi, M., Dehghani, F., and Busse, R. (2000) A gp91phox containing NADPH oxidase selectively expressed in endothelial cells is a major source of oxygen radical generation in the arterial wall. *Circ Res* 87, 26-32
16. Jones, S.A., O'Donnell, V.B., Wood, J.D., Broughton, J.P., Hughes, E.J., and Jones, O.T. (1996) Expression of phagocyte NADPH oxidase components in human endothelial cells. *Am J Physiol Heart Circ Physiol* 271, H1626-H1634
17. Meyer, J .W., Holland, J.A., Ziegler, L.M., Chang, M.M., Beebe, G., and Schmitt, M.E. (1999) Identification of a functional leukocyte-type NADPH oxidase in human endothelial cells: a potential atherogenic source of reactive oxygen species. *Endothelium* 7, 11-22
18. Mohazzab, H., Kaminski, P.M., and Wolin, M.S. (1997) NADH oxidoreductase is a major source of superoxide anion in bovine coronary artery endothelium. *Am J Physiol Heart Circ Physiol* 266, H2568-H2572
19. Griendling, K.K., Sorescu, D., and Ushio-Fukai, M. (2000) NAD(P)H Oxidase: Role in Cardiovascular Disease and Pathology. *Circ Res* 86, 494-501
20. Harrison, D.G. (1997) Endothelial function and oxidant stress. *Clin Cardiol* 20, II-11-7
21. Somers, M.J., Burchfield, J., and Harrison, D.G. (2000) Evidence for a NADH/NADPH oxidase in human umbilical vein endothelial cells using electron spin resonance. *Antioxid Redox Signal* 2, 779-787
22. Zulueta, J.J., Yu, F.S., Hertig, I.A., Thannickal, V.J., and Hassoun, V.J. (1995) Release of hydrogen peroxide in response to hypoxia-reoxygenation: role of and NAD(P)H oxidase-like enzyme in endothelial cell plasma membrane. *Am J Respir Cell Mol Biol* 12, 41-49
23. Duerrschmidt, N., Wippich, N., Goettsch, W., Broemme, H.J., and Morawietz, H. (2000) Endothelin-1 induces NAD(P)H oxidase in human endothelial cells. *Biochem Biophys Res Commun* 269, 713-717
24. Fisher, A.B., Al Mehdi, A.B., and Muzykantov, V. (1999) Activation of endothelial NADPH oxidase as the source of a reactive oxygen species in lung ischemia. *Chest* 116, 25S-26S
25. Hishikawa, K. and Luscher, T.F. (1997) Pulsatile stretch stimulates superoxide production in human aortic endothelial cells. *Circulation* 96, 3610-3616
26. Howard, A.B., Alexander, R.W., Nerem, R.M., Griendling, K.K., and Taylor, W.R. (1997) Cyclic strain induces an oxidative stress in endothelial cells. *Am J Physiol* 272, C421-C427
27. Wei, Z., Costa, K., Al-Mehdi, A.B., Dodia, C., Muzykantov, V., and Fisher, A.B. (1999) Simulated ischemia in flow-adapted endothelial cells leads to generation of reactive oxygen species and cell signaling. *Circ Res* 85, 682-689
28. Zhang, H., Schmeisser, A., Garlichs, C.D., Plotze, K., Damme, U., Mugge, A., and Daniel, W.G. (1999) Angiotensin II-induced superoxide anion generation in human vascular endothelial cells: role of membrane-bound NADH-/NADPH-oxidases. *Cardiovasc Res* 44, 215-222

29. Cross, A.R. (1987) The inhibitory effects of some iodium compounds on the superoxide generating system of neutrophils and their failure to inhibit diaphorase activity. *Biochem Pharm* 36, 489-493
30. Brar, S.S., Kennedy, T.P., Whorton, A.R.,, Sturrock, A.B., Huecksteadt, T., Ghio, A.J., and Hoidal, J.R. (2001) Reactive oxygen species from NAD(P)H:quinone oxido-reductase constitutively activate NF-kappaB in malignant melanoma cells. *Am J Physiol Cell Physiol* 280, C659-C676
31. Dinkova-Kostova, A.T., and Talalay, P. (2001) Persuasive evidence that quinone reductase type I (DT diaphorase) protects cells against the toxicity of electrophiles and reactive forms of oxygen. *Free Rad Biol Med* 29, 231-240
32. Sun, X. and Ross, D. (1996) Quinone-induced apoptosis in human colon adenocarcinoma cells via DT-diaphorase mediated bioactivation. *Chem Biol Interact* 100, 267-276
33. Al Mehdi, A.B., Zhao, G., Dodia, C., Tozawa, K., Costa, K., Muzykantov, V., Ross, C., Blecha, F., Dinauer, M., and Fisher, A.B. (1998) Endothelial NADPH oxidase as the source of oxidants in lungs exposed to ischemia or high K^+. *Circ Res* 83, 730-737
34. Al Mehdi, A.B., Ischiropoulos, H., and Fisher, A.B. (1996) Endothelial cell oxidant generation during K^+ induced membrane depolarization. *J Cell Physiol* 166, 274-282
35. Zhao, G., Al-Mehdi, A.B., and Fisher, A.B. (1997) Anoxia-reoxygenation versus ischemia in isolated rat lungs. *Am J Physiol* 273, L1112-L1117
36. Zulueta, J.J., Sawhney, R., Yu, F.S., Cote, C.C., and Hassoun, P.M. (1997) Intracellular generation of reactive oxygen species in endothelial cells exposed to anoxia-reoxygenation. *Am J Physiol* 272, L897-L902
37. Janiszewski, M., Pedro, M.A., Scheffer, R.C.H., van Asseldonk, J.H., Souza, L.C., Luz, P.L., Augusto, O., and Laurindo, F.R.M. (2000) Inhibition of vascular NADH/NADPH oxidase activity by thiol reagents: lack of correlation with cellular glutathione redox status. *Free Rad Biol Med* 29, 889-899
38. Souza, H.P., Laurindo, F.R.M., Ziegelstein, R.C., Berlowitz, C.O., and Zweier, J.L. (2001) Vascular NAD(P)H oxidase is distinct from phagocytic enzyme and modulates vascular reactivity control. *Am J Physiol Heart Circ Physiol* 280, H658-H667
39. Berridge, M.V., and Tan, A.S. (2000) Cell-surface NAD(P)H oxidase-Relationship to trans-plasma membrane NADH oxidoreductase and a potential source of circulating NADH oxidase. *Antioxid Redox Signal* 2, 277-288, 2000
40. Berridge, M.V., and Tan, A.S. (2000) High-capacity redox control and the plasma membrane of mammalian cells: Trans-membrane, cell surface and serum NADH oxidases. *Antioxid Redox Signal* 2, 231-242, 2000
41. Morre, D.J., and Brightman, A.O. (1991) NADH oxidase of plasma membranes. *J Bioenerg Biomemb* 23, 469-489
42. Terada, L.S., Hybertson, B.M., Connelly, K.G., Weill, D., Piermattei, D., and Repine, J.E. (1997) XO increases neutrophil adherence to endothelial cells by a dual ICAM-1 and P-selectin-mediated mechanism. *J Appl Physiol* 82, 866-873
43. McCord, J.M. (1985) Oxygen derived free radicals in postischemic tissue injury. *N Eng J Med* 312, 159-163
44. McCord, J.M. (1988) Free radicals and myocardial ischemia: Overview and outlook. *Free Rad Biol Med* 4, 9-14
45. Adachi, T., Fukushima, T., Usami, Y., and Hirano, K. (1993) Binding of human xanthine oxidase to sulphated glycosaminoglycans on the endothelial cell surface. *Biochem J* 289, 523-527
46. Houston, M., Estevez, A., Chumley, P., Aslan, M., Marklund, S., Parks, D.A., and Freeman, B.A. (1999) Binding of xanthine oxidase to vascular endothelium. Kinetic characterization and oxidative impairment of nitric oxide-dependent signaling. *J Biol Chem* 274, 4985-4994
47. Rouquette, M., Page, S., Bryant, R., Benboubetra, M., Stevens, C.R., Blake, D.R., Whish, W.D., Harrison, R., and Tosh, D. (1998) Xanthine oxidoreductase is

asymmetrically localized on the outer surface of human endothelial and epithelial cells in culture. *FEBS Lett* 426, 397-401

48. Vickers, S., Schiller, H.J., Hildreth, J.E., and Bulkley, G.B. (1998) Immunoaffinity localization of the enzyme xanthine oxidase on the outside surface of the endothelial cell plasma membrane. *Surgery* 124, 551-560
49. Nielsen, V.G., Tan, S., Baird, M.S., Samuelson, P.N., McCammon, A.T., and Parks, D.A. (1997) Xanthine oxidase mediates myocardial injury after hepatoenteric ischemia-reperfusion. *Crit Care Med* 25, 1044-1050
50. Nielsen, V.G., Weinbroum, A., Tan, S., Samuelson, P.N., Gelman, S., and Parks, D.A. (1994) Xanthine oxidoreductase release after descending thoracic aorta occlusion and reperfusion in rabbits. *J Thorac Cardiovas Sur* 107, 1222-1227
51. Terada, L.S., Dormish, J.J., Shanley, P.F., Leff, J.A., Anderson, B.O., and Repine, J.E. (1992) Circulating xanthine oxidase mediates lung neutrophil sequestration after intestinal ischemia-reperfusion. *Am J Physiol Lung Cell Mol Physiol* 263, L394-L401
52. Galili,Y., Ben Abraham, R., Weinbroum, A., Marmur, S., Iaina, A., Volman, Y., Peer, G., Szold, O., Soffer, D., Klausner, J., Rabau, M., and Kluger, Y. (1998) Methylene blue prevents pulmonary injury after intestinal ischemia-reperfusion. *J Trauma 45*, 222-225
53. Weinbroum, A., Nielsen, V.G., Tan, S., Gelman, S., Matalon, S., Skinner, K.A., Bradley, E. Jr, and Parks, D.A. (1995) Liver ischemia-reperfusion increases pulmonary permeability in rat: role of circulating xanthine oxidase. *Am J Physiol* 268, G988-G996
54. Parks, D.A., and Granger, D.N. (1986) Xanthine oxidase: biochemistry, distribution and physiology. *Acta Physiol Scand* 548, 87-99
55. Walsh, C. (1979) Metalloflavoprotein oxidases and superoxide dismutase. In: *Enzymatic Reaction Mechanisms*, edited by A.C. Bartlett. San Francisco: W.H. Freeman and Co., p. 432-448
56. Zhang, Z., Blake, D.R., Stevens, C.R., Kanczler, J.M., Winyard, P.G., Symons, M.C., Benboubetra, M., and Harrison, R. (1998) A reappraisal of xanthine dehydrogenase and oxidase in hypoxic reperfusion injury: the role of NADH as an electron donor. *Free Rad Res* 28, 151-164
57. Komiyama,T., Kikuchi, T., and Sugiura, Y. (1986) Interactions of anticancer quinone drugs, aclacinomycin A, adriamycin, carbazilquinone, and mitomycin C, with NADPH-cytochrome P-450 reductase, xanthine oxidase and oxygen. *J Pharmacobiodyn* 9, 651-664
58. Sakai, M., Yamagami, K., Kitazawa, Y., Takeyama, N., and Tanaka, T. (1995) Xanthine oxidase mediates paraquat induced toxicity on cultured endothelial cell. *Pharmacol Toxicol* 77, 36-40
59. Yee, S.B. and Pritsos, C.A. (1997) Comparison of oxygen radical generation from the reductive activation of doxorubicin, streptonigrin, and menadione by xanthine oxidase and xanthine dehydrogenase. *Arch Biochem Biophys* 347, 235-241
60. Shaul, P.W., Smart, E.J., Robinson, L.J., German, Z., Yuhanna, I.S., Ying, Y., Anderson, R.G., and Michel, T. (1996) Acylation targets endothelial nitric-oxide synthase to plasmalemmal caveolae. *J Biol Chem* 271, 6518-6522
61. Day, B.J., Patel, M., Calavetta, L., Chang, L.Y., and Stamler, J.S. (1999) A mechanism of paraquat toxicity involving nitric oxide synthase. *Proc Natl Acad Sci* 96, 12760-12765
62. Vasquez-Vivar, J., Martasek, P., Hogg, N., Masters, B.S., Pritchard, K.A.J., and Kalyanaraman, B. (1997) Endothelial nitric oxide synthase-dependent superoxide generation from adriamycin. *Biochem J* 36, 11293-11297
63. Vasquez-Vivar, J., Kalyanaraman, B., Martasek, P., Hogg, N., Masters, B.S., Karoui, H., Tordo, P., and Pritchard, Jr., K.A. (1998) Superoxide generation by endothelial nitric oxide synthase: the influence of cofactors. *Proc Natl Acad Sci* 95, 9220-9225
64. Ignarro, L.J. (1999) Nitric oxide: a unique endogenous signaling molecule in vascular biology. *Biosci Rep* 19, 51-71

65. Goss, S.P., Singh, R.J., Hogg, N., and Kalyanaraman, B. (1999) Reactions of *NO, *NO2 and peroxynitrite in membranes: physiological implications. *Free Radic Res* 31, 597-606
66. Munzel, T., Heitzer, T., and Harrison, D.G. (1997) The physiology and pathophysiology of the nitric oxide/superoxide system. *Herz* 22, 158-172
67. Goss, S.P., Hogg, N., and Kalyanaraman, B. (1995) The antioxidant effect of spermine NONOate in human low density lipoprotein. *Chem Res Toxicol* 8, 800-806
68. Ichinose, M., Sugiura, H., Yamagata, S., Koarai, A., and Shirato, K. (2000) Increase in reactive nitrogen species production in chronic obstructive pulmonary disease airways. *Am J Respir Crit Care Med* 162, 701-706
69. Saleh, D., Barnes, P.J., and Giaid, A. (1997) Increased production of the potent oxidant peroxynitrite in the lungs of patients with idiopathic pulmonary fibrosis. *Am J Respir Crit Care Med* 155, 1763-1769
70. Kooy, N.W., Royall, J.A., Ye, Y.Z., Kelly, D.R., and Beckman, J.S. (1995) Evidence for in vivo peroxynitrite production in human acute lung injury. *Am J Respir Crit Care Med* 151, 1250-1254
71. Crane, F., Sun, I.L., Barr, R., and Low, H. (1991) Electron and proton transport across the plasma membrane. *J Bioenerg Biomemb* 23, 733-803
72. De Grey, A.D.N.J. (2000) Redox 2000: The 5th international conference on plasma membrane redox systems and their role in biological stress and disease. *Antioxid Redox Signal* 2, 373-374
73. Crane, F., Sun, I.L., Clark, M.G., Grebing, C., and Low, H. (1985) Transplasma membrane redox systems in growth and development. *Biochim Biophys Acta* 811, 233-264
74. Kaul, N., Choi, J., and Forman, H.J. (1998) Transmembrane redox signaling activates NF-kappaB in macrophages. *Free Rad Biol Med* 24, 202-207
75. May, J.M., Qu, Z., Morrow, J.D., and Cobb, C.E. (2000) Ascorbate-dependent protection of human erythrocytes against oxidant stress generated by extracellular diazobenzene sulfonate. *Biochem Pharmacol* 60, 47-53
76. Villalba, J. and Navas, P. (2000) Plasma membrane redox system in the control of stress induced apoptosis. *Antioxid Redox Signal* 2, 213-230
77. Villalba, J.M., Navarro, F., Gomez-Diaz, C., Arroyo, A., Bello, R.I., and Navas, P. (1997) Role of cytochrome b5 reductase on the antioxidant function of coenzyme Q in the plasma membrane. *Mol Aspects Med* 18 Suppl, S7-13
78. McKie, A.T., Barrow, D., Latunde-Dada, G., Rolfs, A., Sager, G., Mudaly, E., Mudaly, M., Richardson, C., Barlow, D., Bomord, A., Peters, T., Raja, D., and Shirali, S. (2001) An iron-regulated ferric reductase associated with the absorption of dietary iron. *Science* 291, 1755-1759
79. Pountney, D.J., Raja, K.B., Bottwood, M.J., Wrigglesworth, J.M., and Simpson, R.J. (1996) Mucosal surface ferricyanide reductase activity in mouse duodenum. *Biometals* 9, 15-20
80. Navas, P., Villalba, J.M., and Cordoba, F. (1994) Ascorbate function at the plasma membrane. *Biochim Biophys Acta* 1197, 1-13
81. Mohazzab, K.M., Kaminski, P.M., Agarwal, R., and Wolin, M.S. (1999) Potential role of a membrane-bound NADH oxidoreductase in nitric oxide release and arterial relaxation to nitroprusside. *Circ Res* 84, 220-228
82. Baker, M.A., and Lawen. A. (2000) Plasma membrane NADH -oxidoreductase system: a critical review of the structural and functional data. *Antioxid Redox Signal*, 197-212
83. Bongard, R.D., Krenz, G.S., Linehan, J.H., Roerig, D.L., Merker, M.P., Widell, J.L., and Dawson, C.A. (1994) Reduction and accumulation of methylene blue by the lung. *J Appl Physiol* 77, 1480-1491
84. Bongard, R.D., Merker, M.P., Shundo, R., Okamoto, Y., Roerig, D. L., Linehan, J.H., Dawson, C.A. (1995) Reduction of thiazine dyes by bovine pulmonary arterial endothelial cells in culture. *Am J Physiol Lung Cell Mol Physiol* 269, L78-L84

85. Merker, M.P., Bongard, R.D., Linehan, J.H., Okamoto, Y., Vyprachticky, D., Brantmeier, B.M., Roerig, D.L., and Dawson, C.A. (1997) Pulmonary endothelial thiazine uptake: separation of cell surface reduction from intracellular reoxidation. *Am J Physiol Lung Cell Mol Physiol* 272, L673-L680
86. Dixon, M. and Webb, E.C. (1979) Enzyme Cofactors. In: *Enzymes*, New York: Academic Press, 468-518
87. Bongard, R.D., Merker, M.P., Daum, J.M., and Dawson, C.A. (1999) Quinone reduction by endothelial cells: Potential mechanism for regulating redox status of low density lipoproteins (LDL). *FASEB J* 13, A185
88. Dawson, C.A., Audi, S.H., Bongard, R.D., Okamoto, Y., Olson, L.E., and Merker, M.P. (2000) Transport and reaction at endothelial plasmalemma. Distinguishing intra- from extra-cellular events. *Ann Biomed Eng* 28, 1010-1018
89. De La Fuente, E., Dawson, C.A., Nelin, L.D., Bongard, R.D., McAuliffe, T.L., and Merker, M.P. (1997) Biotinylation of membrane proteins accessible via the pulmonary vasculature in normal and hyperoxic rats. *Am J Physiol Lung Cell Mol Physiol* 272, L461-L470
90. Merker, M.P., Olson, L.E., Bongard, R.D., Patel, M.K., Linehan, J.H., and Dawson, C.A. (1998) Ascorbate-mediated transplasma membrane electron transport in pulmonary arterial endothelial cells. *Am J Physiol Lung Cell Mol Physiol* 274, L685-L693
91. Olson, L.E., Merker, M.P., Patel, M.K., Bongard, R.D., Daum, J.M., Johns, R.A., and Dawson, C.A. (2000) Cyanide increases reduction but decreases sequestration of methylene blue by endothelial cells. *Ann Biomed Eng* 28, 85-93
92. Giulivi, C., and Cadenas, E. (1998) Extracellular activation of fluorinated aziridinylbenzoquinone in HT29 cells EPR studies. *Chem Biol Inter* 113: 191-204, 1998
93. Gillis, C.N., and Roth, J.A. (1976) Pulmonary disposition of circulating vasoactive hormones. *Biochem Pharmacol* 25, 2547-2553
94. Hechtman, H.B. and Shepro, D. (1982) Lung metabolism and systemic organ function. *Circ Shock* 9, 457-467
95. Merker, M.P., Audi, S.H., Brantmeier, B.M., Nithipatikom, K., Goldman, R.S., Roerig, D.L., and Dawson, C.A. (1999) Proline in vasoactive peptides: consequences for peptide hydrolysis in the lung. *Am J Physiol Lung Cell Mol Physiol* 276, L341-L350
96. Claise, C., Edeas, M., Chaouchi, N., Chalas, J., Capel, L., Kalimouttou, S., Vazquez, A., and Lindenbaum, A. (1999) Oxidized-LDL induce apoptosis in HUVEC but not in the endothelial cell line EA.hy 926. *Atherosclerosis* 147, 95-104
97. Drexler, H. and Hornig, B. (1999) Endothelial dysfunction in human disease. *J Mol Cell Cardiol* 31, 51-60
98. Harrison, D.G. (1997) Cellular and molecular mechanisms of endothelial cell dysfunction. *J Clin Invest* 100, 2153-2157
99. Berliner, J.A., and Heinecke, J.W. (1996) The role of oxidized lipoproteins in atherogenesis. *Free Rad Biol Med* 20, 707-727
100. Smalley, D.M., Hogg, N., Kalyanaraman, B., and Pritchard, Jr., K.A. (1997) Endothelial cells prevent accumulation of lipid hydroperoxides in low-density lipoprotein. *Arterio Thromb Vasc Biol* 17, 3469-3474
101. Constantinescu, A., Vink, H., and Spaan, J.E.A. (2001) Elevated capillary tube hematocrit reflects degradation of endothelial cell glycocalyx by oxidized LDL. *Am J Physiol Heart Circ Physiol* 280, H1051-H1057
102. Cominacini, L., Garbin, U., De Santis, A., Campagnola, M., Davoli, A., Pasini, A.F., Faccini, G., Pasqualini, E., Bertozzo, L., Micciolo, R., Pastorino, A.M., and Lo, C. (1996) Mechanisms involved in the in vitro modification of low density lipoprotein by human umbilical vein endothelial cells and copper ions. *J Lipid Media & Cell Signal* 13, 19-33
103. Dugas, T.R., Morel, D.W., and Harrison, E.H. (1998) Impact of LDL carotenoid and alpha-tocopherol content on LDL oxidation by endothelial cells in culture. *J Lipid Res* 39, 999-1007

104. Garner, B., van Reyk, D., Dean, R.T., and Jessup, W. (1997) Direct copper reduction by macrophages. *J Biol Chem* 272, 6927-6935
105. Dawson, C., Bongard, R.D., Merker, M.P., Olson, L.E., and Linehan, J.H. (1998) Pulmonary endothelium reduces copper and ubiquinone: implications for atherosclerosis. *J Vasc Res* 35, 61
106. Schneider, D., and Elstner, E.F. (2001) Coenzyme Q_{10}, vitamin E, and dihydrothioctic acid cooperatively prevent diene conjugation in isolated low density lipoproteins. *Antioxid Redox Signal* 2, 327-333
107. Bowry,V.W., Mohr, D., Cleary, J., and Stocker, R. (1995) Prevention of tocopherol-mediated peroxidation in ubiquinol-10-free human low density lipoprotein. *J Biol Chem* 270, 5756-5763
108. Bowry,V.W., and Stocker R. (1993) Tocopherol mediated peroxidation. The prooxidant effect of Vitamin E on the radical-initiated oxidation of human low density lipoprotein. *J Am Chem Soc* 115, 6029-6044
109. Thomas, S.R., Neuzil, J., Mohr, D., and Stocker, R. (1995) Coantioxidants make alpha-tocopherol an efficient antioxidant for low-density lipoprotein. *Am J Clin Nutr* 62, 1357S-1364S
110. Britigan, B.E., Roeder, T.L., and Shasby, D.M. (1992) Insight into the nature and site of oxygen centered free radical generation by endothelial cell monolayers using a novel spin trapping technique. *Blood* 79, 699-707
111. DeGray, J.A., Rao, D.N., and Mason, R.P. (1991) Reduction of paraquat and related bipyridylium compounds to free radical metabolites by rat hepatocytes. *Arch Biochem Biophys* 289, 145-152

Chapter 10

MECHANISMS OF LUNG EPITHELIAL CELL INJURY BY *PSEUDOMONAS AERUGINOSA*

Jeanine P. Wiener-Kronish, Dara Frank, and Teiji Sawa
Department of Anesthesia and Perioperative Care, University of California at San Francisco, San Francisco, CA, and Department of Microbiology and Molecular Genetics, Medical College of Wisconsin, Milwaukee, WI

INTRODUCTION

The interactions between hosts and pathogens have evolved to select for gene combinations that can confer an advantage to either combatant. Bacteria have evolved several methods of inducing cell injury and are an important cause of acute lung injury (ALI). This review focuses on *Pseudomonas aeruginosa*, a soil bacterium that expresses many toxins that damage epithelial cells and can cause significant lung injury in critically ill patients (1). *P. aeruginosa* is rapidly becoming an important model system to better understand bacterial-host interactions leading to ALI. The recent sequencing and annotation of the *P. aeruginosa* genome will allow for microarray analyses to test the relevance of bacterial gene expression relative to host responses in several kinds of animal models. Moreover, the *Pseudomonas* research community already has a wealth of fundamental information regarding virulence gene and protein expression, and the delivery of toxic substances and enzymes to their cellular targets. Many laboratories are moving rapidly toward a phase where the individual virulence attributes, their expression, function, and regulation will be integrated into the biology of this metabolically and genetically versatile organism.

Lung pathogens, including *P. aeruginosa*, initially contact the respiratory epithelial barrier. It has been shown that *P. aeruginosa* can invade respiratory epithelium (2, 3), and in fact, some propose that epithelial uptake of *P. aeruginosa* is a method utilized by the host to clear the bacteria (4). It appears, however, that in the majority of cases of lung infection, *P. aeruginosa* injures the lung epithelium or enters the lung via pre-existing injuries rather than invading the epithelium (5). This chapter will review the mechanisms of lung epithelial injury mediated by infection with *P. aeruginosa* with a major emphasis on the interaction between bacterial protein toxins and enzymes with host cells.

Bacterial Virulence Factors

P. aeruginosa expresses a number of products, commonly referred to as virulence factors, which have been shown to increase the propensity of the organism to adhere to host tissues, replicate, and damage cells. Adherence factors include exopolysaccharides (alginate), lipopolysaccharides, flagella, and pili (6, 7). *P. aeruginosa* also secretes numerous virulence factors that could potentially injure epithelial cells, including proteases, phospholipases, ADP-ribosylating enzymes, and small toxic molecules such as phenazines, rhamnolipid, and cyanide. Secretion of virulence factors involves three of the four types of known secretion systems in bacteria: type I (8), type II (9), and type III (10). Additionally, *P. aeruginosa* possesses major sensory systems (also called quorum sensing) that detect the density of bacterial populations. As quorum sensing integrates the expression and secretion of virulence determinants in response to bacterial signals, we will review this area of research as it relates to epithelial damage.

Quorum Sensing System

Bacteria signal each other by releasing chemicals to coordinate transcriptional activity, including the production and secretion of virulence factors. *P. aeruginosa* utilize acyl-homoserine lactones as autoinducers. It has been estimated that approximately 4% of the genes encoded by *P. aeruginosa* are controlled directly or indirectly by quorum sensing mechanisms (11). Autoinducers or signaling molecules are produced by specific enzymes and are detected by their ability to activate transcriptional regulatory proteins (12, 13). These signals allow the bacteria to monitor its population density through the induction of transcription. *P. aeruginosa* has at least two quorum-sensing systems. The enzyme that produces the autoinducer is generally a member of the LuxI family of acyl-HSL synthases. *P. aeruginosa* LasI catalyzes the synthesis of N-(3-oxododecanoyl)-HSL (PAI-1) and *P. aeruginosa* RhlI catalyzes the synthesis of N-butyryl-HSL (PAI-2). The intra-bacterial concentration of an acyl-HSL is determined by the environmental concentration, and environmental concentrations increase when the population of the signal-producing bacteria increases (11-13). Recently, a third signaling molecule has been isolated, 2-heptyl-3-hydroxy-4-quinolone, indicating that intercellular communication may not be restricted to acyl-homoserine lactones (14).

The receptors for these acyl-HSL signals are members of the LuxR family of transcriptional regulators (12, 15). LasR is a transcriptional regulator that responds to the LasI-generated signal, PAI-1, and RhlR is the transcriptional regulator that responds to the RhlI-signal, PAI-2. At low population densities *P. aeruginosa* produces low levels of PAI-1. When PAI-1 reaches a critical concentration, it interacts with LasR and this complex activates the transcription of several genes, including, *lasB* (encoding the protease elastase), *toxA* (encoding the ADP-ribosylating toxin, exotoxin A), *xcp* genes (products required for secretion of elastase and exotoxin A), *rhlR*, and *las*I. RhlR responds to PAI-2; activated RhlR leads to the production of rhamnolipid, hydrogen cyanide, and pyocyanin (12, 15, 16). These signal transduction mechanisms and the gene targets are summarized in Figure 1.

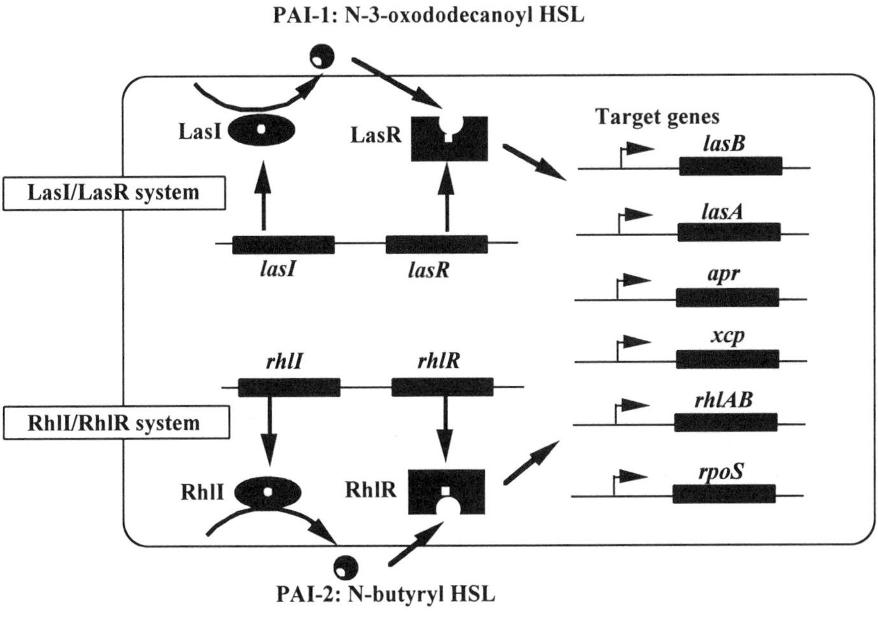

Figure 1. A partial list of gene targets of quorum sensing is shown. The two inducers, N-3-oxododecanoyl HSL (PAI-1, on top of figure) and N-butyryl HSL (PAI-2, on bottom of figure) are produced by different genes and enzyme systems. Once the inducers reach a critical concentration, target protein production is induced as shown.

Neonatal mice infected intratracheally with *lasR* mutant *P. aeruginosa* strains do not develop bacteremia nor die, whereas neonatal mice infected with the wild-type bacteria develop lung injury, bacteremia, and die (6). These data suggest that the proteins governed by the quorum sensing system are involved in lung epithelial injury. Further evidence that the quorum sensing system may be involved with epithelial injury comes from experiments utilizing *Caenorhabditis elegans*. In the nematode *C. elegans*, strain PA14 kills the worms quickly by the production of phenazine compounds (17, 18). Strain PAO1 paralyzes the nematodes so that they cannot eat, defecate, or lay eggs. The paralytic agent appears to be a small molecule regulated by RhlR, either cyanide, pyocyanin, and/or a rhamnolipid (17, 18). Thus, as the bacterial population increases in any given location a number of toxic compounds are released that may cause tissue damage, propagating the spread of *P. aeruginosa* and enhancing further adherence and replication.

Strains whose pathogenesis appears to be primarily due to bacterial products controlled by quorum sensing (e.g. PAO1) appear to be less cytotoxic to cultured epithelial cells and are associated with less mortality in airspace-infected mice when compared to strains whose pathogenesis is associated with type III secretion (5). This phenomenon could be due to genotypic (possession of genes encoding toxins) or phenotypic (ability to induce quorum sensing early in the infection) strain differences, however, higher inoculating doses of strains like PAO1 will lead to epithelial necrosis and death. Further, when strains utilizing the quorum sensing system are administered to neonatal mice, they cause lung inflammation and bacteremia (6). Thus, the number of bacteria causing an infection and the age of the host, appear to be critical to the development of necrotizing lung infections.

Quorum-sensing systems are also involved in biofilm production (12, 19, 20). Biofilms are microbial communities that produce three-dimensional structures that include water-filled channels (20, 21). A number of gene products are involved in biofilm formation and mutations in genes involved in twitching motility, flagellar motility, and swarming behavior prevent the assembly of a normal biofilm (20-25). These functions are postulated to be important for the social behavior required to form higher order bacterial communities and structures. Mutants that cannot produce exopolysaccharide will not form stable biofilms (21, 22). Defects in *lasI* (strains that cannot express the PAI-1 autoinducer) lead to flat and undifferentiated biofilms (22). Biofilm formation has been documented in patients with cystic fibrosis in whom *P. aeruginosa* causes chronic infection of the lungs (23,24). The microcolony formation involved in chronic colonization and the formation of biofilms may damage the epithelium through the localization of high concentrations of bacterial virulence determinants and destructive enzymes or toxins. Biofilms may be an excellent environment for quorum sensing to induce the production of these factors. Moreover, the localization of bacteria within a polysaccharide matrix may prevent phagocytosis and promote chronic inflammatory responses.

Elastase

Elastase expression is controlled by the quorum-sensing system. *Pseudomonas* elastase has been shown to increase paracellular permeability of epithelial cells in animals, as well as *in vitro* (25), and degrade laminin and type III and IV collagens (26). Purified *Pseudomonas* elastase applied to primary cultures of alveolar type II pneumocytes caused redistribution and/or degradation of tight junction proteins (27). *Pseudomonas* elastase is found in the sputa from patients who have cystic fibrosis and chronic *Pseudomonas* infections; elastase concentrations ranging from 3 to 110 μg/ml of sputum were measured (25). *Pseudomonas* elastase has been shown to change lung epithelial cell permeability and to delay airway epithelial wound repair, by alteration of the actin cytoskeleton as well as by inducing epithelial matrix metalloproteinase-2 (25-27).

Type III Secretion System

While most toxins produced by bacteria are secreted into the surrounding extracellular environment, Gram-negative pathogens utilize a specialized secretion system, the type III secretion system, to directly inject toxins into eukaryotic cells (Figure 2). This system is used by pathogenic *Salmonella*, *Shigella*, *E. coli*, *Yersinia* and *Pseudomonas* (28-30). *P. aeruginosa* injects four known toxins via the type III secretion system: ExoS, ExoT, ExoU and ExoY (31). These toxins appear to cause toxicity primarily when they are intracellular; administration of purified toxins to the surface of epithelial cells or other cells has no known effects.

Perhaps the best characterized toxin, to date, is exoenzyme S. Exoenzyme S was first isolated biochemically as an aggregate consisting of two proteins whose masses are 53 and 49 kDa. While early models indicated that the 49 kDa form was a processed product of the 53 kDa molecule, the two proteins are now known to be coordinately expressed by two separate and highly homologous genes termed *exoS* (49 kDa) and *exoT* (53 kDa) (32). Both ExoS and ExoT have ADP-ribosyltransferase activity, particularly when in the presence of the a eukaryotic scaffolding protein, FAS (33). The ADP-ribosyltransferase activity of ExoT, however, is only 0.2% of that expressed by ExoS. The activity of two different Ras-related GTP binding proteins are postulated to be affected by the injection of ExoS into mammalian cells. The C-terminal domain of ExoS inactivates Ras-mediated signal transduction via ADP-ribosylation of an arginine residue important for nucleotide exchange (34). The N-terminus of ExoS inactivates Rho proteins through its GTPase activating protein (GAP) activity. The GAP domain of ExoS down-regulates Rac using an arginine finger to stabilize the transition state of the GTPase (35).

ExoS activity has been measured in pleural fluids of animals infected with *P. aeruginosa* and appears to be expressed during acute infections (36-37). The delivery of ExoS directly into cells and the disruption of cellular signaling may not be the only mechanism of epithelial cell injury. *In vitro* ExoS has ecto-ADP-ribosyltransferase activity and can covalently modify apolipoprotein A1 and immunoglobulin (37). Thus ExoS may have undefined additional functions as an extracellular toxin during infections. Finally, strains isolated from chronic human *Pseudomonas* infections tend to produce larger quantities of ExoS than laboratory strains, suggesting the *in vivo* environment encourages the production of this toxin (38).

ExoU was identified as a major cytotoxin secreted by the *Pseudomonas* type III system (39). Airspace administration of *Pseudomonas* strains producing this toxin kills lung epithelial cells and macrophages, and causes bacteremia, shock, and death (40-41). When these strains are genetically altered not to express ExoU, epithelial cell and macrophage necrosis is significantly decreased, bacterial dissemination does not occur, and the animal survives the lung infection. Comparisons of clinical isolates of *Pseudomonas* suggest that the production of ExoU is associated with increased epithelial toxicity and animal death at lower inoculums (2, 5, 41). The eukaryotic target of ExoU is unknown but microinjection and transfection experiments indicate that only small amounts of intracellular ExoU are required to induce cellular death by necrotic mechanisms (*DW Frank, unpublished observations*).

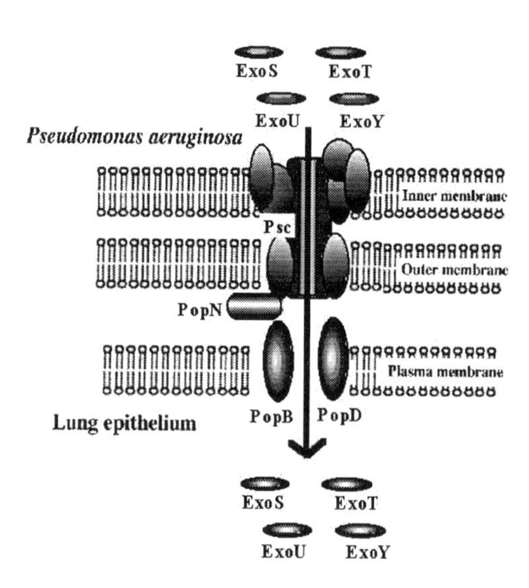

Figure 2. A schematic drawing of the possible structure of the type III secretion system in *P. aeruginosa*. Note that the 4 toxins appear to be directly injected into the eukaryotic cell. Some of the other bacterial proteins (PopN, Psc, PopB and PopD) postulated to participate in the type III system are shown. Our studies (40) indicate that PcrV, like PopB and PopD, participates in the translocation of type III toxins but its localization relative to the bacterial secretion apparatus or the eukaryotic cell has not been established.

Lung Epithelial Permeability and Airspace Infections

The transepithelial efflux of radioactive albumin from the lung airspaces to the bloodstream has been utilized as a measurement of lung epithelial permeability (42). The lung epithelium has been shown to be a tight epithelium, incredibly impervious to various toxins. The airspace instillation of large quantities of *E. coli* endotoxin does not lead to a significant increase in the efflux of radioactive albumin compared to control experiments where only the radioactive albumin is instilled (43). Airspace instillation of live *P. aeruginosa,* however, causes a significant increase in the efflux of the albumin, especially when the bacteria produce type III toxins. Airspace instillation of *P. aeruginosa* strains that do not produce type III toxins still increases the efflux of the albumin; however, the doses of bacteria required to increase the epithelial permeability is 1 to 2 logs greater than the strains producing ExoU (5). The increase in epithelial permeability is also documented by the ability to culture *Pseudomonas* from the systemic circulation in parallel with the albumin efflux (41). Furthermore, the inflammatory mediators produced by the presence of bacteria in the airspace also enter the circulation due to the increased epithelial permeability (41). Thus, the instillation of airspace *P. aeruginosa* strains that produce type III secretory toxins are able to predictably produce lung epithelial cell necrosis and cause septic shock.

Pseudomonas-Induced Epithelial Injury

Utilizing an antibody to a unique protein, rTI_{40}, located on the apical surface of rat epithelial type I cells, we found that release of this protein occurred when type I cells were injured by *P. aeruginosa* instilled into the airspace (44). Instillation of virulent *P. aeruginosa* (containing the type III secretion system and producing ExoU) caused a time-dependent increase in the amount of rTI_{40} in the alveolar fluid obtained at the termination of the experiment. The amount of rTI_{40} was increased 2-fold compared to controls after 2 hours of infection and was increased 80-fold compared to controls after 6 hours of airspace infection (44). In contrast, in animals instilled with the same bacteria that had been genetically modified so that it could not utilize the type III secretion system, the levels of rTI_{40} in the alveolar fluid were the same as in control animals. These experiments verify that the lung epithelial injury was caused by type III secreted toxins.

Electron microscopy of the lungs taken from infected animals documented that after 6 hours of infection, type I cells had vesicles along their basement membranes and many type I cells had been sloughed (44). Furthermore, type II cells had also been injured; type II cells had fewer microvilli and contained large vacuoles. This histology suggested that the injured and sloughed epithelium caused the increased permeability of the epithelial barrier to the radioactive albumin (see above). Permeability experiments were repeated in these animals and again documented that the virulent *P. aeruginosa* increased epithelial permeability significantly, with 28% of the instilled radioactive albumin in the circulation after 6 hours compared to 3% in uninfected animals or animals instilled with the non-toxic bacteria (44).

Apoptosis of the Lung Epithelium

Normal polarized airway epithelial cells in culture are very resistant to *Pseudomonas*-induced apoptosis (45). In contrast, epithelial cells without tight junctions, or treatments that affect tight junctions in epithelial cells, made the cells more susceptible to *Pseudomonas*-induced apoptosis (45-46). Type III secretory products appear to be necessary for *Pseudomonas*-induced epithelial apoptosis and specifically this response appears to be due to the activities of ExoS, ExoT, and perhaps other undefined toxins (47). Exotoxin A may also induce apoptosis (48). *Pseudomonal* contact appears to be required for apoptosis of epithelial cells as bacterial supernatants do not appear to cause apoptosis (47). However, there is one report of *Pseudomonal* supernatants causing apoptosis of macrophages (49). Therefore, specific toxins are linked to the apoptotic response and bacterial contact with cells usually is required for the initiation of apoptosis.

Notably, *in vitro* investigations of epithelial cells with cystic fibrosis transmembrane conductance regulator (CFTR) mutations demonstrated that these epithelia were not more susceptible to apoptosis when exposed to *Pseudomonas* (45). Furthermore, mice that were *cftr-/-* did not have more apoptosis in their *Pseudomonas*-infected lungs compared to wild-type mice. Therefore, apoptosis does not appear to be more frequent in lungs from animals with the genetic defect found in cystic fibrosis.

A recent report suggested that normal lung epithelial cells undergo apoptosis within 4 hours of *Pseudomonas* infection (50). However, the doses of *Pseudomonas* utilized in these experiments were extraordinarily high (1.3 x 10^{10} CFU). This same investigation compared high-dose *Pseudomonal* infections in wild-type mice and mice deficient in CD95 (*lpr* or *gld* mice). These genetically deficient mice have abnormal macrophage function and other abnormalities. Therefore, the clearance of the *Pseudomonas* bacteria could be decreased via several mechanisms. Nonetheless, when the genetically altered mice died more rapidly from their *Pseudomonas* infections, the investigators suggested that the lack of apoptosis led to the decreased clearance of the bacteria (50). The role of apoptosis in bacterial clearance needs to be verified in future experiments. In addition, it remains to be determined whether apoptosis of the lung epithelium commonly occurs in *Pseudomonas*-induced lung infections in patients.

Syndecan-1 Shedding Enhanced by LasA

Cell surface proteins are cleaved by proteinases (sheddases) and ectodomains are released from the cell surface as soluble effectors (51). Shedding is also a regulated host response to tissue injury. Shed syndecan ectodomains are regulators of inflammation (52). Bacterial pathogens can take advantage of this shedding mechanism to enhance their virulence. For example, *E. coli* hemolysin causes the shedding of CD14 and IL-6 receptors, thereby decreasing the inflammatory environment for the bacteria (51). Culture supernatants from *P. aeruginosa* also augment the shedding of the IL-6 receptor (52). LasA, an extracellular *Pseudomonal* protein enhances syndecan-1 shedding by activating the host cell's shedding machinery (51). The extent to which this shedding occurs in human *Pseudomonal*-induced lung infections is unknown.

Surfactant Proteins

The alveolar epithelium plays a role in host defense by producing surfactant proteins. Surfactant protein-A (SP-A) and surfactant protein-D (SP-D) are members of the collectin subgroup of C-type lectins (see Chapter 13). The C-type lectins bind to the carbohydrate surfaces of many microorganisms and thus help phagocytes kill these organisms (53). The role of SP-A in *Pseudomonal*-induced lung infections was investigated in SP-A-deficient (SP-A -/-) mice (54). SP-A -/- mice cleared the *Pseudomonal* airspace infections much more slowly when compared to the rates of bacterial clearance in wild-type mice. Phagocytosis of the *P. aeruginosa* was significantly decreased in the SP-A -/- mice and neutrophil infiltration of the infected lungs in these mice was significantly increased compared to wild-type mice. The cause for the increased neutrophilia is unclear, but could have been caused by the lack of bacterial clearance by the macrophages. SP-A does not apparently bind directly to *P. aeruginosa*, but rather activates macrophages to phagocytose the bacteria (55).

Patients who are at risk for ALI and who actually develop ALI have decreased concentrations of SP-A and SP-D in their bronchoalveolar lavage fluids (56). It is unclear whether patients who have acute lung infections and lung injury due to *P. aeruginosa* develop abnormalities of their SP-A or SP-D due to their infections.

Conclusions

Most clinicians appreciate that the presence of a critical number of *P. aeruginosa* in the lungs of any patient leads to lung epithelial injury. What is less appreciated is that certain strains of *P. aeruginosa* that express the type III secretion system appear to be more toxic to epithelial cells and macrophages, and consequently, infection with a relatively smaller inoculum of these particular strains can lead to lung injury. A larger inoculum of less virulent *P. aeruginosa*, however, can cause epithelial injury as well, by utilizing other mechanisms (induction of biofilm formation, chronic infection and inflammation, production of tissue destructive enzymes, induction of apoptosis, etc.).

In addition to expression of certain bacterial products, the critical inoculum of bacteria necessary to cause lung injury depends on the several host variables. Lack of neutrophils and abnormalities of other host defense mechanisms may allow a smaller inoculum to cause greater cellular injury because of the ability of *P. aeruginosa* to replicate rapidly. An under appreciated point is that injured epithelial cells and non-polarized epithelial cells, appear to be more vulnerable to *P. aeruginosa*-induced injury. Furthermore, the amount or availability of surfactant may also be critical for host defense. *P. aeruginosa* may cripple the ability of the type II cell to produce surfactant in a normal fashion. Thus, the host-pathogen interaction in the lung depends not only on the bacteria and the epithelial cell but also on the surrounding milieu.

REFERENCES

1. Garrouste-Orgeas, M., Chevret, S., Arlet, G., Marie, O., Rouveau, M., Popoff, N., and Schlemmer, B. (1997) Oropharyngeal or gastric colonization and nosocomial pneumonia in adult intensive care unit patients. A prospective study based on genomic DNA analysis. *Am J Respir Crit Care Med* 156, 1647-1655
2. Fleiszig, S.M.J., Wiener-Kronish, J.P., Miyazaki, H., Vallas, V., Mostov, K.E., Kanada, D., Sawa, T., Yen, T.S., and Frank, D.W. (1997) *Pseudomonas aeruginosa*-mediated cytotoxicity and invasion correlate with distinct genotypes at the loci encoding exoenzyme S. *Infect Immun* 65, 579-586
3. Evans, D.J., Frank, D.W., Finck-Barbancon, V., Wu, C., and Fleiszig, S.M. (1998) *Pseudomonas aeruginosa* invasion and cytotoxicity are independent events, both of which involve protein tyrosine kinase activity. *Infect Immun* 66, 1453-1459
4. Pier, G.B., Grout, M., and Zaidi, T.S. (1997) Cystic fibrosis transmembrane conductance regulator is an epithelial cell receptor for clearance of *Pseudomonas aeruginosa* from the lung. *Proc Natl Acad Sci USA* 94, 12088-12093
5. Sawa, T., Ohara, M., Kurahashi, K., Twining, S.S., Frank, D.W., Doroques, D.B., Long, T., Gropper, M.A., and Wiener-Kronish, J.P. (1998) In vitro cellular toxicity predicts *Pseudomonas aeruginosa* virulence in lung infections. *Infect Immun* 66, 3242-3249
6. Pearson, J.P., Feldman, M., Iglewski, B.H., and Prince, A. (2000) *Pseudomonas aeruginosa* cell-to-cell signaling is required for virulence in a model of acute pulmonary infection. *Infect Immun* 68, 4331-4334
7. Stanislavsky, E. S. and Lam, J.S. (1997) *Pseudomonas aeruginosa* antigens as potential vaccines. *FEMS Microbiology Reviews* 21, 243-277
8. Guzzo, J., Duong, F., Wandersoman, C., Murgier, M., and Lazdunski, A. (1991) The secretion genes of *Pseudomonas aeruginosa* alkaline are functionally related to those of Erwinia chrysanthemi proteases and Eschericial coli alpha-haemolysin. *Mol Microbiol* 5, 447-453

9. Filloux, A., Bally, M. Ball, G., Akrim, M., Tommassen, J., and Lazdunski, A. (1990) Protein secretion in Gram-negative bacteria: transport across the outer membrane involves common mechanisms in different bacteria. *EMBO J* 9, 4323-4329
10. Pugsley, A.P. (1993) The general secretory pathway in Gram-negative bacteria. *Microbiol Rev* 57, 50-108
11. Yahr, T.L., Goranson, J., and Frank, D.W. (1996) Exoenzyme S of *Pseudomonas aeruginosa* is secreted by a Type III pathway. *Mol Microbiol* 22, 991-1003
12. Parsek, M.R., Greenberg, E.P. (2000) Acyl-homoserine lactone quorum sensing in Gram-negative bacterial: A signaling mechanism involved in associations with higher organisms. *Proc Natl Acad Sci USA* 97, 8789-8793
13. Pearson, J.P., Gray, K.M., Passador, L., Tucker, K.D., and Eberhard, A. (1994) Structure of the autoinducer required for expression of *Pseudomonas aeruginosa* virulence genes. *Proc Natl Acad Sci USA* 91, 197-201
14. Pesci, E. C., Milbank, J.B.J., Pearson, J.P., McKnight, S., Kende, A.S., Greenberg, E.P., and Iglewski, B.H. (1999) Quinolone signaling in the cell-to-cell communication system of *Pseudomonas aeruginosa. Proc Natl Acad Sci USA* 96, 11229-11234
15. Passador L., Cook, J.M., Gambello, M.J., and Rust, L. (1993) Expression of *Pseudomonas aeruginosa* virulence genes requires cell-to-cell communication. *Science* 260, 1127-1130
16. Brint, J.M., and Ohman, D.E. (1995) Synthesis of multiple exoproducts in *Pseudomonas aeruginosa* is under the control of RHIR-RhlI, another set of regulators in strain PAO1 with homology to the autoinducer-responsive LuxR-LuxI family. *J Bact* 177, 7155-7163
17. Darby, C., Cosma, C.L., Thomas, J.H., and Manoil, C. (1999) Lethal paralysis of *Caenorhabditis elegans* by *Pseudomonas aeruginosa. Proc Natl Acad Sci USA* 96, 15202-15207
18. Mahajan-Miklos, S., Tan, M-W., Rahme, L.G., and Ausubel, F.M. (1999) Molecular mechanisms of bacterial virulence elucidated using a *Pseudomonas aeruginosa-Caenorhabditis elegans* pathogenesis model. *Cell* 96, 47-56
19. Guiney, D.G. (1997) Regulation of bacterial virulence gene expression by the host environment. *J Clin Invest* 99, 565-569
20. Costerton, J.W., Lewandowski, A., Caldwell, D.E., Korber, D.R., and Lappin-Scott, H.M. (1995) Microbial biofilms. *Annu Rev Microbiol* 49, 711-745
21. Costerton, J.W., Stewart, P., and Greenberg, E. (1999) Bacterial biofilms: A common cause of persistent infections. *Science* 284, 1318-1322
22. Davies, D.G., Parsek, M.R., Pearson, J.P., Iglewski, B.H., Costerton, J.W., and Greenberg, E.P. (1998) The involvement of cell-to-cell signals in the development of a bacterial biofilm. *Science* 280, 295-298
23. Singh, P.K., Schaefer, A.L., Parsek, M.R., Moninger, T.O., Welsh, M.J., and Greenberg, E.P. (2000) Quorum-sensing signals indicate that cystic fibrosis lungs are infected with bacterial biofilms. *Nature* 407: 762-764
24. Rashid, M.H., Rumbaugh, K, Passador, L, Davies, D.G., Hamood, A.N., Iglewski, B.H., and Kornberg, A. (2000) Polyphosphate kinase is essential for biofilm development, quorum sensing and virulence of *Pseudomonas aeruginosa. Proc Natl Acad Sci USA* 97, 9636-9641
25. Azghani, A.O., Bedinhaus, T., and Klein, R. (2000) Detection of elastase from *Pseudomonas aeruginosa* in sputum and its potential role in epithelial cell permeability. *Lung* 178, 181-189
26. de Bentzmann, S., Polette , M., Zahm, J-M, Hinnrasky, J., Kileztky, C., Bajolet, O., Klossek, J.M., Filloux, A., Lazdunski, A., and Puchelle, E. (2000) *Pseudomonas aeruginosa* virulence factors delay airway epithelial wound repair by altering the actin cytoskeleton and inducing overactivation of epithelial matrix matalloproteinase-2. *Lab Invest* 80, 209-219
27. Azghani, AO. (1996) *Pseudomonas aeruginosa* and epithelial permeability: role of virulence factors elastase and exotoxin A. *Am J Respir Cell Mol Biol* 15, 132-140

28. Cornelis, G.R., and Van Gijsegem, F. (2000) Assembly and function of type III secretory systems. *Annu Rev Microbiol* 54, 735-74
29. Macnab, R.M. (2000) Type III protein pathway exports *Salmonella* flagella. *ASM News* 66, 738-745
30. Gauthier, A, and Finlay, B.B. (1998) Protein translocation: delivering virulence into the host cell. *Current Biol* 8, R768-R770
31. Yahr, T.L., Vallis, A.J., Hancock, M.K., Barbieri, J.T., and Frank, D.W. (1998) ExoY, an adenylate cyclase secreted by the *Pseudomonas aeruginosa* type III system. *Proc Natl Acad Sci USA* 95, 13899-13904
32. Yahr, T.L., Barbieri, J.T., and Frank, D.W. (1996) Genetic relationship between the 53- and 49-Kilodalton forms of exoenzyme S from *Pseudomonas aeruginosa. J Bacteriol* 178, 1412-1419
33. McGuffie, E.M., Frank, D.W., Vincent T.S., and Olson, J.C. (1998) Modification of Ras in Eukaryotic cells by *Pseudomonas aeruginosa* Exoenzyme S. *Infect Immun* 66, 2607-2613
34. Pederson, K.J., Vallis, A.J., Aktories, K., Frank, D.W., and Barbieri, J.T. (1999) The amino-terminal domain of *Pseudomonas aeruginosa* ExoS disrupts actin filaments via small-molecular-weight GTP-binding proteins. *Molec Micro* 32, 393-401
35. Wurtele, M., Wolf, E., Pederson, K.J., Buchwald, G., Ahmadian, M.R., Barbieri, J.T., and Wittinghofer, A. (2001) How the *Pseudomonas aeruginosa* ExoS toxin down-regulates Rac. *Nat Struct Biol* 8, 23-26
36. Kudoh, I., Wiener-Kronish, J.P., Pittet, J.F., and Frank, D. (1994) Exoproduct secretions of *Pseudomonas aeruginosa* strains influence severity of alveolar epithelial injury. *Am J Physiol (Lung, Cell, Mol Biol)* 267, L551-L556
37. Knight, D.A., and Barbieri, J.T. (1997) Ecto-ADP-Ribosyltransferase activity of *Pseduomonas aeruginosa* exoenzyme S. *Infect Immun* 65, 3304-3309
38. Rumbaugh, K.P., Griswold, J.A., and Hamood, A.N. (1999) *Pseudomonas aeruginosa* strains obtained from patients with tracheal, urinary tract and wound infection: variations in virulence factors and virulence genes. *J Hosp Infect* 43, 211-218
39. Finck-Barbançon, V., Yahr, T.L., and Frank, D.W. (1998) Identification and characterization of SpcU, a chaperone required for efficient secretion of the ExoU cytotoxin. *J Bacteriol* 180, 6224-6231
40. Sawa, T., Yahr, T.L., Ohara, M., Kurashashi, K., Gropper, M.A., Wiener-Kronish, J.P., and Frank, D.W. (1999) Active and passive immunization with the *Pseudomonas* V antigen protects against type III intoxication and lung injury. *Nat Med* 5, 392-398
41. Kurahashi, K., Kajikawa, O., Sawa, T., Ohara, M., Gropper, M.A., Frank, D.W., Martin, T.R., and Wiener-Kronish, J.P. (1999) Pathogenesis of septic shock in *Pseudomonas aeruginosa* pneumonia. *J Clin Invest* 104, 743-750
42. Wiener-Kronish, J.P., Sakuma, T., Kudoh, I., Pittet, J.F., Frank, D., Dobbs, L., Vasil, M.L., and Matthay, M.A. (1993) Alveolar epithelial injury and pleural empyema in acute *P. aeruginosa* pneumonia in anesthetized rabbits. *J App Physiol* 75, 1661-1669
43. Wiener-Kronish, J.P., Albertine, K.H., and Matthay, M.A. (1991) Differential effects of *E. coli* endotoxin on the lung endothelial and epithelial barriers of the lung. *J Clin Invest* 88: 864-875
44. McElroy, M.C., Pittet, J.F., Hashimoto, S., Allen, L., Wiener-Kronish, J.P., and Dobbs, L.G. (1995) A type I cell specific protein is a biochemical marker of epithelial injury in a rat model of pneumonia. *Am J Phys (Lung, Cell, Mol Biol)* 268, L181-L186
45. Rajan, S., Cacalano, G., Bryan, R., Ratner, A.J., Sontich, C.U., van Heerckeren, A., Davis, P., and Prince, A. (2000) *Pseudomonas aeruginosa* induction of apoptosis in respiratory epithelial cells. *Am J Respir Cell Mol Biol* 23, 304-312
46. Lee, A., Chow, D., Haus, B., Tseng, W., Evans, D., Fleiszig, S., Chandy, G., and Machen, T. (1999) Airway epithelial tight junctions and binding and cytotoxicity of *Pseudomonas aeruginosa. Am J Physiol (Lung Cell Mol Physiol)* 277, L204-L217

47. Hauser, A.R., and Engel, J.N. (1999) *Pseudomonas aeruginosa* induces type-III-secretion-mediated apoptosis of macrophages and epithelial cells. *Infect Immun* 67, 5530-5537
48. Hafkemeyer, P., Brinkmann, U., Gottesman, M.M., and Pastan, I. (1999) Apoptosis induced by *Pseudomonas* exotoxin: a sensitive and rapid marker for gene delivery *in vivo*. *Hum Gene Ther* 10, 923-934
49. Zaborina, O., Dhiman, N., Chen, M.L., Kostal, J., Holder, I.A., and Chakrabarty, A.M. (2000) Secreted products of a nonmucoid *Pseudomonas aeruginosa* strain induce two modes of macrophage killing: external-ATP-dependent, P2Z-receptor-mediated necrosis and ATP-independent, caspase-mediated apoptosis. *Microb* 146, 2521-2530
50. Grassme, H., Kirschnek, S., Riethmueller, J., Riehle, A., von Kurthy, G., Lang, F., Weller, M., and Gulbins, E. (2000) CD95/CD95 ligand interactions on epithelial cells in host defense to *Pseudomonas aeruginosa*. *Science* 290, 527-530
51. Park, P.W., Pier, G.B., Preston, M.J., Goldberger, O., Fitzgerald, M.L., and Bernfield, M. (2000) Syndecan-1 shedding is enhanced by LasA, a secreted virulence factor of *Pseudomonas aeruginosa*. *J Biol Chem* 275, 3057-3064
52. Subramanian, S.V., Fitzgerald, M.L., and Bernfield, M. (1997) Regulated shedding of syndecan-1 and -4 ectodomains by thrombin and growth factor receptor activation. *J Biol Chem* 272, 14713-14720
53. Sastry, K., and Ezekowitz, R.A. (1993) Collectins: pattern recognition molecules involved in first line host defense. *Curr Opin Immunol* 5, 59-66
54. LeVine, A.M., Kurak, K.E., Bruno, M.D., Stark, J.M., Whitsett, J.A. and Korfhagen, T.R. (1998) Surfactant protein-A-deficient mice are susceptible to *Pseudomonas aeruginosa* infection. *Am J Respir Cell Mol Biol* 19, 700-708
55. Kabha, K., Schmegner, J., Keisari, Y., Parolis, H., Schlepper-Schaeffer, J., and Ofek, I. (1997) SP-A enhances phagocytosis of *Klebsiella* by interaction with capsular polysaccharides and alveolar macrophages. *Am J Physiol* 272, L344-L352
56. Greene, K.E., Wright, J.R., Steinberg, K.P., Ruzinski, J.T., Caldwell, E., Wong, W.B., Hull, W., Whitsett, J.A., Akino, T., Kuroki, Y., Nagae, H., Hudson, L.D., and Martin, T.R. (1999) Serial changes in surfactant-associated proteins in lung and serum before and after onset of ARDS. *Am J Respir Crit Care Med* 160, 1843-1850

Chapter 11

ALVEOLAR EPITHELIAL REPAIR IN ACUTE LUNG INJURY

Thomas K. Geiser and Michael A. Matthay
Division of Pulmonary Medicine, University Hospital, Bern, Switzerland and Cardiovascular Research Institute, University of California San Francisco, San Francisco, CA

INTRODUCTION

The clinical course of patients with acute lung injury (ALI) or the acute respiratory distress syndrome (ARDS) is variable and influenced by a variety of factors. One of the most important mechanisms that determines the severity of lung dysfunction is the magnitude of injury to the alveolar epithelial barrier. The capacity to repair the epithelial injury is a major determinant of recovery. Specific treatments for accelerating alveolar epithelial repair do not exist although progress in studies with experimental models of ALI suggests that specific treatment may be possible in the future. Most of the treatment modalities that have been tested recently were based on diminishing the inflammatory response in the lung in order to minimize the initial injury. However, an alternative therapeutic approach is to accelerate the repair process of the alveolar epithelium in order to enhance the resolution of pulmonary edema and improve outcomes in patients with ALI. In this chapter, we will focus on the cellular and molecular mechanisms that mediate alveolar epithelial repair, since they may represent potential new targets for treatment of patients with ALI.

Injury of the Alveolar Epithelium in ALI

The normal alveolar barrier is composed of three different structures: 1) the capillary endothelium, 2) the interstitial space including the basement membrane and the extracellular matrix, and 3) the alveolar epithelium. The alveolar epithelium consists of alveolar type 1 and alveolar type 2 cells. The flat alveolar type 1 cells line more than 90% of the alveolar surface area (1). The attenuated cytoplasm provides for close approximation of the alveolar lumen and the bloodstream, optimizing the exchange of respiratory gases. The cuboidal alveolar type 2 cells are multifunctional cells. They produce surfactant, are important for active alveolar ion and liquid clearance, and represent the progenitor cells that regenerate the alveolar epithelium after injury (2).

Under normal conditions, the epithelial barrier is much less permeable than the endothelial barrier (3), and prevents cells and plasma from flooding the air spaces, thereby maintaining normal gas exchange. Several studies have demonstrated the critical importance of the alveolar epithelium in the pathogenesis and recovery from severe ALI (3-6). Efficient alveolar epithelial repair is therefore crucial for the recovery of patients with ALI (7).

Pathology of ALI

While there are many different causes of ALI, the response of the lung to injury is limited, resulting in a similar pathological pattern largely independent of the type of injury. Diffuse alveolar damage is a hallmark in patients with ALI, whether ALI is caused directly (i.e. pneumonia or acid-aspiration) or indirectly (i.e. sepsis or severe trauma) (8). In histological sections from patients who died with ALI, the first lesions appear to be interstitial edema, followed by severe alveolar epithelial damage (6). Based on the pathological findings, different phases have been defined, although they may overlap to varying degrees (9).

The exudative phase is characterized by interstitial and intra-alveolar edema, accumulation of neutrophils in the alveolar space and extensive endothelial and epithelial injury. The alveolar epithelium usually demonstrates extensive necrosis of alveolar type 1 cells, leaving a denuded, but mainly intact basement membrane with overlying hyaline membranes (6). The type 1 alveolar epithelial cell is highly vulnerable to injury, whereas the alveolar type 2 cell is more resistant and can therefore function as progenitor cells for the regeneration of the alveolar epithelium after injury (10). The loss of the integrity of the alveolar epithelium has several pathological and functional consequences. There is influx of protein-rich edema fluid into the air spaces with deposition of hyaline membranes on the denuded basement membranes.

Hyperplastic alveolar type 2 cells characterize the proliferative phase of ALI. Alveolar type 2 cells migrate and begin to proliferate along the alveolar septa in an attempt to cover the denuded basement membrane and re-establish the continuity of the alveolar epithelium. Within the alveolar interstitium, fibroblasts proliferate and migrate through the basement membrane into the fibrinous intra-alveolar exudate. If the fibrinous exudates can be resolved, restoration of the normal lung architecture may be achieved. However, if alveolar type 2 cells migrate over the surface of the organizing granulation tissue, thereby transforming the intra-alveolar exudate into interstitial tissue, interstitial fibrosis of the lung may develop (9 and Chapter 14).

The factors which determine if pulmonary fibrosis or restoration of the normal pulmonary architecture occur after ALI are unknown. Efficient alveolar epithelial repair can reduce the development of fibrosing alveolitis, since the presence of an intact pulmonary epithelial layer suppresses fibroblast proliferation and matrix deposition (11). This property of the epithelium has been confirmed *in vivo* and *ex vivo* using several animal models, which demonstrate that delaying alveolar epithelialization after lung injury leads to an enhanced fibrotic response (12). Fibrosing alveolitis on histologic analysis from patients with ALI correlates with an increased risk of death (13). Moreover, the early appearance of procollagen III, a precursor for collagen synthesis and marker of a fibroproliferative response in the alveolar space, is associated with an increased risk of death (14-16). In summary, efficient restoration of the alveolar epithelium in the early phase of ALI may be a mechanism to accelerate recovery and to prevent the development of pulmonary fibrosis after ALI (7, 17).

REPAIR OF THE ALVEOLAR EPITHELIUM

Clinical Studies of ALI

Two major studies have demonstrated that many patients with ALI have impaired alveolar fluid clearance. In the first study published in 1990, the patients with impaired alveolar fluid clearance appeared to have a worse outcome with a higher mortality than patients with intact alveolar fluid clearance (18). However, that study included only 16 patients with ALI. A more recent study has examined 69 patients with ALI and demonstrated that patients with maximal alveolar fluid clearance have a shorter duration of mechanical ventilation and a lower mortality (4), thus providing strong support for the hypothesis that a functional alveolar epithelial barrier is important to the resolution of clinical lung injury. Also, surfactant production and composition is not normal in many patients with ALI, suggesting alterations in the function of alveolar epithelial type 2 cells (19 and Chapter 12). Taken together, these observations provide evidence that the alveolar epithelial barrier is functionally impaired in most patients with ALI, and that efficient repair of the epithelial lining is critical to recovery and survival in ALI.

Cellular and molecular mechanisms of alveolar epithelial repair

Alveolar epithelial repair is a complex process involving multiple steps including cell spreading and migration, proliferation, production of the underlying alveolar basement membrane and differentiation of some of the type 2 cells into type 1 cells. In the case of ALI with extensive alveolar epithelial damage, the alveolar type 2 cell is responsible for efficient repair and the restoration of a functional alveolar epithelium. It is widely accepted that alveolar type 2 cells are the progenitor cells of the alveolar epithelium (10, 20, 21). The turnover rate of alveolar type 2 cells is low in the normal lung, but can be dramatically increased following lung injury.

Several mechanisms play a major role in alveolar epithelial repair *in vivo* and *in vitro*. Proliferation of alveolar type 2 cells is the most obvious and easily measurable event in epithelial repair *in vivo*, however it needs one or two days to be significant (7). Therefore, other mechanisms may contribute to early alveolar epithelial wound repair. *In vitro* studies using primary alveolar type 2 epithelial cells showed that cell spreading and migration are primarily responsible for efficient epithelial wound repair (22, 23). In accordance with the *in vitro* results from studies using primary alveolar epithelial cells, the most important and earliest events after denudation of the airway epithelium *in vivo* was shown to be cell spreading and migration (24). It is therefore reasonable to speculate that *in vivo* cell spreading and migration are the primary mechanisms during the early phase of alveolar epithelial repair, followed by cell proliferation leading to alveolar type 2 cell hyperplasia (7).

Studies using lung tissue from patients with ALI have suggested that programmed cell death (apoptosis) may play a role in alveolar epithelial repair. Apoptosis is a form of genetically regulated cell death characterized by a distinct sequence of morphologic rearrangements including condensation of the nuclear chromatin and fragmentation into membrane-bound vesicles (see Chapter 15). Typically, apoptosis does not lead to an inflammatory response and thereby provides a biological mechanism for inflammatory cell elimination and tissue remodeling after injury. While some recent studies demonstrated the potential importance of apoptosis in granulocyte removal after ALI (25), there is evidence that epithelial cell apoptosis may be a mechanism to restore the normal cell number of alveolar epithelial cells after type 2 hyperplasia during the fibroproliferative phase of ALI (26). In human lung specimens from patients in the acute and organizing phases of diffuse alveolar damage, a marked increase of alveolar epithelial cell apoptosis was detected, accompanied by an upregulation of the proapoptotic proteins p53, WAF-1 (wild-type p53-activated fragment), and BAX (27, 28). These data are supported by studies showing that alveolar epithelial cell apoptosis is an important mechanism in restoration of KGF-induced alveolar type 2 cell hyperplasia in rats (29, 30).

Finally, the Fas-Fas Ligand interaction, one of the major systems that modulates apoptosis in alveolar epithelial cells (31), is involved in the development of bleomycin-induced lung fibrosis (32-34). Initial reports indicate that the Fas-Fas Ligand system may also play a role in patients with ALI (34, 35). In addition, apoptosis also appears to be an important mechanism for eliminating granulation tissue from the alveolar space in patients with ALI. Bronchoalveolar lavage fluid obtained from patients with ALI during the reparative phase induced fibroblast and endothelial cell apoptosis *in vitro* and apoptotic cells were detected in the granulation tissue *in vivo* (36). These data support the hypothesis that apoptosis may be a major mechanism in the resolution of granulation tissue into scar during wound repair (37).

Mediators of alveolar epithelial repair

The repair process includes interactions between cells, as well as between the alveolar type 2 cell and the extracellular matrix that are coordinated by a variety of soluble mediators released into the alveolar space during ALI (Table 1). Many of these mediators are increased in the lung during ALI, indicating that they may play a significant role in the pathogenesis or in the resolution of ALI (5). Identification of molecules that induce alveolar epithelial repair is therefore important, as they might be useful as a potential therapeutic targets in patients with ALI. Though a variety of proteins, including growth factors, cytokines, and extracellular matrix proteins, are increased in pulmonary edema or bronchoalveolar lavage fluid from patients with ALI (5), we will focus on proteins which were shown to play an important role in alveolar epithelial repair *in vitro* or *in vivo*.

Table 1. Mediators inducing alveolar epithelial repair.

Mediator	Activity on alveolar type 2 Epithelial Cells (AEC)	Reference
Epidermal Growth Factor (EGF) and Transforming Growth Factor-α (TGFα)	Alveolar fluid removal, AEC proliferation, *In vitro* epithelial wound repair.	22,48
Hepatocyte Growth Factor (HGF)	Proliferation and migration of AEC.	49, 50
Keratinocyte Growth Factor (KGF)	Proliferation of AEC, *In vitro* epithelial wound repair.	38, 39, 47
Interleukin-1β (IL-1β)	Cell spreading, cell migration, *In vitro* epithelial wound repair.	23, 84
Fibronectin	Cell spreading, cell migration, *In vitro* epithelial wound repair.	66
Collagenases	Cell spreading, cell migration Decreased cell attachment, *In vitro* epithelial wound repair	70

Growth factors

Keratinocyte growth factor (KGF), also known as fibroblast growth factor-7, is one of the most potent mitogens observed for adult alveolar type 2 cells. KGF, which is produced by mesenchymal cells, was shown to induce alveolar type 2 proliferation *in vitro* (38) and *in vivo* (39). These studies indicated that KGF may play a role in the repair phase of ALI. Administration of KGF had no beneficial effect given after bleomycin-induced or acid-induced lung injury (40, 41), whereas a protective effect was shown if KGF was given before lung injury was induced by bleomycin (40, 42), hyperoxia (43), or *Pseudomonas aeruginosa* (44). Moreover, the protective effect of KGF was reported in an *ex vivo* model of ventilator-induced lung injury (45). However, the mechanism of the protective effect of KGF in lung injury is unclear. In a recent study, increased KGF levels were found to correlate with alveolar epithelial cell proliferation after bleomycin-induced lung injury in rats (46), indicating that the increased levels of KGF induce efficient epithelial repair. This hypothesis is supported by *in vitro* wound healing studies showing that KGF increases alveolar epithelial repair (47, and *d'Ortho et al unpublished data*). Also, we measured biologically active KGF in pulmonary edema fluid from patients with ALI (48). Although KGF may play an important role in recovery from ALI, further studies are needed to elucidate the mechanism of protection from lung injury induced by KGF.

In contrast to KGF, hepatocyte growth factor (HGF), another important growth factor for alveolar epithelial cells, was increased in the pulmonary edema fluid from patients with ALI (48). HGF is a potent mitogen for alveolar type 2 cells *in vivo* (49) and *in vitro* (50) and is upregulated after acid-instillation in rat lungs (51). Furthermore, DNA synthesis was stimulated after intravenous administration of HGF in acid-induced lung injury. HGF is mainly produced by mesenchymal cells and upregulated by proinflammatory cytokines including IL-1β and TNF-α, with the potential benefit to stimulate proliferation of alveolar type 2 cells and restore the alveolar epithelial barrier. Since biologically active HGF is present in high concentrations in pulmonary edema fluid from patients with ALI (48), it can be hypothesized that HGF may play an important role in alveolar epithelial repair after injury.

There is increasing evidence that EGF, TGF-α, and their common receptor, EGFR, can regulate epithelial repair *in vivo* (52, 53) and *in vitro* (22, 54). EGFR is overexpressed and activated in response to lung epithelial injury and plays a crucial role in epithelial repair (55). An increase of both TGF-α and EGFR have been shown in bleomycin-injured rat lungs when compared to controls (52). In addition, TGF-α has been identified in the bronchoalveolar lavage fluid and edema fluid from patients with ALI (56, 57). These data indicate that EGF and TGF-α may be beneficial in the resolution and repair phase of patients with ALI. TGF-α was shown to upregulate the rate of alveolar fluid clearance in rats (58), and over-expression of TGF-α in mice attenuated polytetrafluoroethylene fumes-induced lung injury (59), suggesting a beneficial role of TGF-α in ALI. However, more work is needed to better define the therapeutic value of administration of growth factors like EGF and TGF-α in ALI.

The role of TGF-β in alveolar epithelial repair has yet to be defined. In bleomycin- and hyperoxia-induced lung injury, there is a significant decrease in the secretion of biologically active TGF-β. Since TGF-β inhibits epithelial cell growth, it is possible that low levels of TGF-β support epithelial cell proliferation, allowing efficient alveolar epithelial repair. Moreover, TGF-β was shown to induce the production of matrix components and the upregulation of integrins on airway epithelial cells (60), two major components that facilitate cell migration and spreading during alveolar epithelial repair (22 and Chapter 14).

The role of other growth factors such as platelet-derived growth factor (PDGF), granulocyte-colony stimulating factor (G-CSF), or insulin-like growth factor (IGF) are not well defined in alveolar epithelial repair. PDGF-BB and IGF-1 are increased in injured lungs (61). Biologically active PDGF is present in bronchoalveolar lavage fluids of patients with ALI and was shown to induce fibroblast proliferation and migration (62). However, the effect on alveolar epithelial repair has not been studied.

Extracellular matrix proteins and metalloproteinases

The interaction of the alveolar epithelial cells with the underlying extracellular matrix is important for efficient epithelial repair, since cell spreading and migration are major mechanisms of alveolar epithelial repair. Alveolar epithelial cells are capable of producing their own extracellular matrix (63-65) and to recreate their own basement membrane. For example, the production of fibronectin is upregulated in alveolar epithelial cells, thus contributing to the provisional matrix in the injured lung. Scanning electron micrographs from patients with ALI showed that alveolar epithelial repair occurs over a provisional matrix containing fibrin and fibronectin. The hypothesis that fibronectin plays a key role in epithelial repair was supported by the results of Garat et al (66), who reported that both soluble and insoluble fibronectin stimulated *in vitro* alveolar epithelial repair.

Cell migration during epithelial repair is accompanied by digestion of the provisional matrix allowing its replacement by normal basement membranes containing type IV collagen, laminin, and nidogen. Matrix metalloproteinases (MMP), which cleave extracellular matrix components, may therefore represent important proteins in epithelial repair, although these enzymes can also destroy the normal architecture of the lung and therefore contribute to the pathogenesis of ALI. Collagenases (MMP-1 and MMP-13) and gelatinases (MMP-2 and MMP-9) were found to be increased in bronchoalveolar lavage fluid from patients with ALI (67-69). These levels correlated to the disruption of the basement membrane structures, leading to the increased permeability of the alveolocapillary barrier in patients with ALI. However, the collagenases MMP-1 and MMP-13 induce alveolar epithelial repair *in vitro* (70) and may therefore also have beneficial effects in the injured alveolus. Alveolar type 2 cells *in vitro* also produce gelatinase A (MMP-2) and gelatinase B (MMP-9) (71). MMP-9 (Gelatinase B) was shown to induce repair of the airway epithelium (72). Moreover, the expression of matrilysin (MMP-7) is increased in airway epithelium and induced in alveolar type 2 cells in lungs from patients with cystic fibrosis (73), indicating that MMP-7 is involved in the remodeling process of the injured lung. Repair of the injured tracheal epithelium was impaired in matrilysin-null mice (73). These data suggest that MMP-7 may be a predominant MMP controlling epithelial repair.

Inflammatory Cytokines

During the early phase of ALI, a variety of inflammatory mediators are released into the alveolar space (5 and Chapter 2). Interleukin-1β (IL-1β) and tumor necrosis factor-α (TNF-α) are markedly increased in edema fluid or bronchoalveolar lavage fluid in patients at an early stage of ALI. In both bronchoalveolar lavage fluid and pulmonary edema fluid IL-1β was biologically active and primarily responsible for the inflammatory response induced by the pulmonary edema fluid from ALI patients, whereas very little biologically active TNF-α was found. Interestingly, pulmonary edema fluid obtained from patients with ALI within the first 12 hours after endotracheal intubation had an increased alveolar epithelial repair activity compared to control patients with hydrostatic edema, as determined in our *in vitro* wound healing system (Figure 1) (74).

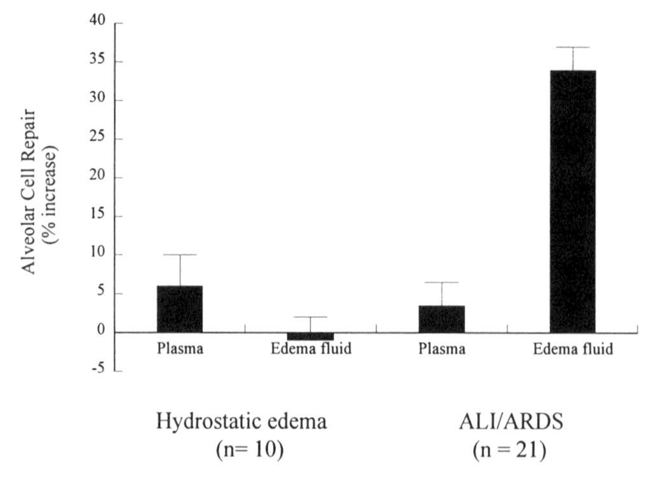

Figure 1. Effect of pulmonary edema fluid and plasma from patients with ALI or hydrostatic pulmonary edema on alveolar epithelial repair *in vitro*. The epithelial repair activity of pulmonary edema fluid or plasma was determined for each sample using an *in vitro* epithelial wound repair assay with A549 alveolar epithelial cells and expressed as percent increase of alveolar epithelial repair compared to pooled plasma from healthy donors (control). Results are reported as mean ± SEM, *p<0.01 (compared to control) (74).

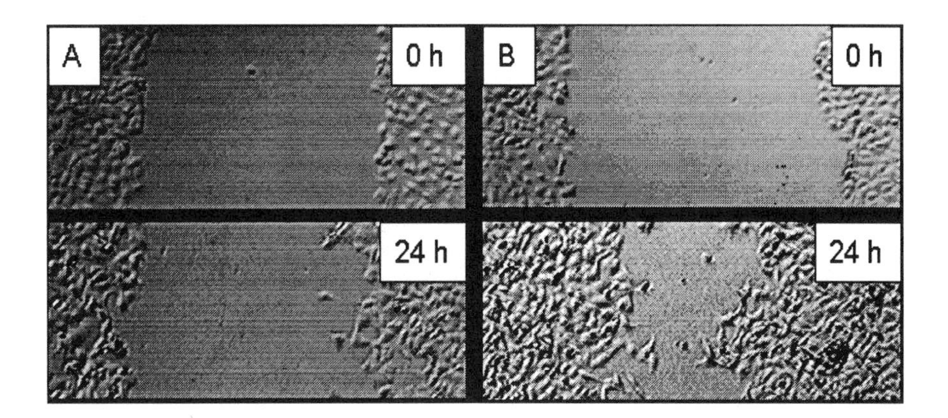

Figure 2. An *in vitro* alveolar epithelial wound repair model was established in our laboratory to study the mechanisms of alveolar epithelial repair and to quantify the epithelial repair activity of pulmonary edema fluid or purified reagents. Panel A: Epithelial wound without closure in serum-free medium (control), Panel B: Promotion of wound closure in the presence of Interleukin-1β (10 ng/mL) (23, 74).

Summary and Conclusions

Specific treatment to hasten recovery from ALI is currently limited to lung protective ventilation strategies (8). Since severe alveolar epithelial injury is a crucial component of the pathogenesis of ALI, acceleration of alveolar epithelial repair may be a promising and novel treatment strategy in patients with ALI (7). As discussed in this chapter, substantial progress has been made in understanding some of the cellular and molecular mechanisms that promote alveolar epithelial repair. Efficient alveolar epithelial repair may restore the functions of the alveolar epithelium, decrease the duration of mechanical ventilation, and therefore improve clinical outcomes. Strategies to accelerate alveolar epithelial repair might include administration of epithelial specific growth factors (e.g. KGF or HGF). Moreover, modulation of apoptosis may represent a possible future strategy to improve resolution of inflammation and promotion of alveolar remodeling after ALI. However, since strategies to promote alveolar epithelial repair are still in their embryonic stage of development, further studies are needed to understand the important mechanisms involved in alveolar epithelial repair before designing clinical trials.

REFERENCES

1. Crapo, J.D., Barry, B.E., Gehr, P., Bachofen, M., and Weibel, E. (1982) Cell number and cell characteristics of the normal human lung. *Am Rev Respir Dis* 126, 332-337
2. Mason, R.J., and Shannon, J.M. (1997) Alveolar type 2 cells. *In* The Lung. R.G. Crystal, J.B. West, P.J. Barnes, and E.R. Weibel, editors. Lippincott - Raven Publishers, Philadelphia - New York. 543-555
3. Wiener-Kronish, J.P., Albertine, K.H., and M.A. Matthay, M.A. (1991) Differential responses of the endothelial and epithelial barriers of the lung in sheep to Escherichia coli endotoxin. *J Clin Invest* 88, 864-75
4. Ware, L.B., and Matthay, M.A. (2001) The majority of patients with acute lung injury and the acute respiratory distress syndrome have impaired alveolar fluid clearance. *Am J Respir Crit Care Med* (In press).
5. Pittet, J.F., Mackersie, R.C., Martin, T.R., and Matthay, M.A. (1997) Biological markers of acute lung injury: prognostic and pathogenetic significance. *Am J Respir Crit Care Med* 155, 1187-205
6. Bachofen, M., and Weibel, E.R. (1977) Alterations of the gas exchange apparatus in adult respiratory insufficiency associated with septicemia. *Am Rev Respir Dis* 116, 589-615
7. Berthiaume, Y., Lesur, O., and Dagenais, A. (1999) Treatment of adult respiratory distress syndrome: plea for rescue therapy of the alveolar epithelium. *Thorax* 54, 150-160
8. Ware, L.B., and Matthay, M.A. (2000) The acute respiratory distress syndrome. *N Engl J Med* 342, 1334-1349
9. Tomashefski, J.F.J. (2000) Pulmonary pathology of acute respiratory distress syndrome. *In* Acute Respiratory Distress Syndrome. Vol. 21 (3). M. Wiedmann, M.A., editor. 435-466.
10. Uhal, B. (1997) Cell cycle kinetics in the alveolar epithelium. *Am J Physiol* 272, L1031-L1045
11. Adamson, I.Y., Young, L., and Bowden, D.H. (1988) Relationship of alveolar epithelial injury and repair to the induction of pulmonary fibrosis. *Am J Pathol* 130, 377-383
12. Adamson, I.Y., Hedgecock, C., and Bowden, D.H. (1990) Epithelial cell-fibroblast interactions in lung injury and repair. *Am J Pathol* 137, 385-392
13. Martin, C., Papazian, L., Payan, M.J., Saux, P., and Gouin, F. (1995) Pulmonary fibrosis correlates with the outcome in the adult respiratory distress syndrome: a study in mechanically ventilated patients. *Chest* 107, 196-200
14. Chesnutt, A.N., Matthay, M.A., Tibayan, F.A., and Clark, J.G. (1997) Early detection of type III procollagen peptide in acute lung injury. Pathogenetic and prognostic significance. *Am J Respir Crit Care Med* 156, 840-845
15. Clark, J.G., J.A. Milberg, K.P. Steinberg, and L.D. Hudson. 1995. Type III procollagen peptide in the adult respiratory distress syndrome: association of increased peptide levels in bronchoalveolar lavage fluid with increased risk for death. *Ann Intern Med.* 122, 17-23
16. Marshall, R.P., Bellingan, G., Webb, S., Puddicombe, A., Goldsack, N., McAnulty, R., and Laurent, G.J. (2000) Fibroproliferation occurs early in the acute respiratory distress syndrome and impacts on outcome. *Am J Respir Crit Care Med* 162, 1783-1788
17. Matthay, M.A. (1995) Fibrosing alveolitis in the adult respiratory distress syndrome. *Ann Intern Med* 122, 65-66
18. Matthay, M.A., and Wiener-Kronish, J.P. (1990) Intact epithelial barrier function is critical for the resolution of alveolar edema in humans [see comments]. *Am Rev Respir Dis* 142, 1250-1257
19. Gregory, T.J., Longmore, W.J., and Moxley, M.A. (1991) Surfactant chemical composition and biophysical activity in acute respiratory distress syndrome. *J Clin Invest* 88, 1976-1981

20. Adamson, I.Y., and Bowden, D.H. (1974) The type 2 cell as progenitor of alveolar epithelial regeneration. A cytodynamic study in mice after exposure to oxygen. *Lab Invest* 30, 35-42
21. Fehrenbach, H. (2001) Alveolar epithelial type 2 cell: defender of the alveolus revisited. *Respir Res* 2, 33-46
22. Kheradmand, F., Folkesson, H.G., Shum, L., Derynk, R., Pytela, R., and M.A. Matthay, M.A. (1994) Transforming growth factor-α enhances alveolar epithelial cell repair in a new *in vitro* model. *Am J Physiol* 267, L728-L738
23. Geiser, T., Jarreau, P.H., Atabai, K., and Matthay, M.A. (2000) Interleukin-1β increases *in vitro* alveolar epithelial repair. *Am J Physiol* 279, 1184-1190
24. Lane, B.P., and Gordon, R. (1974) Regeneration of the rat tracheal epithelium after mechanical injury. The relationship between mitotic activity and cellular differentiation. *Proc Soc Exp Biol Med* 145, 1139-1144
25. Matute-Bello, G., Liles, W.C., Radella, F.N., Steinberg, K.P., Ruzinski, J., Jonas, M., Chi, E.Y., Hudson, L.D., and Martin, T.R. (1997) Neutrophil apoptosis in the acute respiratory distress syndrome. *Am J Respir Crit Care Med* 156, 1969-1977
26. Bardales, R.H., Xie, S.S., Schaefer, R.R., and Hsu, S.M. (1996) Apoptosis is a major pathway responsible for the resolution of type 2 pneumocytes in acute lung injury. *Am J Pathol* 149, 845-852
27. Guinee, D., Fleming, M., Hayashi, T., Woodward, M., Zhang, J., Walls, J., Koss, M., Ferrans, V., and Travis, W. (1996) Association of p53 and WAF1 expression with apoptosis in diffuse alveolar damage. *Am J Pathol* 149, 531-538
28. Guinee, D.J., Brambilla, E., Fleming, M., Hayashi, T., Rahn, M., Koss, M., Ferrans, V., and Travis, W. (1997) The potential role of BAX and BCL-2 expression in diffuse alveolar damage. *Am J Pathol* 151, 999-1007
29. Fehrenbach, H., Kasper, M., Tschernig, T., Pan, T., Schuh, D., Shannon, J., Mueller, M., and Mason, R. (1999) Keratinocyte growth factor-induced hyperplasia of rat alveolar type 2 cells is resolved by differentiation into type I cells and by apoptosis. *Eur Respir J* 14, 534-544
30. Fehrenbach, H., Kasper, M., Koslowski, R., Pan, T., Schuh, D., Mueller, M., and Mason, R.J. (2000) Alveolar epithelial type 2 cell apoptosis in vivo during resolution of keratinocyte growth factor-induced hyperplasia in the rat. *Histochem Cell Biol* 114, 49-61
31. Fine, A., Anderson, N.L., Rothstein, T.L., Williams, M.C., and Gochuico, B.R., (1997) Fas expression in pulmonary alveolar epithelial type 2 cells. *Am J Physiol* 273, L64-L71
32. Hagimoto, N., Kuwano, K., Nomoto, Y., Kunitake, R., and Hara, N. (1997) Apoptosis and expression of Fas/Fas ligand mRNA in bleomycin-induced pulmonary fibrosis in mice. *Am J Respir Cell Mol Biol* 16, 91-101
33. Kuwano, K., Miyazaki, H., Hagimoto, N., Kawasaki, M., Fujita, M., Kunitake, R., Kaneko, Y., and H. N. 1999. The involvement of Fas-Fas Ligand pathway in fibrosing lung disease. *Am J Respir Cell Mol Biol* 20, 53-60
34. Hashimoto, S., Kobayashi, A., Kooguchi, K., Kitamura, Y., Onodera, H., and Nakajima, H. (2000) Upregulation of two death pathways of perforin/granzyme and FasL/Fas in septic acute respiratory distress syndrome. *Am J Respir Crit Care Med* 161, 237-243
35. Matute-Bello, G., Liles, W.C., Steinberg, K.P., Kiener, P.A., Mongovin, S., Chi, E.Y., Jonas, M., and Martin, T.R. (1999) Soluble Fas-Ligand induces epithelial cell apoptosis in humans with acute lung injury (ARDS). *J Immunol* 163, 2217-2225
36. Polunovsky, V., Baruch, C., Henke, C., Snover, D., Wendt, C., Ingbar, D., and Bittermann P. (1993) Role of mesenchymal cell death in lung remodeling after injury. *J Clin Invest* 92, 388-397
37. Desmouliere, A., Redard, M., Darby, I., and Gabbiani, G. (1995) Apoptosis mediates the cellularity during the transition between granulation tissue and scar. *Am J Pathol* 146, 56-66

38. Panos, R.J., Rubin, J.S., Aaronson, S.A., and Mason, R.J. (1993) Keratinocyte growth factor and hepatocyte growth factor/scatter factor are heparin-binding growth factors for alveolar type 2 cells in fibroblast conditioned medium. *J Clin Invest* 92, 969-977
39. Ulich, T.R., Yi, E.S., Longmuir, K., Yin, S., Biltz, R., Morris, C.F., R. Housley, R., and Pierce, G.F. (1994) Keratinocyte growth factor is a growth factor for type 2 pneumocytes *in vivo*. *J Clin Invest* 93, 1298-1306
40. Deterding, R.R., Havill, A.M., Toshiyuki, Y., Middleton, S.C., Jacoby, C.R., Shannon, J.M., Simonet, W.S., and Mason, R.J. (1997) Prevention of bleomycin-induced lung injury in rats by keratinocyte growth factor. *Proc Assoc Am Phys* 109, 254-268
41. Yano, T., Deterding, R.R., Simonet, W.S., Shannon, J.M., and Mason, R.J. (1996) Keratinocyte growth factor reduces lung damage due to acid instillation in rats. *Am J Respir Cell Mol Biol* 15, 433-442
42. Guo, J., Yi, E.S., Havill, A.M., Sarosi, I., Whitcomb, L., Yin, S., Middleton, S.C., Piguet, P., and Ulich, T.R. (1998) Intravenous keratinocyte growth factor protects against experimental pulmonary injury. *Am J Physiol* 275, L800-L805
43. Barazzone, C., Donati, Y.R., Rochat, A.F., Vesin, C., Kan, C.D., Pache, J.C., and Piguet, P.F. (1999) Keratinocyte growth factor protects the alveolar epithelium and endothelium from oxygen-induced injury in mice. *Am J Pathol* 154, 1479-1487
44. Viget, N.B., Guery, B.P.H., Ader, F., Neviere, R., Alfandari, S., Creuzy, C., Roussel-Delvallez, M., Foucher, C., Mason, C.M., Beaucaire, G., and Pittet, J.F. (2001) Keratinocyte growth factor protects against Pseudomonas aeruginosa-induced lung injury. *Am J Physiol* 279, L1199-L1209
45. Welsh, D.A., Summer, W.R., Dobard, E.P., Nelson, S., and Mason, C.M. (2000) Keratinocyte growth factor prevents ventilator-induced lung injury in an *ex vivo* rat model. *Am J Respir Crit Care Med* 162, 1081-1086
46. Adamson, I.Y.R., and Bakowska, J. (1999) Relationship of keratinocyte growth factor and hepatocyte growth factor levels in rat lung lavage fluid to epithelial cell regeneration after bleomycin. *Am J Pathol* 155, 949-954
47. Ware, L.B., Folkesson, H.G., and Matthay, M.A. (1998) Keratinocyte growth factor increases alveolar epithelial wound healing in vitro. *FASEB* 12, A778
48. Verghese, G.M., McCormick-Shannon, K., Mason, R.J., and Matthay, M.A. (1998) Hepatocyte growth factor and keratinocyte growth factor in the pulmonary edema fluid of patients with acute lung injury. Biologic and clinical significance. *Am J Respir Crit Care Med* 158, 386-394
49. Ohmichi, H., Matsumoto, K., and Nakamura. T. (1996) *In vivo* mitogenic action of HGF on lung epithelial cells: pulmotrophic role in lung regeneration. *Am J Physiol* 270, L1031-L1039
50. Shiratori, M., Michalopoulos, G., Shinozuka, H., Singh, G., Ogasawara, D., and Katyal, S.L. (1995) Hepatocyte growth factor stimulates DNA synthesis in alveolar type 2 cells *in vitro*. *Am J Respir Cell Mol Biol* 12, 171-180
51. Yanagita, K., Matsumoto, K., Sekiguchi, K., Ishibashi, H., Niho, Y., and Nakamura, T. (1993) Hepatocyte growth factor may act as a pulmotrophic factor on lung regeneration after acute lung injury. *J Biol Chem* 268, 21212-21217
52. Madtes, D.K., Busby, H.K., Strandjord, T.P., and Clark. J.G. (1994) Expression of transforming growth factor-α and epidermal growth factor receptor is increased following bleomycin-induced lung injury in rats. *Am J Respir Cell Mol Biol* 11, 540-551
53. Pai, R., Ohta, M., Itani, R.M., Sarfeh, I.J., and Tarnawski, A.S. (1998) Induction of mitogen-activated protein kinase signal transduction pathway during gastric ulcer healing in rats. *Gastroenterology* 114, 706-713
54. Nici, L., Medina, M., and Frackelton, A.R. (1996) The epidermal growth factor receptor network in type 2 pneumocytes exposed to hyperoxia *in vitro*. *Am J Physiol* 270, L242-L250
55. Van Winkle, L.S., Isaac, J.M., and Plopper, C.G. (1997) Distribution of epidermal growth factor receptor and ligands during bronchiolar epithelial repair from naphthalene-induced Clara cell injury in the mouse. *Am J Pathol* 151, 443-459

56. Chesnutt, A.N., Kheradmand, F., Folkesson, H.G., Alberts, M., and Matthay, M.A. (1997) Soluble transforming growth factor-alpha is present in the pulmonary edema fluid of patients with acute lung injury. *Chest* 111, 652-656
57. Madtes, D.K., Rubenfeld, G., Klima, L.D., Milberg, J.A., Steinberg, K.P., Martin, T.R., Raghu, G., Hudson, L.D., and Clark. J.G. (1998) Elevated transforming growth factor-alpha levels in bronchoalveolar lavage fluid of patients with acute respiratory distress syndrome. *Am J Respir Crit Care Med* 158, 424-430
58. Folkesson, H.G., Pittet, J.F., Nitenberg, G., and Matthay, M.A. (1996) Transforming growth factor-alpha increases alveolar liquid clearance in anesthetized ventilated rats. *Am J Physiol* 271, L236-L244
59. Hardie, W.D., Prows, D.R., Leikauf, G.D., and Korfhagen, T.R. (1999) Attenuation of acute lung injury in transgenic mice expressing human transforming growth factor-α. *Am J Physiol* 277, L1045-L1050
60. Sheppard, D., D.S. Cohen, A. Wang, and M. Busk. (1992) Transforming growth factor β differentially regulates expression of integrin subunits in guinea pig airway epithelial cells. *J Biol Chem*. 267:17409-17414.
61. Homma, S., Nagaoka, I., Abe, H., Takahashi, K., Seyama, K., and Nukiwa. T.S. (1995) Localization of platelet-derived growth factor and insulin-like growth factor 1 in the fibrotic lung. *Am J Respir Crit Care Med* 152, 2084-2089
62. Snyder, L.S., Hertz, M.I., Peterson, M.S., Harmon, K.R., Marinelli, W.A., Henke, C. A., Greenheck, J.R., Chen, B., and Bitterman. P.B. (1991) Acute lung injury: Pathogenesis of intraalveolar fibrosis. *J Clin Invest* 88, 663-673
63. Dunsmore, S.E., Martinez-Williams, C., Goodman, R.A., and Rannels, D.E. (1995) Composition of extracellular matrix of type 2 pulmonary epithelial cells in primary culture. *Am J Physiol*. 269, L754-L765
64. Dunsmore, S.E., Martinez-Williams, C., Goodman, R.A., and Rannels, D.E. (1995) Turnover of fibronectin and laminin by alveolar epithelial cells. *Am J Physiol* 269, L766-L775
65. Dunsmore, S.E., and Rannels, D.E. (1995) Turnover of extracellular matrix by type 2 pulmonary epithelial cells. *Am J Physiol* 268, L336-L346
66. Garat, C., Kheradmand, F., Albertine, K.H., Folkesson, H.G., and Matthay, M.A. (1996) Soluble and insoluble fibronectin increases alveolar epithelial wound healing in vitro. *Am J Physiol* 271, L844-L853
67. Torii, K., Iida, K., Miyazaki, Y., Saga, S., Kondho, Y., Taniguchi, H., Taki, F., Takagi, K., Matsuyama, M., and Suzuki. R. (1997) Higher concentrations of matrix metalloproteinases in bronchoalveolar lavage fluid of patients with adult respiratory distress syndrome. *Am J Respir Crit Care Med* 155, 43-46
68. Delclaux, C., d'Ortho, M.P., Delacourt, C., Lebargy, F., Brun-Buisson, C., Brochard, L., Lemaire, F., Lafuma, C., and Harf, A. (1997) Gelatinases in epithelial lining fluid of patients with adult respiratory distress syndrome. *Am J Physiol* 272, L442-L451
69. Ricou, B., Nicod, L., Lacraz, S., Welgus, H.G., Suter, P.M., and Dayer, J.M. (1996) Matrix metalloproteinases and TIMP in acute respiratory distress syndrome. *Am J Respir Crit Care Med* 154, 346-352
70. Planus, E., Galiacy, S., Matthay, M.A., Laurent, V., Gavrilovic, J., G. Murphy, G., Clerici, C., Isabey, D., Lafuma, C., and d'Ortho, M.P. (1999) Role of collagenase in mediating in vitro alveolar epithelial wound repair. *J Cell Sci* 112, 243-252
71. d'Ortho, M.P., Clerici, C., Yao, P.M., Delacourt, C., Delclaux, C., Franco-Montoya, M.L., Harf, A., and Lafuma, C. (1997) Alveolar epithelial cells in vitro produce gelatinases and tissue inhibitor of matrix metalloproteinase-2. *Am J Physiol* 273, L663-L675
72. Buisson, A.C., Zahm, J.M., Polette, M., Pierrot, D., Bellon, G., Puchelle, E., Birembaut, P., and Tournier, J.M. (1996) Gelatinase B is involved in the in vitro wound repair of human respiratory epithelium. *J Cell Physiol* 166, 413-426

73. Dunsmore, S.E., Saarialho-Kere, U.K., Roby, J.D., Wilson, C.L., Matrisian, L.M., Welgus, H.G., and Parks, W.C. (1998) Matrilysin expression and function in airway epithelium. *J Clin Invest* 102, 1321-1331
74. Geiser, T., Atabai, K., Jarreau, P.H., Ware, L.B., Pugin, J., and Matthay, M.A. (2001) Pulmonary edema fluid from patients with acute lung injury augments alveolar epithelial repair by an interleukin-1β dependent mechanism. *Am J Respir Crit Care Med* (In press)

Chapter 12

MOLECULAR BASIS FOR SURFACTANT FUNCTION IN ARDS

Liqian Zhang and Jeffrey A. Whitsett
Division of Pulmonary Biology, Children's Hospital Medical Center and Children's Hospital Research Foundation, Cincinnati, OH

INTRODUCTION

While knowledge regarding the molecular and physiologic basis of surfactant function has been expanded primarily through study of the preterm lung, the pathogenesis of adult respiratory distress syndrome (ARDS) and the precise roles played by the surfactant system in this complex disorder have remained less well defined. Since ARDS is initiated by various systemic or lung injuries, the heterogeneity of mechanisms involved in the pathogenesis of ARDS are likely to be complex and involve numerous cellular pathways critical for lung function.

This chapter will discuss the role of the surfactant system in the pathogenesis of ARDS. Increasing knowledge of surfactant function derived from experience with preterm neonates, and the successful application of surfactant replacement therapy for neonatal RDS and meconium aspiration provides a framework with which to approach the complicated clinical issues presented by ARDS.

Structure and Function of Pulmonary Surfactant

Pulmonary surfactant is required for the reduction of surface tension at the air-liquid interface in the alveoli, maintaining alveolar stability at low lung volumes at end expiration. The lack of surfactant activity, typified by premature infants with RDS, causes cyanosis and respiratory distress secondary to atelectesis and V-Q mismatch. In the alveoli, surface tension is generated by the unequal forces of water molecules at the air-liquid interface, the attractive forces of water producing surface tension of approximately 70 dynes/cm^2. In ARDS the presence of abnormal serum and cellular proteins in the airspace inhibits the relatively low surface tension created by phospholipids in pulmonary surfactant, the surface tension of most proteins generally being greater than 20 dynes/cm^2. The evolutionary solution to the problem of surface tension reduction in the alveoli consisted of the production of a distinct physical-chemical phase generated by the organization of phospholipids at the air-liquid interface.

Surfactant phospholipids are produced by Type II epithelial cells and are packaged with surfactant proteins B and C (SP-B and SP-C) in the form of lamellar bodies that are secreted into the airspace by exocytosis (Figure 1). Extracellularly, phospholipids and lamellar bodies interact with SP-A and calcium to produce tubular myelin forms, a highly organized, lipid-rich material from which phospholipid films consisting of monolayers and multilayers are generated at the air-liquid interface. Formation and the stability of the lipid films are facilitated by the interactions of SP-B and SP-C with surfactant phospholipids. Surfactant phospholipids are particularly enriched in dipalmitoyl phosphatidylcholine and phosphatidylglycerol, and are structurally suited for the formation of stable surface films. The presence of hydrophobic SP-B and SP-C facilitates the spreading and stability of phospholipid films in the alveolus, serving to enhance surfactant function throughout the respiratory cycle, recruiting phospholipids to the interface, and maintaining the lipid film as the alveolar surfaces expand and contract (Table 1, reviewed in references 1-4).

Surfactant Pool and Forms

Surfactant lipids and proteins are maintained at precise concentrations by regulation of synthesis, storage, secretion, reutilization, and catabolism. The pool size of surfactant phospholipids is approximately 4 mg/kg in the adult lung, but are much higher (75-100 mg/kg) in the newborn infant. Surfactant lipids and proteins are efficiently reutilized, and surfactant phospholipids and proteins are cleared from the airspaces of adult animals with a half-life of approximately 6-8 hours (2). In the normal lung, alveolar surfactant consists primarily of dense, large aggregate forms that are enriched in surfactant proteins and tubular myelin. Large aggregate fractions are surface active, spreading rapidly to reduce surface tension at the air-liquid interface. In contrast, smaller, lighter vesicular fractions or small aggregate forms, are relatively depleted of surfactant proteins and do not reduce surface tension actively at the air-liquid interface. Small aggregate phospholipids may represent forms destined for catabolism or recycling. The ratio of small to large aggregate forms is critical to maintaining surfactant function in health and disease. Decreased content of these surface active, large aggregate forms is associated with lung injury. Large aggregate forms of surfactant lipids are enriched in SP-A, -B, and -C, the latter two proteins contributing to the surface activity, spreading and stability of surfactant phospholipids in the alveoli.

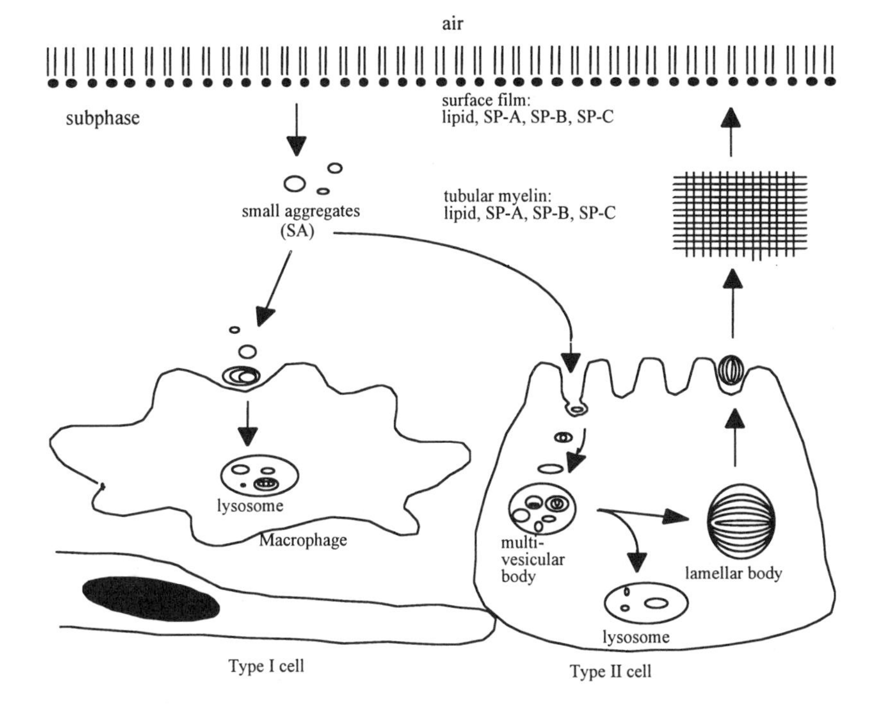

Figure 1. Surfactant phospholipid and surfactant proteins are synthesized by Type II epithelial cells lining the surface of the alveoli. Surfactant lipids and hydrophobic surfactant proteins, SP-B and SP-C, are co-packaged with lamellar bodies and secreted into the airspace. In the presence of calcium, SP-A interacts with the unraveling lamellar bodies to form tubular myelin, a highly organized, large aggregate surfactant form that comprises most of the active surfactant fraction present in the subphase. Monolayers and multilayers of phospholipids are generated at the air-liquid interface. Formation of this surface film is enhanced by SP-B and SP-C, which maintain stability of the lipid film during the respiratory cycle. Surfactant is recycled and degraded by the Type II epithelial cell. Surfactant lipids and proteins are catabolized by alveolar macrophages.

Table 1. Hydrophobic proteins in surfactant.

	Functions
SP-B (79 aa)	Lamellar body, tubular myelin structure PL adsorption, spreading, stability Processing of proSP-C Protection from inactivation Protein, lipid recycling
SP-C (33-34 aa)	Surface activity Spreading, stability of PL film Protection from inactivation

Mutations in SP-B and SP-C Reveal their Role in Surfactant Function and Homeostasis *In Vivo*

Addition of SP-B or SP-C to surfactant phospholipid mixtures enhances their surface properties *in vitro* and *in vivo* (5). Genetic experiments in which surfactant protein were decreased or deleted in transgenic mice provided direct insight into their function *in vivo*. SP-B is required for lung formation after birth. Mutations in the SP-B gene cause severe respiratory distress in newborn infants with hereditary SP-B deficiency (6). Likewise, absence of or mutations in the SP-C gene are associated with severe lung disease, including ARDS, in infants and adults (7). Studies in transgenic mice, in which either the SP-B or SP-C gene was deleted, demonstrated the critical role of the surfactant proteins in lung function (8, 9).

Deletion of the SP-B gene caused lethal respiratory distress in both newborn mice and infants. SP-B deficiency was associated with numerous abnormalities in Type II cell morphology and function. SP-B was required for formation of lamellar bodies, secretion of surfactant lipids and proteins, as well as the generation of tubular myelin (8). Lung volumes were reduced and hysteresis absent in newborn SP-B -/- mice, consistent with the inability to reduce surface tension in the airspaces. Furthermore, the processing of proSP-C and the reutilization of SP-A and lipids was disrupted in the lungs of mice and humans with hereditary SP-B deficiency, consistent with the requirement of SP-B for processing of proSP-C by the Type II cell. Partially processed proSP-C accumulated in the airspaces of SP-B -/- mice and infants, its presence being diagnostic of hereditary SP-B deficiency (10). The active SP-C peptide was absent from the airspace of mice and infants with hereditary SP-B deficiency. This deletion of proSP-B results in the failure to synthesize both active peptides, SP-B and SP-C.

Reduction of SP-B renders mice susceptible to lung injury *in vivo*. SP-B concentrations were reduced by 50% in SP-B +/- mice, and lung compliance was mildly decreased at steady state in SP-B +/- mice; however, these mice were highly sensitive to lung injury when placed in hyperoxia (11). Treatment of SP-B +/- with SP-B containing surfactant rescued them from lung injury and alveolar capillary leak during exposure of oxygen (12).

Mice deficient in SP-C survive, but have abnormal surfactant function, with instability of film formation (9). Furthermore, patients with mutations in SP-C were recently identified. Mutations in human SP-C caused familial interstitial lung disease. Individuals in families with mutations in the SP-C gene developed ARDS (7). Taken together, these genetic experiments demonstrate that both SP-B and SP-C play important roles in surfactant homeostasis and lung function *in vivo*. Since expression of SP-B and SP-C is reduced in preterm infants with RDS and in patients with ARDS, the reduction of SP-B or SP-C, secondary to prematurity, injury, or infection, likely renders the subject susceptible to surfactant dysfunction.

Changes in Surfactant Function and Homeostasis in ARDS

Surfactant function can be impaired by processes that 1) decrease the surfactant lipid protein composition or concentrations in the airspace, or 2) impair surfactant function mediated by the inhibitory effects of non-surfactant proteins that leak into the airspace. The hydrophobic proteins, SP-B and SP-C, enhance surfactant function and protect surfactant from inhibition.

Not surprisingly, the lung injuries involved in the pathogenesis of ARDS may influence all aspects of surfactant homeostasis, including synthesis, recycling, catabolism, and inactivation (2). Figure 2 represents the pathophysiology that leads to disruption of surfactant function in ARDS derived from both human and animal studies.

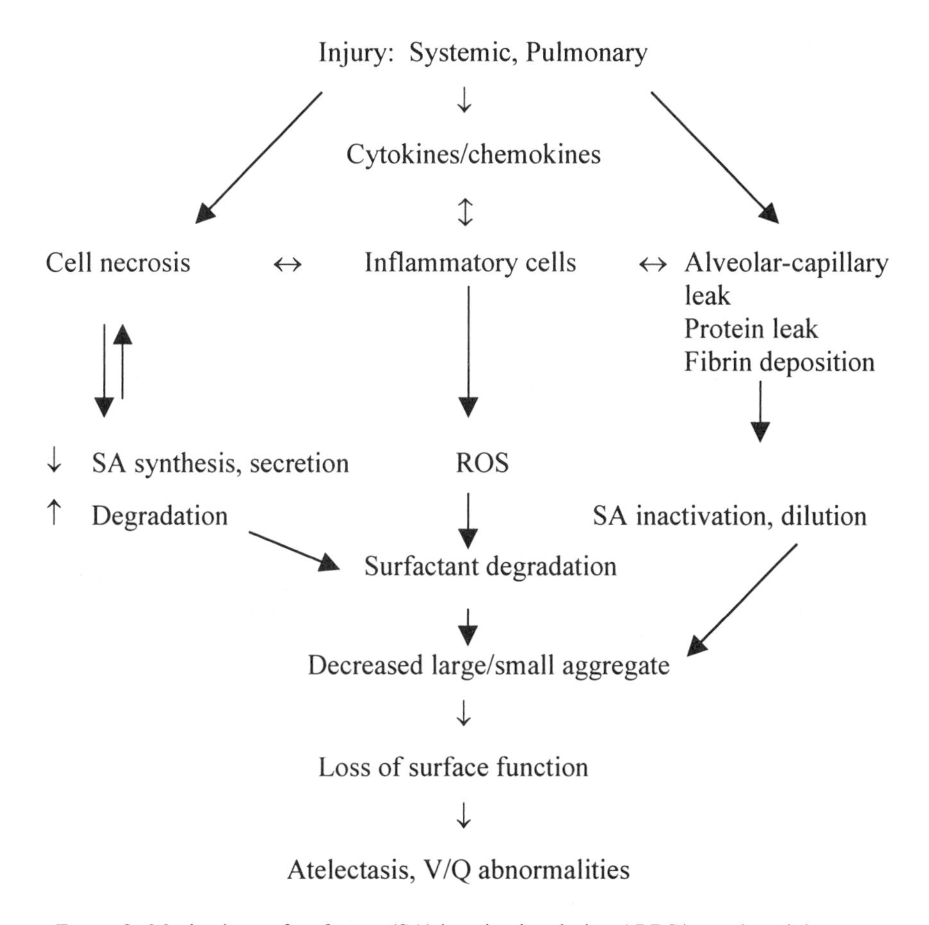

Figure 2. Mechanisms of surfactant (SA) inactivation during ARDS/acute lung injury.

Changes in Surfactant Composition During ARDS

Changes in surfactant structure and composition accompany the course of ARDS. ARDS is caused by various systemic or pulmonary injuries that alter permeability of the alveolar-capillary cell membranes, causing high-permeability pulmonary edema that dilutes and inactivates the pulmonary surfactant. Leakage of serum or cellular proteins rapidly inactivates surfactant function *in vitro* and *in vivo* (Table 2).

Table 2. Factors involved in surfactant inactivation.

Alveolar-capillary leak
Serum and tissue proteins
Edema - dilution
Fibrin deposition, hemorrhage
Reactive oxygen species - proteins, lipids
Proteases, phospholipases
Decreased large aggregate to small aggregate surfactant

Analysis of broncho-alveolar lavage (BAL) fluid obtained from patients with sepsis-induced ARDS demonstrated changes in phospholipid (PL) and surfactant protein composition, decreased proportion of large aggregate (LA) fraction within the air space, and abnormal surface activity of the isolated surfactant (13). In patients with inflammatory lung injury, PL concentrations and LA content in BAL fluid decreased in association with increased alveolar protein content and neutrophilic infiltration. Likewise, composition of surfactant phospholipids was altered in ARDS with decreased phosphatidylcholine (PC) and phosphatidylglycerol (PG) and increased phosphatidylinositol (PI), sphingomyelin (SPH), phosphatidylethanolamine, and phosphatydylserine (14, 15, 16). The lecithin/sphingomyelin ratio and SP-A concentrations in BAL fluid from children with ARDS, bacterial pneumonia, and viral pneumonitis were significantly reduced (17). In adult patients with ARDS, SP-A and SP-B concentration in BALF decreased (18), and the proportion of the large aggregate fraction of surfactant decreased (13). Direct damage to SP-A may occur in lungs of patients with ARDS, perhaps related to increased proteolysis by neutrophil elastase (19). The concentrations of SP-A and SP-B were decreased in the BAL fluids of patients at risk for ARDS, being detected before the onset of clinically defined lung injury (18). Immunoreactive SP-A, SP-B, and SP-D were higher in the serum of patients with ARDS and acute cardiogenic pulmonary edema (16, 18, 20). In serum, immunoreactive SP-A and SP-B concentrations were inversely related to blood oxygenation and static respiratory system compliance (20, 21), suggesting that increased alveolar-capillary leak resulted in leakage of the alveolar proteins into the systemic circulation. Thus, clinical ARDS has been associated with modest changes in surfactant phospholipid content and composition, decreased surfactant protein concentration, and loss of surface active large aggregate forms, findings that may contribute, at least in part, to decreased pulmonary function in ARDS.

Animal Models Used for Study of ARDS

A number of animal models have been used to assess surfactant homeostasis and the potential utility of surfactant replacement for treatment of ARDS. Endogenous surfactant has been depleted by repetitive lung lavage (22-31). ARDS-like injury has been induced by instillation of bacteria, endotoxin (27, 32-39), and excessive mechanical ventilation (40). Exposure to 100% oxygen (41), detergent (42), paraquat (43), HCl (27, 30), N-nitroso-N-methylurethane (NNMU) (30), and oleic acid (27, 44, 45) have been used to study the pathophysiology of ARDS-like injury in animals. These animal models vary in the extent of injury and may create distinct abnormalities in physiology.

Surfactant depletion and HCl instillation models generally produce acute hypoxemia in an otherwise hemodynamically stable animal (27). Variations in lavage volume and repetition produce distinct severities of ARDS (46). Pathophysiology in NNMU-injured rabbits and sepsis-induced ARDS was similar, including changes in surfactant protein and PL composition, decreased proportion of large aggregate surfactant, and abnormal surface activity of the isolated surfactant (30). HCl-induced injury in rabbits represents a model of aspiration of gastric contents or inhalation of toxic fumes (30).

A number of models create lung injury in association with shock-like syndromes that may be relevant to ARDS in some clinical settings. Oleic acid infusion creates a model of marked cardiovascular instability, pulmonary hypertension, and profound hypoxemia (27, 46, 47). In rats, systemic endotoxin infusion creates a model of cardiovascular instability and pulmonary hypertension but fails to produce hypoxemia (27). However, following intratracheal treatment with aerosolized lipopolysaccharide (LPS) in rats, hypoxemia, decreased lung compliance, and increased microvascular permeability are observed and associated with increased TNF-α and polymorphonuclear cell infiltration (35). Lung elastance, tissue resistance, and arterial-alveolar oxygen gradients were increased in association with decreased surfactant function, large aggregates ratio, and PL concentrations (39). In rats, a short period of mechanical ventilation caused decreased lung compliance associated with influx of proteins into the alveolar space and decreased lavage aggregate surfactant (48). Excessive mechanical ventilation of dogs caused decreased oxygen tension and increased BAL neutrophil count, protein concentrations, and the ratio of small surfactant aggregate to large aggregate surfactant (40). Sepsis caused by cecal ligation and puncture in rats lowered blood oxygenation and surfactant pool sizes, increased the small surfactant to large surfactant aggregate ratio, and increased levels of SP-A in BAL fluid (36).

Taken together, these various animal models have been created for study of lung injury and ARDS, creating models with abnormalities of lung physiology that may share some features with clinical ARDS. The models may be of considerable use in understanding the pathophysiology of lung injury and in testing of various therapies that maybe useful in ARDS.

Variability of Surfactant Replacement in ARDS Models

Several factors influence the efficacy of surfactant replacement therapy in ARDS, including the differences in the function of various surfactant preparations, delivery methods, surfactant doses, timing of surfactant treatment during the course of the disease, and differences in ventilation strategies (49, 50).

Surfactant Preparations. A number of mammalian-based and synthetic surfactant preparations have been used effectively in the treatment or prevention of neonatal RDS and surfactant replacement is now standard practice for infants with RDS. The composition and properties of these preparations vary, and all have been considered for use in ARDS (Table 3). In general, natural surfactants or synthetic surfactants with SP-C- or SP-B-like activity are more rapid and effective in improving pulmonary function in ARDS models. The hydrophobic proteins SP-B and SP-C enhance the rate of improvement in lung function, and surfactant preparations containing them are more resistant to protein inactivation associated with lung injury. In rat a lung lavage model, instillation of bovine surfactant (Survanta and Alveofact; recombinant surfactant protein C (rSP-C) based surfactant) were more effective than Exosurf in preventing deterioration of lung function (22). Similarly, administration of KL4-Surfactant (an artificial preparation containing a synthetic 21 amino-acid peptide with SP-B like activity), or Survanta, was more effective than Exosurf in improving oxygenation in a lung lavage model in neonatal piglets (34). Treatment of LPS challenged rats with Curosurf enhanced survival, improved respiratory frequency, decreased lung wet weight, total protein, and the numbers of polymorphonuclear cells in the BAL fluid (51). In an endotoxin-induced ARDS model in rats, synthetically reconstituted surfactant with porcine SP-B and SP-C was as effective as natural surfactant in improving oxygenation (52). Similarly, in a rat lung lavage model of ARDS, protein containing surfactants (rSP-C surfactant, bovine lung surfactant extract, Infasurf and Survanta) were more effective than the synthetic surfactants, ALEC (a synthetic phospholipid mixture) or Exosurf. Addition of rSP-C to phospholipid mixtures improved activity to that of the natural surfactants (53). The exogenously applied rSP-C surfactant was distributed homogeneously along the alveolar lining, improved oxygenation, and prevented hyaline membrane formation (54). In a lavage model of ARDS in adult sheep, surfactant based on a rSP-C was effective in improving gas exchange, although higher doses (100-200 mg PL/kg) were required for optimal responses (29). Taken together, the increased speed of physiologic action and resistance to protein inhibition typical of mammalian-based surfactants containing SP-B, SP-C, or both support their preferential utility in ARDS models.

Table 3. Surfactant preparations studied in ARDS models or clinical ARDS.

Synthetic	
	Exosurf: Phospholipid (PL), tyloxapol
	ALEC: PL
	KL4: Synthetic peptide, PL
Mammalian	
	Survanta: SP-C, SP-B, PL
	Curosurf: SP-C, SP-B, PL
	Infasurf: SP-C, SP-B, PL
	Alveofact: SP-C, SP-B, PL
	Venticute: Recombinant SP-C, PL

Delivery Methods. Surfactant has been delivered intratracheally by bolus, drip, aerosol, and bronchoscopic spray. In neonates and adults, distribution is more effective and homogenous when given by using relatively large volumes (2). For example, in surfactant-depleted rats with respiratory failure, instillation of four fractional surfactant doses was not as effective in improving ventilation as the same total dose given by a single bolus (23). The tracheal instillation of a rSP-C surfactant produced results similar to those of the bronchoscopic administration technique, but was easier and shorter (29).

Aerosolization was generally inefficient for delivery of surfactant. Furthermore, aerosolized surfactant was distributed preferentially to ventilated regions, and was not delivered to atelectatic regions. In endotoxin injured rats, bolus instillation of a modified natural surfactant increased oxygenation more rapidly than after aerosolization (32). Ultrasonically delivered natural surfactant (8.8 mg/kg) was as effective as bolus administration of surfactant (80 mg/kg) in reducing shunt flow in saline lavage model of ARDS in rabbits (31). In endotoxin-induced lung injury in pigs, aerosolized surfactant decreased leukocyte sequestration and improved oxygenation (38). Overall, bolus methods of delivery appear to be more effective than aerosolization in animal models of ARDS.

Lung Lavage. Exogenous surfactant has been delivered in dilute preparations during lung lavage, the lung wash being utilized to remove proteins that inhibit surfactant function. Sequential segmental lavage was performed in ARDS patients with a dilute synthetic surfactant containing the KL4 peptide (Surfaxin) in 12 adults with ARDS, with reductions of $PaCO_2$ and positive end-expiratory pressure (PEEP) (55). Pre-inflation of the lung, to reduce atelectasis, resulted in more uniform distribution of dilute surfactant and greater improvement in pulmonary function than when administered by bolus alone (56). The use of this lavage/dilution technique improved oxygenation and pulmonary function using various surfactants, including Infasurf, KL4-surfactant, and Exosurf (28, 56, 57).

Impact of Ventilatory Strategy. Mechanical ventilation contributes to lung injury and influences surfactant function. Lung injury caused by repeated alveolar collapse and subsequent reexpansion can be prevented by maintaining alveolar volumes at end-expiration with PEEP (58). In lavage and exdotoxin models of ARDS in rabbits, surfactant treatment followed by ventilation with the low-V_T modes (5 cm H_2O and 9 cm H_2O) led to higher PaO_2 values and higher fraction of LA surfactant than did the normal V_T group. Recent clinical trials demonstrated improved survival of ARDS patients using a low tidal volume strategy, supporting the concept that the mode of ventilation is an important factor in the pathogenesis of lung injury in ARDS (59).

Clinical Experience with Surfactant Replacement for ARDS. Success in the treatment of ARDS with surfactant replacement is confined primarily to anecdotal reports, and several clinical trials in which relatively few patients have been studied. Initial successful case reports, and small clinical trials, have utilized animal-based surfactant preparations containing SP-B and -C (e.g. Curosurf, Survanta, and Infasurf). In contrast, relative large, controlled studies with synthetic surfactant (Exosurf), failed to show benefit in ARDS. These experiences are not directly comparable. Differences in the modes of delivery, the surfactants used, and the selection of patients for treatment, complicate interpretation of the data.

All available mammalian-based surfactants contain SP-B and SP-C, and are generally more resistant to inactivation by serum or cellular proteins that accumulate in the lungs of patients with ARDS. However, the utility of surfactant replacement in ARDS remains unproven. Methods for surfactant treatments have varied in the published studies, and have included aerosol-ization, bolus, and drip instillation. In animal studies of ARDS, bolus instillation resulted in more homogenous delivery of surfactant into the lung. In contrast, aerosolization is generally inefficient and delivery generally favors the aerated, rather than collapsed airspaces, limiting delivery to the most injured regions of the lung. In a relatively small number of patients, porcine surfactant was delivered by selective bronchial instillation with improvement in oxygenation (60). Surfactant has also been used in association with lung lavage, that may remove or dilute inhibitory substances, prior to replacement of surfactant by bolus or bronchoscopic administration. Whether these therapeutic modalities will improve respiratory physiology, decrease the requirements for oxygen and mechanical ventilation or result in decreased morbidity and mortality in patients, remain to be validated.

Administration of replacement surfactant to pediatric patients with ARDS did not improve survival, but improved dynamic compliance and stabilized gas exchange earlier in the disease (61). Rapid improvement in oxygenation and moderation of ventilatory support was observed following surfactant replacement in children with acute hypoxemic respiratory failure (62). Mortality was decreased in a relative small study group of ARDS patients receiving bovine surfactant (Survanta) by endotracheal instillation (63). Likewise, surfactant replacement (bovine- and porcine-derived preparations) enhanced oxygenation in children and adults (64, 65). After the limits of mechanical ventilation had been reached in four severely burned patients with inhalation injury complicated by ARDS, bronchoscopic intrabronchial administration of bovine surfactant (Alveofact) was followed by temporarily improved gas exchange with an increase in PaO_2, accompanied by a reduction in FiO_2, and improved lung compliance (66). In summary, the use of mammalian-based replacement surfactants for treatment of ARDS generally improved lung function and was not harmful. Randomized, controlled studies that demonstrate the efficacy of surfactant replacement in reducing morbidity or mortality in ARDS are lacking. Thus at present, surfactant replacement for ARDS remains anecdotal and experimental.

Summary

It is increasingly clear that surfactant deficiency and dysfunction accompanies acute lung injury/ARDS and contributes to the disease. The salutatory effects of surfactant replacement in reducing morbidity and mortality in treatment of RDS in infants has been clearly established (3). Likewise, the pathophysiology and biochemistry of surfactant dysfunction in ARDS and ARDS models supports the potential utility of surfactant replacement. While various surfactant replacement strategies improve pulmonary function in experimental models, it remains unclear whether surfactant replacement will improve morbidity and mortality in the setting of multiorgan disease often characteristic of ARDS. Since the pathophysiologic abnormalities in lung function are often well established and severe by the time ARDS is diagnosed, the success of surfactant replacement therapies may depend on early diagnosis (pre-ARDS) and treatment. Successful therapy of ARDS will often depend on appropriate treatment of the underlying disease process associated with ARDS.

ACKNOWLEDGEMENT

Supported by grants from the National Institutes of Health: PPG HL61646 and HL38859.

REFERENCES

1. Goerke, J. (1998) Pulmonary surfactant: functions and molecular composition. *Biochim Biophys Acta* 1408, 79-89
2. Jobe, A.H., and Ikegami, M. (1997) Surfactant for acute respiratory distress syndrome. *Adv Intern Med* 42, 203-230
3. Jobe, A.H. (1993) Pulmonary surfactant therapy. *N Engl J Med* 328, 861-868
4. Evans, D.A., Wilmott, R.W., and Whitsett, J.A. (1996) Surfactant replacement therapy for adult respiratory distress syndrome in children. *Pediatr Pulmonol* 21, 328-336
5. Weaver, T.E., and Conkright, J.J. (2001) Functions of surfactant proteins B and C. *Annu Rev Physiol* 63, 555-578
6. Nogee, L.M., de Mello, D.E., Dehner, L.P., and Colten, H.R. (1993) Brief report: deficiency of pulmonary surfactant protein B in congenital alveolar proteinosis. *N Engl J Med* 328, 406-410
7. Nogee, L.M., Dunbar, A.E., Wert, S.E., Askin, F., Hamvas, A., and Whitsett, J.A. (2001) A mutation in the surfactant protein C gene associated with familial interstitial lung disease. *N Engl J Med* 344, 573-579
8. Clark, J.C., Wert, S.E., Bachurski, C.J., Stahlman, M.T., Stripp, B.R., Weaver, T.E., and Whitsett, J.A. (1995) Targeted disruption of the surfactant protein B gene disrupts surfactant homeostasis, causing respiratory failure in newborn mice. *Proc Natl Acad Sci USA* 92, 7794-7798
9. Glasser, S.W., Burhans, M.S., Korfhagen, T.R., Na, C-L., Sly, P.D., Ross, G.F., Ikegami, M. and Whitsett, J.A. (2001) Altered stability of pulmonary surfactant in SP-C deficient mice. *Proc Natl Acad Sci USA*, (in press)
10. Vorbroker, D.K., Profitt, S.A., Nogee, L.M., and Whitsett, J.A. (1995) Aberrant processing of surfactant protein C (SP-C) in hereditary SP-B deficiency. *Am J Physiol (Lung Cell Mol Physiol 12)* 268, L647-L656
11. Tokieda, K., Whitsett, J.A., Clark, J.C., Weaver, T.E., Ikeda, K., Jobe, A.H., Ikegami, M., and Iwamoto, H.S. (1997) Pulmonary dysfunction in neonatal SP-B deficient mice. *Am J Physiol (Lung Cell Mol Physiol)* 273, L875-L882
12. Tokieda, K., Iwamoto, H.S., Bachurski, C., Wert, S.E., Hull, W.M., Ikeda, K., and Whitsett, J.A. (1999) Surfactant protein-B-deficient mice are susceptible to hyperoxic lung injury. *Am J Respir Cell Mol Biol* 21, 463-472
13. Veldhuizen, R.A.W., McCaig, L.A., Akino, T., and Lewis, J.F. (1995) Pulmonary surfactant subfractions in patients with the acute respiratory distress syndrome. *Am J Respir Crit Care Med* 152, 1867-1871
14. Gunther, A., Siebert, C., Schmidt, R., Ziegler, S., Grimminger, F., Yabut, M., Temmesfild, B., Walmrath, D., Morr, H., and Seeger, W. (1996) Surfactant alterations in severe pneumonia, acute respiratory distress syndrome, and cardiogenic lung edema. *Am J Respir Crit Care Med* 153, 176-184

15. Nakos, G., Kitsiouli, E.I., Tsangaris, I., and Lekka, M.E. (1998) Bronchoalveolar lavage fluid characteristics of early intermediate and late phases of ARDS. Alterations in leukocytes, proteins, PAF and surfactant components. *Intensive Care Med.* 24, 296-303
16. Bersten, A.D., Doyle, I.R., Davidson, K.G., Barr, H.A., Nicholas, T.E., and Kermeen, F. (1998) Surfactant composition reflects lung over-inflation and arterial oxygenation in patients with acute lung injury. *Eur Respir J* 12, 301-308
17. LeVine, A.M., Lotze, A., Stanley, S., Stroud, C., O'Donnell, R., Whitsett, J., and Pollack, M.M. (1996) Surfactant content in children with inflammatory lung disease. *Crit Care Med* 24, 1062-1067
18. Greene, K.E., Wright, J.A., Steinberg, K.P., Ruzinski, J.T., Caldwell, E., Wong, W.B., Hull, W., Whitsett, J.A., Akino, T., Kuroki, Y., Nagae, H., Hudson, L.D., and Martin, T.R. (1999) Serial changes in surfactant-associated proteins in lung and serum before and after onset of ARDS. *Am J Respir Crit Care Med* 160, 1843-1850
19. Baker, C.S., Evans, T.W., Randle, B.J., and Haslam, P.L. (1999) Damage to surfactant-specific protein in acute respiratory distress syndrome. *Lancet* 353, 1232-1237
20. Doyle, I.R., Bersten A.D., and Nicholas, T.E. (1997) Surfactant proteins-A and -B are elevated in plasma of patients with acute respiratory failure. *Am J Respir Crit Care Med* 156, 1217-1229
21. Doyle, I.R., Nicholas, T.E., Bersten, A.D. (1995) Serum surfactant protein-A levels in patients with acute cardiogenic pulmonary edema and adult respiratory distress syndrome. *Am J Respir Crit Care Med* 152, 307-317
22. Hafner, D., Beume, R., Killian, U., Krasznai, G., and Lachmann, B. (1995) Dose-response comparisons of five lung surfactant factor (LSF) preparations in an animal model of adult respiratory distress syndrome (ARDS). *Br J Pharmacol* 115, 451-458
23. Alvarez, F.J., Alfonso, L.F., Gastiasoro, E., Lopez-Heredia, J., Arnaiz, A., and Valls-Soler, A. (1995) The effects of multiple small doses of exogenous surfactant on experimental respiratory failure induced by lung lavage in rats. *Acta Anaesthesiol Scand* 39, 970-974
24. Makhoul, I.R., Kugelman, A., Bui, K.C., Berkeland, J.E., Saiki, K., Lew, C.D., and Garg, M. (1996) Reduction of respiratory system resistance of rabbits with surfactant deficiency using a novel ultra thin walled endotracheal tube. *ASAIO J* 42, 1000-1005
25. Sood, S.L., Balaraman, V., Finn, K.C., Wilderson, S.Y., Mundie, T.F., and Easa, D. (1996) Exogenous surfactant decreases oxygenation in Escherichia coli endotoxin-treated neonatal piglets. *Pediatr Pulmonol* 22, 376-386
26. Ito, Y., Manwell, S.E.E., Kerr, C.L., Veldhuizen, R.A.W., Yao, L.J., Bjarneson, D., McCaig, L.A., Bartlett, A.F., and Lewis, J.F. (1998) Effects of ventilation strategies on the efficacy of exogenous surfactant therapy in a rabbit model of acute lung injury. *Am J Respir Crit Care Med* 157, 145-155
27. Rosenthal, C., Caronia, C., Quinn, C., Lugo, N., and Sagy, M. (1998) A comparison among animal models of acute lung injury. *Crit Care Med* 26, 912-916
28. Gommers, D., Eijking, E.P., So, K.L., van't Veen, A., and Lachmann, B. (1998) Bronchoalveolar lavage with a diluted surfactant suspension prior to surfactant instillation improves the effectiveness of surfactant therapy in experimental acute respiratory distress syndrome (ARDS). *Intensive Care Med* 24, 494-500
29. Lewis, J., McCaig, L., Hafner, D., Spragg, R., Veldhuizen, R., and Kerr, C. (1999) Dosing and delivery of a recombinant surfactant in lung-injured adult sheep. *Am J Respir Crit Care Med* 159, 741-747
30. Puligandla, P.S., Gill, T., McCaig, L.A., Yao, L-J., Veldhuizen, R.A.W., Possmayer, F., and Lewis, J.F. (2000) Alveolar environment influences the metabolic and biophysical properties of exogenous surfactants. *J Appl Physiol* 88, 1061-1071
31. Schermuly, R.T., Gunther, A., Weissmann, N., Ghofrani, H.A., Seeger, W., Grimminger, F., and Walmrath, D. (2000) Differential impact of ultrasonically nebulized versus tracheal-instilled surfactant on ventilation-perfusion (V_A/Q) mismatch in a model of acute lung injury. *Am J Respir Crit Care Med* 161, 152-159

32. Tashiro, K., Yamada, K., Li, W.Z., Matsumoto, Y., and Kobayashi. T. (1996) Aerosolized and instilled surfactant therapies for acute lung injury caused by intratracheal endotoxin in rats. *Crit Care Med* 24, 488-494
33. Picone, A., Gatto, L.A., Nieman, G.F., Paskanik, A.M., and Lutz, C. (1996) Pulmonary surfactant function following endotoxin: effects of exogenous surfactant treatment. *Shock* 5, 304-310
34. Sood, S.L., Balaraman, V., Finn, K.C., Britton, B., Uyehara, C.F., and Easa, D. (1996) Exogenous surfactants in a piglet model of acute respiratory distress syndrome. *Am J Respir Crit Care Med* 153, 820-828
35. van Helden, H.P., Kuijpers, W.C., Steenvoorden, D., Go, C., Bruijnzeel, P.L., van Eijk, M., and Haagsman, H.P. (1997) Intratracheal aerosolization of endotoxin (LPS) in the rat: a comprehensive animal model to study adult (acute) respiratory distress syndrome. *Exp Lung Res* 23, 297-316
36. Molloy, J., McCaig, L., Veldhuizen, R., Yao, L.J., Joseph, M., Whitsett, J., and Lewis, J. (1997) Alterations of the endogenous surfactant system in septic adult rats. *Am J Respir Crit Care Med* 156, 617-623
37. Holguin, F., Moss, I.M., Brown, L.A.S., and Guidot, D.M. (1998) Chronic ethanol ingestion impairs alveolar type II cell glutathione homeostasis and function and predisposes to endotoxin-mediated acute edematous lung injury in rats. *J Clin Invest* 101, 761-768
38. Lutz, C., Carney, D., Finck, C., Picone, A., Gatto, L.A., Paskanik, A., Langenback, E., and Nieman, G. (1998) Aerosolized surfactant improves pulmonary function in endotoxin-induced lung injury. *Am J Respir Crit Care Med* 158, 840-845
39. Mora, R., Arold, S., Marzan, Y., Suki, B., and Ingenito., E.P. (2000) Determinants of surfactant function in acute lung injury and early recovery. *Am J Physiol Lung Cell Mol Physiol* 279, L342-L349
40. Novick, R.J., Gilpin, A.A., Gehman, K.E., Ali, I.S., Veldhuizen, R.A., Duplan, J., Denning, L., Possmayer, F., Bjarneson, D., and Lewis, J.F. (1997) Mitigation of injury in canine lung grafts by exogenous surfactant therapy. *J Thorac Cardiovasc Surg* 113, 342-353
41. Novotny, W. (1995) Hyperoxic lung injury reduces exogenous surfactant clearance in vivo. *Am J Respir Crit Care Med* 151, 1843-1847
42. Shermuly, R., Schmehl, T., Gunther, A., Grimminger, F., Seeger, W., Walmrath, D. (1997) Ultrasonic nebulization for efficient delivery of surfactant in a model of acute lung injury. *Am J Respir Crit Care Med* 156, 445-453
43. Silva, M.F., and Saldiva, P.H. (1998) Paraquat poisoning: an experimental model of dose-dependent acute lung injury due to surfactant dysfunction. *Braz J Mel Biol Res* 31, 445-450
44. Zhu, G.F., Sun, B., Niu, S.F., Cai, U.Y., Lin, K., Lindwall, R., and Robertson, B. (1998) Combined surfactant therapy and inhaled nitric oxide in rabbits with oleic acid-induced acute respiratory distress syndrome. *Am J Respir Crit Care Med* 158, 437-443
45. Jacobs, B.R., Smith, D.J., Zingarelli, B., Passerini, D.J., Ballard, E.T., and Brilli, R.J. (2000) Soluble nitric oxide donor and surfactant improve oxygenation and pulmonary hypertension in porcine lung injury. *Nitric Oxide* 4, 412-422
46. Hafner, D., and Germann, P.G. (1999) A rat model of acute respiratory distress syndrome (ARDS) Part 2, influence of lavage volume, lavage repetition, and therapeutic treatment with rSP-C surfactant. *J Pharmacol Toxicol Methods* 41, 97-106
47. Zhou, Z.-H., Sun, B., Lin, K., and Zhu, L-W (2000) Prevention of rabbit acute lung injury by surfactant, inhaled nitric oxide, and pressure support ventilation. *Am J Respir Crit Care Med* 161, 581-588
48. Veldhuizen, R.A., Tremblay, L.N., Govindarajan, A., van Rozendaal, B.A., Haagsman, H.P., and Slutsky, A.S. (2000) Pulmonary surfactant is altered during mechanical ventilation of isolated rat lung. *Crit Care Med* 28, 2545-2551
49. Lewis, J.F., and Veldhuizen, R.A. (1995) Factors influencing efficacy of exogenous surfactant in acute lung injury. *Biol Neonate* 67, 48-60

50. Robertson, B. (1998) Surfactant inactivation and surfactant therapy in acute respiratory distress syndrome (ARDS). *Monaldi Arch Chest Dis* 53, 64-69
51. van Helden, H.P., Kuijpers, W.C., Langerwerf, P.E., Langen, R.C., Haagsman, H.P., and Bruijnzeel, P.L. (1998) Efficacy of Curosurf in a rat model of acute respiratory distress syndrome. *Eur Respir J* 12, 533-539
52. Tashiro, K., Nishizuka, K., Matsumoto, Y., Ohta, K., Suzuki, Y., and Kobagashi, T. (1999) Modified natural and synthetically reconstituted surfactant therapies for acute lung injury caused by endotoxin in rats. *Acta Anaesthesiol Scand* 43, 821-828
53. Hafner, D., Germann, P-G., and Hauschke, D. (1998) Comparison of rSP-C surfactant with natural and synthetic surfactants after late treatment in a rat model of the acute respiratory distress syndrome. *Br J Pharmacol* 124, 1083-1090
54. Hafner, D., Germann, P.G., Hauschke, D., and Killian, U. (1999) Effects of early treatment with rSP-C surfactant on oxygenation and histology in rats with acute lung injury. *Pulm Pharmacol Ther* 12, 193-201
55. Wiswell, T.E., Smith, R.M., Katz, L.B., Mastroianni, L., Wong, D.Y., Willms, D., Heard, S., Wilson, M., Hite, R.D., Anzueto, A., Revak, S.D., and Cochrane, C.G. (1999) Bronchopulmonary segmental lavage with surfactant (KL_4-surfactant) for acute respiratory distress syndrome. *Am J Respir Crit Care Med* 160, 1188-1195
56. Balaraman, V., Meister, J., Ku, T.L., Sood, S.L., Tam, E., Killeen, J., Uyehara, C.F.T., Egan, E., and Easa, D. (1998) Lavage administration of dilute surfactants after acute lung injury in neonatal piglets. *Am J Respir Crit Care Med* 158, 12-17
57. Cochrane, C.G., and Revak, S.D. (1999) Surfactant lavage treatment in a model of respiratory distress syndrome. *Chest* 116, 85S-87S
58. Verbrugge, S.J., Sorm, V., and Lachmann, B. (1997) Mechanisms of acute respiratory distress syndrome: role of surfactant changes and mechanical ventilation. *J Physiol Pharmacol* 48, 537-557
59. The Acute Respiratory Distress Syndrome Network. Ventilation with lower tidal volumes as compared with traditional tidal volumes for acute lung injury and the acute respiratory distress syndrome. N Engl J Med. 342:1301-1308, 2000
60. Spragg, R.G., Gilliard, N., Richman, P., Smith, R.M., Hite, R.D., Pappert, D., Robertson, B., Curstedt, T., and Strayer, D. (1994) Acute effects of a single dose of porcine surfactant on patients with the adult respiratory distress syndrome. *Chest* 105, 195-202
61. Perez-Benavides, F., Riff, E., and Franks, C. (1995) Adult respiratory distress syndrome and artificial surfactant replacement in the pediatric patient. *Pediatr Emerg Care* 11, 153-155
62. Willson, D.F., Jiao, J.H., Bauman, L.A. Zaritsky, A., Craft, H., Dockery, K., Conrad, D., and Dalton, H. (1996) Calf's lung surfactant extract in acute hypoxemic respiratory failure in children. *Crit Care Med* 24, 1316-1322
63. Lopez-Herce, J., de Lucas, N., Carrillo, A., Bustinza, A., and Moral, R. (1999) Surfactant treatment for acute respiratory distress syndrome. *Arch Dis Child* 80, 248-252
64. Walmrath, D., Gunther, A., Ghofrani, H.A., Schermuly, R., Schneider, T., Grimminger, F., and Seeger, W. (1996) Bronchoscopic surfactant administration in patients with severe adult respiratory distress syndrome and sepsis. *Am J Respir Crit Care Med* 154, 57-62
65. Gregory, T.J., Steinberg, K.P., Spragg, R., Gadek, J.E., Hyers, T.M., Longmore, W.J., Moxley, M.A., Cai, G.Z., Hite, R.D., Smith, R.M., Hudson, L.D., Crim, C., Newton, P., Mitchell, B.R., and Gold, A.J. (1997) Bovine surfactant therapy for patients with acute respiratory distress syndrome. *Am J Respir Crit Care Med* 155, 1309-1315
66. Pallua, N., Warbanow, K., Noah, E.M., Machens, H.G., Poets, C., Bernhard, W., and Berger, A. (1998) Intrabronchial surfactant application in cases of inhalation injury: first results from patients with severe burns and ARDS. *Burns* 24, 197-206

Chapter 13

PULMONARY COLLECTINS AND DEFENSINS

Ann Marie LeVine
Division of Pulmonary Biology and Critical Care Medicine, Children's Hospital Medical Center and Children's Hospital Research Foundation, Cincinnati, OH

INTRODUCTION

Maintaining a sterile respiratory tract presents a unique host defense challenge determined, in part, by the large surface area of lung tissue that comes in direct contact with inhaled pathogens, particles, and gases. The constant exposure of the respiratory tract to microbial pathogens and associated inflammatory molecules is accommodated by a complex innate and acquired immune system that enhances clearance and killing of pathogens, while simultaneously attempting to minimize systemic acquired immune responses and local inflammation. Therefore, it is not surprising that a complex and multifaceted innate immune system has evolved to protect the lung against a large variety of pathogens. Pulmonary cells synthesize a repertoire of host defense molecules that bind, opsonize, or kill various pathogenic organisms. In addition to the contributions of lung parenchymal cells to host defense, phagocytes, leukocytes, mast cells, eosinophils, and lymphocytes also synthesize mediators of innate host defense. Lung parenchymal cells produce numerous small molecules and proteins with antimicrobial activities, including reactive oxygen and nitrogen species, lysozyme, lactoferrin, defensins, phospholipases, complement components, proteinase inhibitor, and secretory IgA. Respiratory epithelial cells also express two members of the collectin family of mammalian lectins, surfactant protein-A (SP-A) and surfactant protein-D (SP-D) that contribute to distinct aspects of innate host defense in the lung (1). This chapter will discuss the critical roles played by both SP-A and SP-D in the orchestration of innate defense, alveolar macrophage function, and modulation of inflammatory responses in the respiratory tract. Lung cells also produce antimicrobial peptides that play key roles in the response to respiratory epithelial compromise and microbial invasion. This chapter will also discuss the antimicrobial mechanisms of these peptides and specifically the role of defensins in acute lung injury (ALI).

SP-A and SP-D: Structure and Synthesis

SP-A and SP-D are members of the collectin family of the mammalian C-type lectins that also includes mannose-binding protein (MBP) and conglutinin (1). MBP is produced by the liver and secreted into the serum. Children carrying mutations in the MBP gene are more susceptible to recurrent infections (2), supporting the concept that the collectins are an important part of innate immunity against microbial pathogens. The collectins, including SP-A and SP-D, are involved in innate host defense against a variety of bacterial and viral pathogens. Collectins form multimeric structures resembling C1q (the first component of the complement cascade), consisting of multimeric collagenous amino-terminal domains and globular carboxy-terminal, carbohydrate binding domains. The C-type lectins bind carbohydrate surfaces of many microorganisms mediating phagocytosis and killing by phagocytic cells (1).

SP-A is an abundant pulmonary surfactant-associated protein encoded by two genes located on chromosome 10 (3). The primary translation product generated from the SP-A mRNAs encodes a leader sequence that is cleaved during processing, a distinct N-terminal collagen-like, hydrophobic neck, and C-terminal C-type lectin domains (4). The mature peptide forms trimers that assemble to form an octadecamer that is the predominant airway form (5). SP-A binds various lipids and glycolipids including dipalmitoylphosphatidylcholine (DPPC) and binds to surface receptors on alveolar type II (ATII) cells and macrophages (1). In the human, SP-A synthesis is initiated in the developing respiratory epithelium in the second trimester of gestation, and is expressed by ATII cells, Clara cells, and subsets of cells in tracheo-bronchial glands in the adult lung (6).

SP-D was first identified as a component of pulmonary surfactant. It is now well known that SP-D is also expressed by a variety of non-pulmonary tissues in humans and other mammals (7). SP-D is encoded by a single gene located in close proximity to SP-A and other members of the collectin family of mammalian lectins, located on chromosome 10. SP-D mRNAs encode a complex polypeptide, that contains a signal peptide, a distinct amino-terminal collagen-like, α-helical coiled coil neck, and C-terminal C-type lectin domains (8). In the human lung, SP-D is expressed primarily in ATII cells and in serous cells in tracheal-bronchial glands. SP-D binds to phosphatidylinositol of pulmonary surfactant lipids (1).

SP-A and SP-D: Phagocytosis and Killing of Microorganisms

In vitro studies support a role for SP-A in enhancing microbial phagocytosis by macrophages. SP-A acts as opsonin and can directly stimulate macrophage activity (1). SP-A binds a variety of microorganisms *in vitro* (Table 1). SP-A also enhances calcium-dependent neutrophil uptake of *E. coli*, *S. pneumoniae* and *S. aureus* (9). In addition to binding microorganisms, SP-A is chemotactic for alveolar macrophages and peritoneal macrophages (10). The structure and activity of SP-A *in vitro* strongly supports a role for SP-A in clearance of microbial pathogens from the lung as part of the innate immunity. As will be discussed below, an *in vivo* role for SP-A-mediated innate immunity has now been established by the use of SP-A deficient mice.

Table 1. Pathogens bound by SP-A (References 1, 11-17)

Gram-positive bacteria	Gram-negative bacteria
Group A *Streptococcus*	E. *coli* J5
Group B *Streptococcus*	*H. influenzae* (type a)
S. aureus	*K. pneumoniae*
S. pneumoniae	*P. aeruginosa*
Viruses	Fungi
Adenovirus	*A. fumigatus*
Cytomegalovirus	*C. neoformans*
Herpes simplex virus, Type 1	
Influenza virus A (H3N2)	Parasites
Respiratory syncytial virus	*P. carinii*

In vitro studies also support a role for SP-D in microbial phagocytosis by alveolar macrophages. Similar to SP-A, SP-D can bind Gram-negative and Gram-positive bacteria, viruses, fungi and parasites in a calcium-dependent manner (Table 2), and also enhances calcium-dependent uptake of *E. coli*, *S. pneumoniae* and *S. aureus* by neutrophils (9). In addition to binding microorganisms, SP-D is chemotactic for monocytes and neutrophils at lower concentrations than required for SP-A-dependent chemotactic activity (18). These *in vitro* findings strongly support the role of SP-D in binding and opsonization of microbial pathogens in the lung.

Table 2. Pathogens bound by SP-D (References 1, 17, 19, 20)

Gram-positive bacteria	Gram-negative bacteria
Group *B Streptococcus*	*E. coli*
	H. influenzae
	K. pneumoniae
	P. aeruginosa
	S. minnesota
Viruses	Fungi
Influenza virus A	*A. fumigatus*
Respiratory syncytial virus	*C. neoformans*
	Parasites
	P. carinii

SP-A and SP-D: Production of Free Radicals

Reactive oxygen and nitrogen species are generated by phagocytic cells and have an important role in killing of ingested and extracellular pathogens, and also have a role in both intracellular and intercellular signaling pathways. *In vitro* studies provide conflicting results regarding the role of SP-A in regulating oxygen radical production. In two different studies, SP-A stimulated the release of oxygen radicals from alveolar macrophages (21, 22). In contrast, other studies have found that SP-A decreases PMA-stimulated superoxide production in canine neutrophils and alveolar macrophages (23), and in rat alveolar macrophages (24). The conformation of SP-A may be important in the regulation of oxygen radical production by phagocytic cells and may explain, in part, the differences found in these *in vitro* studies. Stimulation of oxygen radicals by SP-A was not seen with rat peritoneal macrophages, rat or human neutrophils, or human peripheral blood monocytes (21) suggesting that SP-A regulation of oxygen radical production is specific for alveolar macrophages.

Nitric oxide (NO) is a reactive nitrogen species, which also has a role in phagocytic cell-mediated host defense (see Chapters 6 and 7). A variety of inflammatory stimuli increase alveolar macrophage NO production through the induction of inducible NO synthase. SP-A has variable effects on NO production. Similar to the observations regarding reactive oxygen species, SP-A does not stimulate NO production by peritoneal macrophages or ATII cells (25). Purified SP-A is often contaminated with lipopolysaccharide (LPS), which may explain differing results between *in vitro* studies, since LPS can stimulate immune cell function. SP-A treated to remove LPS does not stimulate NO production by macrophages (26), however, SP-A inhibits production of NO in macrophages stimulated with LPS (27). Alveolar macrophages stimulated with interferon-γ (IFN-γ), IFN-γ plus LPS (27), or IFN-γ followed by *Mycoplasma pulmonis* infection (28) generate more NO in the presence of SP-A. In contrast, SP-A inhibits NO production by IFN-γ activated macrophages incubated with *Mycobacterium tuberculosis* (29). These studies suggest that SP-A exerts differential effects on the production of reactive nitrogen species by alveolar macrophages depending on the stimulus and the activation state of the immune cells.

SP-D, similar to SP-A, enhances the production of oxygen radicals as assessed by monitoring lucigenin-dependent chemiluminescence by rat alveolar macrophages (30). This response appears to be specific for alveolar macrophages, as SP-D does not stimulate the production of oxygen radicals by neutrophils (31). SP-D, treated to remove endotoxin, does not stimulate production of NO by alveolar macrophages (26). In addition, SP-D does not affect the production of NO by macrophages stimulated with either LPS or IFN-γ (27). Further studies with SP-D are necessary to determine its role in regulating production of reactive oxygen and nitrogen species.

SP-A and SP-D: Lymphocyte Activation

T cells are active participants and regulators of acute pulmonary inflammation due to the release of cytokines. SP-A enhances proliferation of concanavalin A (Con A) stimulated spleen lymphocytes (32, 33). In contrast, SP-A suppresses phytohemagglutinin (PHA)-driven T cell proliferation and anti-CD-3 stimulated T cell proliferation (33). The three mitogens used in these studies have slightly different binding sites, but similar biochemical pathways for T-cell activation and proliferation (34). PHA and Con A bind to T-cells via the T-cell receptor, whereas anti-CD-3 binds to the T cell via the CD3-complex. These studies suggest that SP-A disrupts co-stimulatory signals unique or dominant in PHA- and anti-CD3-mediated T cell activation, and support a role for SP-A in regulating T cell responses depending on the mitogen encountered in the lung.

SP-D, similar to SP-A, modulates activation of T cells. SP-D has an inhibitory effect on Con-A, PHA, and anti-CD3 T cell proliferation (35). In addition, SP-D has a suppressive effect on allergen-stimulated lymphocyte proliferation of lymphocytes obtained from asthmatic and healthy children (36). These results suggest that SP-D plays an important role in regulating T cell responses in the lung.

SP-A Dependent Cytokine Production

Cytokines are a diverse group of biologically active proteins, many of which are thought to contribute to the pathogenesis of ALI by increasing the production of substances that promote local and systemic inflammatory processes. SP-A stimulates production of the cytokines TNF-α, IL-1α, IL-1β, and IL-6 by human peripheral blood mononuclear cells. SP-A also stimulates the production of TNF-α by rat peripheral blood mononuclear cells, alveolar macrophages, and splenocytes (1). Other studies have found that SP-A inhibits TNF-α release by alveolar (37) and interstitial (38) macrophages stimulated with LPS, inhibits the production of IL-2 by stimulated human lymphocytes (33), and inhibits IL-8 production by stimulated human eosinophils (39). Alveolar macrophages infected with *Candida albicans* generate less proinflammatory cytokines (TNF-α, IL-1β, macrophage inflammatory protein-1α, and monocyte chemoattractant protein-1) in the presence of SP-A (40). The reported variable effects of SP-A on cytokine release *in vitro* may be due to differences in study conditions and purification methods of SP-A. Despite these limitations, however, the studies support a role for SP-A in modulating cytokine expression in the lung during ALI. Relatively less is known regarding SP-D-dependent regulation of cytokine production.

SP-A Gene Deficient Mice

To investigate the role of SP-A in the regulation of immune function *in vivo*, our laboratory generated mice in which the SP-A gene was ablated by homologous recombination (SP-A deficient mice). Lung function is unchanged in SP-A deficient mice, thereby providing a powerful model for assessing the *in vivo* role of SP-A in host defense. SP-A deficient mice have increased susceptibility to infection by various pathogens including Group B *Streptococcus* (11), *Pseudomonas aeruginosa* (41), *Haemophilus influenza* (19), *Klebsiella pneumoniae* (42), *Mycoplasma pulmonis* (28), respiratory syncytial virus (43), adenovirus (14), influenza A virus (*LeVine et al unpublished data*), and *Pneumocystis carinii* (*Linke et al unpublished data*). Alveolar macrophages from the SP-A deficient mice display impaired phagocytosis and have an impaired ability to generate reactive oxygen species (11). In addition, inflammation is increased in the lungs of SP-A deficient mice following bacterial and viral infection (11, 43). These data, together with many *in vitro* studies, suggest that SP-A may opsonize bacteria and viruses and enhance their phagocytosis and clearance from the airspaces, thereby providing an early line of defense against pulmonary infection (Figure 1).

SP-D Gene Deficient Mice

To investigate the role of SP-D in the regulation of immune function *in vivo*, our laboratory also generated mice in which the SP-D gene was ablated by homologous recombination (SP-D deficient mice). Surfactant lipid pools and alveolar macrophages are increased in SP-D deficient mice. In the absence of SP-D the mice develop progressive emphysema. In addition, SP-D deficient mice have increased lung production of reactive oxygen species and metalloproteinases, both of which are postulated to play important roles in the development of emphysema (44). The phenotype observed in SP-D deficient mice supports an important role for SP-D in surfactant and lung architectural homeostasis.

With regard to SP-D and *in vivo* immune function, SP-D deficient mice efficiently cleared both Group B *Streptococcus* and *Haemophilus influenzae* from the lung (19). Despite the ability to clear bacteria from the lung, deficiency of SP-D is associated with increased inflammatory cell recruitment to the lung after infection. Alveolar macrophages from the SP-D deficient mice display impaired phagocytosis compared to macrophages from wild type mice. Oxygen radical production by the alveolar macrophages was markedly increased in SP-D deficient mice. In contrast to the efficient bacterial clearance, SP-D deficient mice are susceptible to both influenzae A (*LeVine et al, unpublished data*) and respiratory syncytial viral pneumonia, including increased expression of proinflammatory cytokines and increased neutrophilic infiltrates (45). These finding suggest that SP-D plays a role in modulating cytokine production and inflammatory responses during viral and bacterial infections in the lung (Figure 1).

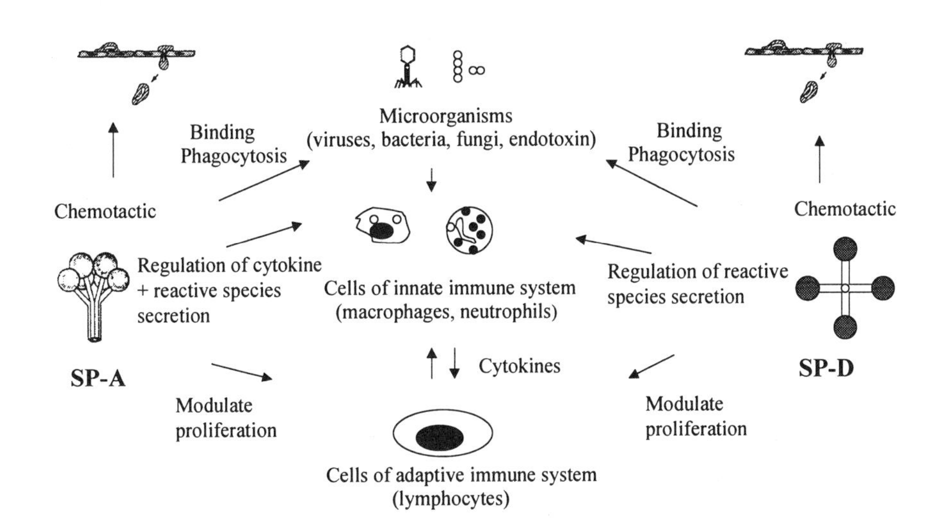

Figure 1. Model for SP-A and SP-D involvement in host defense in the lung. The surfactant-associated proteins SP-A and SP-D are members of a family of host defense lectins, designated collectins. There is increasing evidence that these pulmonary proteins are important components of the innate immune response to microbial challenge. SP-A and SP-D can bind to a broad spectrum of pathogens, including bacteria, viruses, fungi, parasites, as well as lipopolysaccharides (LPS, endotoxin) and allergens. In addition, SP-A and SP-D enhance the clearance of various pathogens by macrophages and neutrophils by aggregation of the pathogens or by direct interaction of the collectins with receptors on the phagocytic cells. The surfactant proteins have also been shown to have important roles in modulating the immune response including cytokine production, reactive species production, and lymphocyte proliferation.

SP-A and SP-D: Role in Lung Disease

SP-A levels are altered in a variety of pulmonary diseases. SP-A levels are lower than normal in both Gram-positive and Gram-negative bacterial pneumonia (1). SP-A levels relative to total protein are decreased in tracheal aspirates of children with bacterial pneumonia, viral pneumonitis, and ALI (46). In the lungs of patients with ALI, both lipid and SP-A levels are decreased and SP-A levels correlate with disease severity. Patients with severe lung injury have low levels of SP-A in the lung, however, with moderate lung injury SP-A levels return to more normal levels and actually increase to greater than control values 3 to 6 days after trauma (47). Similarly, SP-A concentrations are reduced in the lung of patients at risk for ALI who subsequently develop clinically apparent ALI and remain low for as long as 14 days in patients with sustained ALI (48). Bronchoalveolar lavage (BAL) fluid from patients with ALI contains SP-A that is cleaved, possibly due to proteolytic cleavage by neutrophil elastase (49). In contrast, serum concentrations of SP-A increase in patients with ALI and acute cardiogenic pulmonary edema (1, 48). Increased serum levels of SP-A are thought to be due to SP-A leakage through the damaged pulmonary capillary endothelial/epithelial barrier into the serum.

Other conditions associated with reduced SP-A in the lung include infant respiratory distress syndrome, bronchopulmonary dysplasia, idiopathic pulmonary fibrosis, asthma, and cystic fibrosis (1, 50). In contrast, some pulmonary conditions are associated with increased SP-A levels in the lung including hyperoxia-related lung injury, AIDS-related pneumonia with *P. carinii*, silicosis, alveolar proteinosis, and hypersensitivity pneumonitis (1). Intratracheal instillation of LPS increases surfactant proteins in the lung associated with a proliferation of ATII cells (51) suggesting that increased protein synthesis and possibly changes in protein turnover may contribute to the increased SP-A levels secondary to LPS stimulation.

The regulation of SP-D production in the setting of lung disease has not been investigated as thoroughly as the regulation of SP-A. SP-D levels in BAL fluid are normal in patients at risk for ALI and those that develop ALI, however, serum SP-D levels increase in patients with ALI (48). These finding are in contrast to those with SP-A suggesting that SP-D may be regulated differently in the lungs of patients with ALI. In animal models, silica, LPS, and *P. carinii* infection increase SP-D levels in the lung. SP-D levels are increase in BAL fluids of patients with alveolar proteinosis, but not with idiopathic pulmonary fibrosis, interstitial pneumonia secondary to collagen-vascular diseases, or pulmonary sarcoidosis (1).

SP-A and SP-D: Summary Comments

The lung is constantly exposed to toxins, antigens, and microbes that can be efficiently cleared without the generation of destructive inflammatory responses. The pulmonary collectins SP-A and SP-D, traditionally associated with pulmonary surfactant function (i.e. lowering alveolar surface tension), are now also recognized as central modulators of innate host defense and inflammation in the lung. SP-A and SP-D are mammalian lectins capable of recognizing pathogens, leading to binding and enhancing phagocytosis of microbes. Furthermore, SP-A and SP-D modulate production of reactive oxygen species involved in microbial killing, and orchestrate cellular inflammatory responses to diverse types of respiratory pathogens, including bacteria and viruses.

Defensins: Structure and Synthesis

Antimicrobial substances used by host cells range from simple inorganic compounds (e.g. hydrogen peroxide, hypochlorous acid, NO) to relatively complex antimicrobial peptides. Some antimicrobial peptides are enzymes (proteases, and hydrolytic enzymes such as phospholipase, glycosidases, and lysozyme) that can digest the protective layers of microbes, while other antimicrobial peptides directly disrupt biological membranes. Antimicrobial peptides have been identified in organisms as diverse as humans, frogs, insects, plants, and protozoa. The antimicrobial peptides are microbicidal at micromolar concentrations against a wide range of target organisms. One family of antimicrobial peptides, the defensins, were named in 1985. Apart from antimicrobial activity, defensins also play a role in inflammation, wound repair, and specific immune responses (acquired immunity).

Mammalian defensins are cationic antimicrobial peptides characterized by the presence of six cysteine residues that form three intramolecular disulfide bonds. Defensins can be subdivided into two classes, the α-defensins and the β-defensins, based on alternative spacing of their six cysteine residues and differences in the alignment of the disulfide bridges. Despite their differences in sequence and disulfide bond pattern, the α- and β-defensins share a similar three-dimensional structure in solution (52). They also exhibit comparable antimicrobial activity, have genes closely clustered and are expressed in similar anatomical locations. Some of the defensin encoding genes are constitutively expressed, while others are inducible by infections and/or inflammatory agents.

The α-defensins are 29 to 35 amino acids in length and contain three disulfide bridges. α-Defensins are found in great abundance in the azurophilic granules of circulating neutrophils (5-18% of total protein) and in granule-containing Paneth cells of the small intestine (53-56). The human α-defensin family consists of four defensins isolated from neutrophils, HD-1 through HD-4, and two defensins expressed in the secretory granules of Paneth cells of the small intestine and epithelial cells of the female genital tract, HD-5 (55, 57) and HD-6 (58). Low level expression of HD-5 has been found in human nasal and bronchial epithelial cells (59).

The β-defensins are 36 to 42 amino acids in length with six cysteines in a spacing pattern and a disulfide alignment (60). In most tissues the constitutive level of β-defensin expression seems to be low, however, expression is increased during inflammation *in vivo* and *in vitro* in response to LPS (61). Two human β-defensins have been identified, HBD-1 (62) and HBD-2 (62). The gene encoding HBD-1 has been shown to be expressed in epithelial tissues including kidney, lung (both upper respiratory tract and parenchyma), pancreas, testis, gingival tissue, and vagina. HBD-2 is expressed in skin and tracheal mucosa (64). Expression of HBD-2 is increased in cultured keratinocytes exposed to pro-inflammatory cytokines, bacteria, and fungi, suggesting that HBD-2 synthesis is induced during inflammation (65).

A cluster of genes on chromosome 8p22-23.1 encodes HBD-1 and HBD-2 immediately adjacent to the α-defensin locus (63,66). β-defensin molecules consist of a structurally rigid triple-stranded, anti-parallel β-sheet stabilized by intramolecular disulfide bonds, as demonstrated by NMR spectroscopy and x-ray crystallography (52). The mature β-defensin sequence consists of the carboxy terminus of a pre-pro-peptide (93 to 95 amino acids) containing an amino-terminal signal sequence followed by a 40 to 45 amino acid anionic pro-piece, which may be important for neutralization, processing, and folding of the cationic c-terminal peptide (67,68).

Antimicrobial Activity of Defensins

Defensins were originally identified based on their antimicrobial activity against Gram-negative and Gram-positive bacteria, fungi, and enveloped viruses (60). Microorganisms susceptible to human defensin-mediated killing are presented in Table 3. Human defensins 1 to 3 seem to be the major bacterial peptides in human neutrophils constituting 5-7% of total protein content in neutrophils and 30-50% of total protein content of the azurophilic granules. Human defensin 4 has a different antibacterial spectrum compared to HD 1-3 and is present in neutrophils at low concentrations (69). α-Defensins are microbicidal at 1-100 μg/mL (60) and are thought to have antimicrobial activity primarily in the phagolysosome where they reach high concentrations. Increasing salt concentrations competitively inhibit defensin activity (52). Epithelial cells have initial contact with microbes providing the first line of host defense prior to neutrophil recruitment. The β-defensin are expressed by epithelial cells and also have antimicrobicidal activity against Gram-positive and Gram-negative bacteria and fungi at concentrations of 10-100 μg/mL and HBD-2 is 10 times more potent than HBD-1 (60).

Defensins are microbicidal rather than static agents. Bacterial killing occurs in minutes and in most cases requires bacterial cell growth. Defensins interact with bacterial membranes by disrupting the order of the phospholipid bilayer causing loss of membrane integrity. The energy gradient across the membrane is destroyed, membrane damage occurs, and lysis of the bacteria follows. Studies supporting the interaction of defensins at the plasma membranes have demonstrated that defensins induce leakage of potassium and other cellular constituents (75). Defensins also form voltage-dependent, weakly anion-selective channels in planar lipid bilayer membranes (76).

Table 3. Antimicrobial Spectrum of Human Defensins (References 52, 60, 70-74)

Gram-positive bacteria	Gram-negative bacteria
Staphylococcus aureus	*Acinetobacter calcoaceticus*
	Capnocytophaga spp. strains
	Escherichia coli
	Pseudomonas aeruginosa
Viruses	Fungi
Cytomegalovirus	*Chlamydia trachomatis*
Herpes simplex virus, Type 1 and 2	*Cryptococcus neoformans*
Influenza virus A/WSN	*Giardia lamblia*
Vesicular stomatitis virus	

Role of Defensins in Inflammation

In addition to the antimicrobial activity, defensins may also play a role in inflammation and regulation of specific immune responses. The modulating effects of α-defensins on inflammation include chemotaxis, histamine release by mast cells, stimulation of wound repair, and apoptosis (77-79). Defensins are chemotactic for monocytes (78) and T lymphocytes (79), and stimulate cytokine production by T lymphocytes (80). Recent studies demonstrated expression of defensins (HD 1-3) in human T and natural killer cells (NK) (81). β-defensins attract immature dendritic cells and memory T cells, which initiate a primary and recall immune response, respectively (82). These findings provide evidence for a role of defensins as a link between the innate and adaptive immune responses.

α-Defensins are increased in disease states characterized by neutrophilic infiltrates. Defensin levels are increased in airway secretions with inflammatory lung diseases such as ALI, chronic bronchitis, α-1 antitrypsin deficiency, and cystic fibrosis (83). In addition, defensin levels are increase in the plasma of patients with sepsis or meningitis (84). In patients with diffuse panbronchiolitis, BAL fluid levels of defensins are increased in parallel with the neutrophil chemoattractant, IL-8 (85). *In vitro*, defensins induce IL-8 secretion in airway epithelial cells (86). Collectively, these data suggest that the defensins may stimulate chemokine expression and mediate the recruitment of neutrophils into the airways. In addition, defensins stimulate the release of leukotriene B_4 by neutrophils and IL-8 by macrophages (83), both neutrophil chemoattractants. *In vivo*, subcutaneous injection of defensins in mice produces a neutrophil and macrophage infiltrate (79) further supporting a role for defensins in regulating the inflammatory response that is typically seen during ALI.

Defensins in ALI

ALI is characterized by epithelial/endothelial injury and increased permeability, neutrophil accumulation in the airspaces, and increased pro-inflammatory cytokines within the lung compartment. Neutrophil-mediated lung injury is due, in part, to the serine proteases and neutrophil defensins which both affect the integrity of the epithelial layer, decrease the frequency of ciliary beating, increase secretion of mucus, and induce the synthesis of epithelial-derived mediators that may influence the amplification and resolution of inflammation. Neutrophil products that mediate tissue injury at sites of inflammation include the neutrophil serine proteinases, elastase, cathepsin G, proteinase 3, and the nonenzymatic defensins (87). To protect against the extracellular activity of serine proteinases, the lung is equipped with serine proteinase inhibitors. Defensins bind to members of the serine proteinase inhibitor (Serpin) family such as α-1 proteinase inhibitor (α1PI), alpha 1- antichymotrypsin, alpha 2-antiplasmin, and antithrombin III. When α-1PI, an inhibitor of neutrophil elastase, is bound to the defensins it is unable to bind and inactivate neutrophil elastase (88), which can injure lung tissue. *In vitro*, defensins cause lysis of airway epithelial cells. Co-incubation of elastase with defensin *in vitro* reduces defensin-induced cell lysis and prevents defensin-induced IL-8 synthesis (89).

In addition to the release of proteases by neutrophils, the production of reactive oxygen species can cause damage to the lung. Oxidative damage follows increased oxidative stress from reactive oxygen species from activated neutrophils and macrophages, and from the use of high oxygen concentrations during treatment. Glutathione, a potent antioxidant in the lung, protects against oxidant injury. Alterations in alveolar and lung glutathione metabolism are recognized as a central feature in ALI (87). *In vitro*, defensins decrease glutathione levels in airway epithelial cells (90), which may result in an imbalance in antioxidant versus pro-inflammatory responses to oxidative stress resulting in lung damage. In addition, the combination of defensins with hydrogen peroxide, both secreted by activated neutrophils, act synergistically to cause cell lysis (91) suggesting that defensins may have a pro-inflammatory role during ALI.

Complement activation, either within the lung or within the circulation, is associated with lung injury (see Chapter 5). C5 fragments (C5a and C5a-des-Arg) are especially potent PMN chemotaxins and activators. If complement activation is extensive, endothelial and type II epithelial damage can occur, accompanied by edema, hemorrhage, and fibrin deposition (87). Defensins have anti-inflammatory activities by inhibiting activation of the classical complement pathway (92) by binding to C1q (the first component of complement).

Extravascular fibrin deposition in ALI occurs, in part, due to depression of the fibrinolytic activity in the distal airspaces of the lung (87). Plasminogen concentrations in BAL fluid of patients with ALI are increased compared with values in control patients, however, most of the plasminogen activator in BAL fluid is coupled to inhibitors (93). Defensins have been reported to inhibit fibrinolysis possibly by shielding of fibrin-bound plasminogen from activation by tissue type plasminogen activation (94).

LPS (endotoxin) has been implicated as an important agent in the induction of ALI. Endotoxin levels in the plasma of patients with ALI, or patients at risk that subsequently develop ALI, are increased compared with values measured in patients who do not develop ALI (95). *E. coli* endotoxin enhances complement-mediated neutrophil activation and superoxide production, and tissue injury (87). *In vitro*, LPS and TNF-α induce expression of defensins in tracheal epithelial cells (61). The inducible response of the defensins may represent a mechanism to recognize Gram-negative bacteria at epithelial surfaces and mount a host defense response important in preventing colonization and/or subsequent infection and inflammation.

Data regarding BAL levels of β-defensins is limited. In healthy volunteers, HBD-1 and HBD-2 are present in BAL fluid. Both HBD-1 and HBD-2 are present in BAL fluid of patients with cystic fibrosis or idiopathic pulmonary fibrosis (83). Plasma from patients with bacterial pneumonia contains greater concentrations of HBD-2 compared to plasma from healthy controls (96). *In vitro*, HBD-2 is inducible in airway epithelial cells either upon contact with *Pseudomonas aeruginosa*, and with stimulation by LPS, TNF-α, or IL-1β (97).

The bactericidal activity of the β-defensins is affected by local salt concentrations. In this regard, the airway surface fluid from patients with cystic fibrosis lacks antimicrobial activity believed to be due to the increased salt concentrations, which may decrease the activity of hBD-1 and hBD-2 (98). Recently the effect of salt concentration effect on the β-defensins has been questioned and other factors such as mucin or DNA have been proposed to inhibit bactericidal activity and could be of equal importance in inactivation of defensins in lung disease associated with cystic fibrosis. Future studies are necessary to further characterize the role of the β-defensins in ALI.

Summary

ALI may be the result of direct damage to the lung by the inciting disease or agent, or alternatively, injury may be the result of indirect damage resulting from the formation of oxidants, release of proteases, or release of mediators as part of the host response to the inciting agent. Since the lung is the usual site of entry of respiratory pathogens the pulmonary collectins and defensins play an important role in initial host defense and regulation of inflammation in the lung. Insufficient expression of collectins and defensins may predispose to disease and may be markers for prediction of disease. In addition, further investigation of the role of collectins and defensins in ALI may provide novel therapeutic agents for prevention and treatment of pulmonary infections.

ACKNOWLEGEMENTS

The author's published and unpublished work presented in this chapter was supported by grants from the National Institutes of Health (HL03905, HL58794, HL61646, and HL56387) and the Children's Hospital Research Foundation.

REFERENCES

1. Wright, J.R. (1997) Immunomodulatory functions of surfactant *Physiol Rev* 77, 931-962
2. Sumiya, M., Super, M., Tabona, P., Levinsky, R.J., Arai, T., Turner, M.W., and Summerfield, J.A. (1991) Molecular basis of opsonic defect in immunodeficient children. *Lancet* 337, 1660-1670
3. Katyal, S.L., Singh, G., and Locker, J. (1992) Characterization of a second human pulmonary surfactant-associated protein SP-A gene. *Am J Respir Cell Mol Biol* 6, 446-452
4. White, R.T., Damm, D., Miller, J., Spratt, K., Schilling, J., Hawgood, S., Benson, B., and Cordell, B. (1985) Isolation and characterization of the human pulmonary surfactant apoprotein gene. *Nature* 317, 361-363
5. Weaver, T.E., and Whitsett, J.A. (1988) Structure and function of pulmonary surfactant proteins. *Semin Perinatol* 12, 213-220
6. Khoor, A., Gray, M.E., Hull, W.M., Whitsett, J.A., and Stahlman, M.T. (1993) Developmental expression of SP-A and SP-A mRNA in the proximal and distal respiratory epithelium in the human fetus and newborn. *J Histochem Cytochem* 41, 1311-1319
7. Madsen, J., Kliem, A., Tornoe, I., Skjodt, K., Koch, C., and Holmskov, U., (2000) Localization of lung surfactant protein D on mucosal surfaces in human tissue. *J Immunol* 164, 5866-5870
8. Crouch, E., Rust, K., Veile, R., Donis-Keller, H., and Grosso, L. (1993) Genomic organization of human surfactant protein D (SP-D). *J Biol Chem* 268, 2976-2983
9. Hartshorn, K.L., Crouch, E., White, M.R., Colamussi, M.L., Kakkanatt, A., Tauber, B., Shepherd, V., and Sastry, K.N. (1998) Pulmonary surfactant proteins A and D enhance neutrophil uptake of bacteria. *Am J Physiol* 274, L958-L969
10. Wright, J.R., and Youmans, D.C. (1993) Pulmonary surfactant protein-A stimulates chemotaxis of alveolar macrophages. *Am J Physiol* 264, L338-L344
11. LeVine, A.M., Kurak, K.E., Wright, J.R., Watford, W.T., Bruno, M.D., Ross G.F., Whitsett, and J.A., Korfhagen, T.R. (1999) Surfactant protein-A binds Group B streptococcus, enhancing phagocytosis and clearance from lungs of surfactant protein-A-deficient mice. *Am J Respir Cell Mol Biol* 20, 279-286
12. Kabha, K., Schmegner J., Keisari, Y., Parolis, H., Schlepper-Schaefer, J., and Ofek, I. (1997) SP-A enhances phagocytosis of *Klebsiella* by interaction with capsular polysaccharides and alveolar macrophages. *Am J Physiol* 272, L344-L352

13. Mariencheck, W.I., Savov, J., Dong, Q., Tino, M.J., and Wright, J.R. (1999) Surfactant protein A enhances alveolar macrophage phagocytosis of a live, mucoid strain of *P. aeruginosa. Am J Physiol* 277, L777-L786
14. Harrod, K.S., Trapnell, B.C., Otake, K., Korfhagen, T.R., and Whitsett, J.A. (1999) SP-A enhances viral clearance and inhibits inflammation after pulmonary adenoviral infection. *Am J Physiol* 277, L580-L588
15. Weyer, C., Sabat, R., Wissel, H., Kruger, D.H., Stevens, P.A., and Prosch, S. (2000) Surfactant protein A binding to cytomegalovirus proteins enhances virus entry into rat lung cells. *Am J Respir Cell Mol Biol* 23, 71-78
16. Ghildyal, R., Hartley, C., Varrasso, A., Meanger, J., Voelker, D.R., Anders, E.M., and Mills, J. (1999) Surfactant protein A binds to the fusion glycoprotein of respiratory syncytial virus and neutralizes virion infectivity. *J Infect Dis* 180, 2009-2013
17. Madan, T., Eggleton, P., Kishore, U., Strong, P., Aggrawal, S.S., Sarma, P.U., and Reid, K.B.M. (1997) Binding of pulmonary surfactant proteins A and D to *Aspergillus fumigatus* conidia enhances phagocytosis and killing by human neutrophils and alveolar macrophages. *Infect Immun* 65, 3171-3179
18. Crouch, E.C., Persson, A., Griffin, G.L., Chang, D., and Senior, R.M. (1995) Interactions of pulmonary surfactant protein D (SP-D) with human blood leukocytes. *Am J Resp Cell Mol Biol* 12, 410-415
19. LeVine, A.M., Whitsett, J.A., Gwozdz J.A., Richardson, T.R., Fisher, J.H., Burhans, M.S., and Korfhagen, T.R. (2000) Distinct effects of surfactant protein A and D deficiency during bacterial infection on the lung. *J Immunol* 165, 3934-3940
20. Hickling, T.P., Bright, H., Wing, K., Gower, D., Martin, S.L., Sim, R.B. and Malhotra, R. (1999) A recombinant trimeric surfactant protein D carbohydrate recognition domain inhibits respiratory syncytial virus infection *in vitro* and *in vivo*. *Eur J Immunol* 29, 3478-3484
21. van Iwaarden, F., Welmers, B., Verhoef, J., Haagsman, H.P., and van Golde, L.MG. 1990) Pulmonary surfactant protein A enhances the host-defense mechanism of rat alveolar macrophages. *Am J Respir Cell Mol Biol* 2, 91-98
22. Weissbach, S., Neuendank, A., Pettersson, M., Schaberg T., and Pison, U. (1994) Surfactant protein A modulates release of reactive oxygen species from alveolar macrophages. *Am J Physiol* 267, L660-L666
23. Weber, H., Heilmann, P., Meyer, B., and Maier, K.L. (1990) Effect of canine surfactant protein (SP-A) on the respiratory burst of phagocytic cells. *FEBS Lett* 270, 90-94
24. Katsura, H., Kawada, H., and Konno, K. (1993) Rat surfactant apo-protein A (SP-A) exhibits antioxidant effects on alveolar macrophages. *Am J Respir Cell Mol Biol* 6, 446-452
25. Blau, H., Riklis, S., van Iwaarden, J.F., McCormack, R.X., and Kalina, M. (1997) Nitric oxide production by rat alveolar macrophages can be modulated *in vitro* by surfactant protein A. *Am J Physiol* 272, L1198-L1204
26. Wright, J.R., Zlogar, D.F., Taylor, J.C., Zlogar, T.M., and Restrepo, C.I. (1999) Effects of endotoxin on surfactant protein A and D stimulation of NO production by alveolar macrophages. *Am J Physiol* 276, L650-L658
27. Stamme, C., Walsh, E., and Wright, J.R. (2000) Surfactant protein A differentially regulates IFN-γ and LPS-induced nitrite production by rat alveolar macrophages. *m J Respir Cell Mol Biol* 23, 772-779
28. Hickman-Davis, J., Gibbs-Erwin, J., Lindsey, J.R., and Matalon, S. (1999) Surfactant protein A mediates mycoplasmacidal activity of alveolar macrophages by production of peroxynitrite. *Proc Natl Acad Sci USA* 96, 4953-4958
29. Pasula, R., Wright, J.R., Kachel, D.L., and Martin, II, W.J. (1999) Surfactant protein A suppresses reactive nitrogen intermediates by alveolar macrophages in response to *Mycobacterium tuberculosis. J Clin Invest* 103, 483-490
30. van Iwaarden, J. F., Shimizu, H., van Golde, P. H. M, Voelker, D. R., and van Golde, L.M.G. (1992) Rat surfactant protein D enhances the production of oxygen radicals by rat alveolar macrophages. *Biochem J* 286, 5-8

31. Hartshorn, K.L., Crouch, E.C.,. White, M.R, Eggleton, P., Tauber, A.I., Chang, D., and Sastry, K. (1994) Evidence for a protective role of pulmonary surfactant protein D (SP-D) against influenza A viruses. *J Clin Invest* 94, 311-319
32. Kremlev, S.G., Umstead, T.M., and Phelps, D.S. (1994) Effects of surfactant protein A and surfactant lipids on lymphocyte proliferation *in vitro*. *Am J Physiol* 267, L357-L364
33. Borron, P., Veldhuizen, R.A.W., Lewis, J.F., Possmayer, F., Caveney, A., Inchley, K., McFadden, R.G., and Fraher, L.J. (1996) Surfactant associated protein-A inhibits human lymphocyte proliferation and IL-2 production. *Am J Respir Cell Mol Biol.* 15, 115-121
34. Kanellopoulos, J.M., DePetris, S., Leca, G., and Crumpton, M.J. (1985) The mitogenic lectin from *Phaseolus vulgaris* does not recognize the T3 antigen of human T lymphocytes. *Eur J Immunol* 15, 479-486
35. Borron, P.J., Crouch, E.C, Lewis, J.F., Wright, J.R., Possmayer, F., and Fraher, L.J. (1998) Recombinant rat surfactant-associated protein D inhibits human T lymphocyte proliferation and IL-2 production. *J Immunol* 161, 4599-4603
36. Wang, J.Y., Shieh, C.C., You, P.F., Lei, H.Y., and Reid, K.B.M. (1998) Inhibitory effect of pulmonary surfactant protein A and D on allergen-induced lymphocyte proliferation and histamine release in children with asthma. *Am J Respir Crit Care Med* 158, 510-518
37. McIntosh, J.C., Mervin-Blake, S., Conner, E., and Wright, J.R. (1996) Surfactant protein A protects growing cells and reduces TNF-α activity from LPS-stimulated macrophages. *Am J Physiol* 271, L310-L319
38. Aria-Diaz, J., Garcia-Verdugo, I., Casals, C., Sanchez-Rico, N., Vara, E., and Balibrea, J.L. (2000) Effect of surfactant protein A (SP-A) on the production of cytokines by human pulmonary macrophages. *Shock* 14, 300-306
39. Cheng, G, Ueda, T., Nakajima, H., Nakajima, A., Arima, M., Kinjyo, S., and Fukuda, T. (2000) Surfactant protein A exhibits inhibitory effect on eosinophils IL-8 production. *Biochem Biophy Res Comm* 270, 831-835
40. Rosseau, S., Hammerl, P., Maus, U., Gunther, A., Seeger, W., Grimminger, F., and Lohmeyer, J. (1999) Surfactant protein A down-regulates proinflammatory cytokine production evoked by *Candida albicans* in human alveolar macrophages and monocytes. *J Immun* 163, 4495-4502
41. LeVine, A.M., Kurak, K.E., Bruno, M.D., Stark, J.M., Whitsett, J.A., and Korfhagen, T.R. (1998) Surfactant protein-A deficient mice are susceptible to *Pseudomonas aeruginosa* infection. *Am J Resp Cell Mol Biol* 19, 700-708
42. Korfhagen, T.R., Bruno, M.D., Silver, J.A., Whitsett, J.A., and LeVine, A.M. Enhanced *K. pneumoniae* pulmonary infection in mice lacking SP-A. *Am J Respir Care Med* 161, A514 (Abstract)
43. LeVine, A.M., Gwozdz, J., Stark, J., Bruno, M., Whitsett, J., and Korfhagen, T. (1999) Surfactant protein-A enhances respiratory syncytial virus clearance *in vivo*. *J Clin Invest* 103, 1015-1021
44. Wert, S.E., Yoshida, M., LeVine, A.M., Ikegami, M., Jones, T., Ross, G.F., Fisher, J.H., Korfhagen, T.R., and Whitsett, J.A. (2000) Increased metalloproteinase activity, oxidant production, and emphysema in surfactant protein D gene-inactivated mice. *Proc Natl Acad Sci USA* 97, 5972-5977
45. LeVine, A.M., Gwozdz, J., Fisher, J., Whitsett, J., and Korfhagen, T. (2000) Surfactant protein-D modulates lung inflammation with respiratory syncytial virus infection *in vivo*. *Am J Respir Crit Care* Med 161, A515
46. LeVine, A.M., Lotze, A., Stanley, S., Stroud, C., O'Donnell, R., Whitsett, J. and Pollack, M.M. (1996) Surfactant content in children with inflammatory lung disease. *Crit Care Med* 24, 1062-1067
47. Pison, U., Obertacke, U., Seeger, W., and Hawgood, S. (1992) Surfactant protein A (SP-A) is decreased in acute parenchymal lung injury associated with polytrauma. *Eur J Clin Invest* 22, 712-718
48. Green, K.E., Wright, J.R., Steinberg, K.P., Ruzinski, J.T., Caldwell, E., Wong, W.B., Hull, W., Whitsett, J.A., Akino, T., Kuroki, Y., Nagae, H., Hudson, L.D., and Martin,

T.R. (1999) Serial changes in surfactant-associated proteins in lung and serum before and after onset of ARDS. *Am J Respir Crit Care Med* 160, 1843-1850

49. Baker, C.S., Evans, T.W., Randle, B.J., and Haslam, P.L. (1999) Damage to surfactant-specific protein in acute respiratory distress syndrome. *Lancet* 353, 1232-1237
50. Postle, A.D., Mander, A., Reid, K.B.M., Wang, J.Y., Wright, S.M., Moustake, M., and Warner, J.O. (1999) Deficient hydrophilic lung surfactant protein A and D with normal surfactant phospholipid molecular species in cystic fibrosis. *Am J Respir Cell Mol Biol* 20, 90-98
51. Viviano, C.J., Bakewell, W.E.., Dixon, D., Dethloff, L.A., and Hook, G.E.R. (1995) Altered regulation of surfactant phospholipid and protein during acute pulmonary inflammation. *Biochim Biophy Acta* 1259, 235-244
52. Lehrer, R.I., Lichtenstein, A.K., and Ganz, T. (1993) Defensins: antimicrobial and cytotoxic peptides of mammalian cells. *Annu Rev Immunol* 11, 105-128
53. Rice, W.G., Ganz, T., Kinkade, J.M., Selsted, M.E., Lehrer, R.I., and Parmely, R.T. (1987) Defensin-rich dense granules of human neutrophils. *Blood* 70, 757-765
54. Gabay, J.E., Scott, R.W., Campanelli, D., Griffith, J., Wilde, C., Marra, M.N., Seeger, M., and Nathan, C.F. (1989) Antibiotic proteins of human polymorphonuclear leukocytes. *Proc Natl Acad Sci USA* 86, 5610-5614
55. Selsted, M.E., Miller, S.I., Henschen, A.H., and Ouellette, A.J. (1992) Enteric defensins: antibiotic peptide components of intestine host defense *J Cell Biol* 118, 929-936
56. Porter, E., Liu, L., Oren, A., Anton, P., and Ganz, T. (1997) Localization of human intestinal defensin 5 in Paneth cell granules. *Infect Immun* 65, 2389-2395
57. Porter, E.M., Poles, M.A., Lee, J.S., Naitoh, J., Bevins, C.L. and Ganz, T. (1998) Isolation of human intestinal defensins from ileal neobladder urine. *FEBS Lett* 434, 272-276
58. Jones, D.E., and Bevins, C.L. (1993) Defensin-6 mRNA in human Paneth cells: implications for antimicrobial peptides in host defense of the human bowel. *FEBS Lett* 315, 187-192
59. Frye, M., Bargon, J., Dauletbaev, N., Weber, A., Wagner, T.O.F., and Gropp, R. (2000) Expression of human-α-defensin 5 (HD5) mRNA in nasal and bronchial epithelial cells. *J Clin Pathol* 53, 770-773
60. Ganz, T. and Lehrer, R.I. (1995) Defensins. *Pharmac Ther* 66:191-205

61 Russell, J.P., Diamond, G., Tarver, A.P., Scanlin, T.F. and Bevins, C.L. (1996) Coordinate induction of two antibiotic genes in tracheal epithelial cells exposed to the inflammatory mediators lipopolysaccharide and tumor necrosis factor alpha. *Infect Immun* 64, 1565-1568

61. Bensch, K.W., Raida, M., Magert, H.J., Schulz-Knappe, and P., Forssmann, W.G. ((1995) hBD-1: a novel β-defensin from human plasma. *FEBS Lett* 368, 331-335.
62. Harder, J., Bartels, J., Christophers, E., and Schroder, J.M. (1997) A peptide antibiotic from human skin. *Nature* 387, 861
63. Huttner, K.M. and Bevins, C.L. (1999) Antimicrobial peptides as mediators of epithelial host defense. *Pediatr Res* 45, 785-794
64. Singh, P.K., Kia, H.P., Wiles, K., Hesselberth, J., Liu, L., and Conway, B.D. (1998) Production of β-defensins by human airway epithelia. *Proc Natl Acad Sci USA* 95, 14961-14966
65. Liu, L., Zhao, C., Heng, H.H.Q. and Ganz, T. (1997) The human β-defensin-1 and α-defensins are encoded by adjacent genes: Two peptide families with differing disulfide topology share a common ancestry. *Genomics* 43, 316-320
66. Michaeklson, D., Rayner, J., Couto, M., and Ganz, T. (1992) Cationic defensins arise from charge neutralized propeptides: a mechanism for avoiding leukocyte autotoxicity? *J Leuk Biol* 51, 634-639
67. Valore, E.V., and Ganz, T. (1992) Posttranslational processing of defensins in immature human myeloid cells. *Blood* 79, 1538-1544

68. Wilde, C.G., Griffith, J.E., Marra, M.N., Snable, J.L., and Scott, R.W. (1989) Purification and characterization of human neutrophil peptide 4, a novel member of the defensin family. *J Biol Chem* 264, 11200-11203
69. Greenwald, G.I., and Ganz, T. (1987) Defensins mediate the microbicidal activity of human neutrophil granule extract against *Acinetobacter calcoaceticus*. *Infect Immun* 55, 1365-1368
70. Miyasaki, K.T., Bodeau, A.L., Ganz, T., Selsted, M.E., and Lehrer, R.I. (1990) *In vitro* sensitivity of oral gram-negative, facultative bacteria to the bactericidal activity of human neutrophil defensins. *Infect Immun* 58, 3934-3940
71. Ogata, K., Linzer, B.A., Zuberi, R.I., Ganz, T., Lehrer, R.I., and Catanzaro, A. (1992) Activity of defensins from human neutrophilic granulocytes against *Mycobacterium avium-Mycobacterium intracellulare*. *Infect Immun* 60, 4720-4725
72. Yasin, B., Harwig, S.S., Lehrer, R.I., and Wagar, E.A. (1996) Susceptibility of *Chlamydia trachomatis* to protegrins and defensins. *Infect Immun* 64, 709-713
73. Aley, S.B., Zimmerman, M., Hetsko, M., Selsted, M.E., and Gillin, F.D. (1994) Killing of *Giardia lamblia* by cryptdins and cationic neutrophil peptides. *Infect Immun* 62, 5397-5403
74. Lehrer, R.I., Barton, A., Daher, K.A., Harwig, S.S., Ganz, T., and Selsted, M.E. (1989) Interaction of human defensins with *Escherichia coli*. *J Clin Invest* 84, 553-561
75. Kagan, B.L., Selsted, M.E., Ganz, T., and Lehrer, R.I. (1990) Antimicrobial defensin peptides form voltage-dependent ion-permeable channels in planar lipid bilayer membranes. *Proc Natl Acad Sci USA* 87, 210-214
76. Befus, A.D., Mowat, C., Gilchrist, M., Hu J., Solomon, S., and Bateman, A. (1999) Neutrophil defensins induce histamine secretion from mast cells: mechanisms of action. *J Immunol* 163, 947-953
77. Territo, M.C., Ganz, T., Selsted, M.E., and Lehrer, R. (1989) Monocyte-chemotactic activity of defensins from human neutrophils. *J Clin Invest* 84, 2017-2020
78. Chertov, O., Michiel, D.F., Xu, L., Wang, J.M., Tani, K., Murphy, W.J., Longo, D.L., Taub, D.D., and Oppenheim, J.J. (1996) Identification of defensin-1, defensin-2, and cap37/azurocidin as T-cell chemoattractant proteins released from interleukin-8-stimulated neutrophils. *J Biol Chem* 271, 2935-2940
79. Lillard, J.W., Jr., Boyaka, P.N., Chertov, O., Oppenheim, J.J., and McGhee, J.R. (1999) Mechanisms for induction of acquired host immunity by neutrophil peptide defensins. *Proc Natl Acad Sci USA* 96, 651-656
80. Agerberth, B., Charo, J., Werr, J., Olsson, B., Idali, F., Lindbom, L., Kiessling, R., Jornvall, H., Wigzell, H., and Gudmundsson, G.H. (2000) The human antimicrobial and chemotactic peptide Ll-37 and α-defensins are expressed by specific lymphocyte and monocyte populations. *Blood* 96, 3086-3093
81. Yang, D., Chertov, O., Bykovskaia, S.N., Chen, Q., Buffo, M.J., Shogan, J., Anderson, M., Schroder, J.M., Wang, J.M. Howard, O.M.Z., and Oppenheim, J.J. (1999) β-Defensins: linking innate and adaptive immunity through dendritic and T cell CCR6. *Science* 286, 525-528
82. van Wetering, S., Sterk, P.J., Rabe, K.F., and Hiemstra, P.S. (1999) Defensins: key players or bystanders in infection , injury and repair in the lung? *J Allergy Clin Immunol* 104, 1131-1138
83. Panyutich, A.V., Panyutich, E.A., Krapivin, V.A., Baturevich, E.A., and Ganz, T. (1993) Plasma defensin concentrations are elevated in patients with septicemia or bacterial meningitis. *J Lab Clin Med* 122, 202-207
84. Ashitani, J., Mukae, H., Nakazato, M., Ihi, T., Mashimoto, H., Kadota, J., Kohno, S., and Matsukura S. (1998) Elevated concentrations of defensins in bronchoalveolar lavage fluid in diffuse panbronchiolitis. *Eur Respir J* 11, 104-111
85. van Wetering, S., Mannesse-Lazeroms, S.P.G., van Sterkenburg, M.A.J.A., Daha M.R., Dijkman J.H., Hiemstra P.S. (1997) Effect of defensins on IL-8 synthesis in airway epithelial cells. *Am J Physiol* 272, L888-L896

86. Pittet, J.F., Mackersie, R.C., Martin, T.R., and Matthay, M.A. (1997) Biological markers of acute lung injury: prognostic and pathogenetic significance. *Am J Respir Crit Care Med* 155, 1187-1205
87. Panyutich, A.V., Hiemstra, P.S., van Wetering, S., and Ganz, T. (1995) Human neutrophil defensin and serpins form complexes and inactivate each other. *Am J Respir Cell Mol Biol* 12, 351-357
88. van Wetering, S., Mannesse-Lazeroms, S.P.G., Dijkman, J.H., and Hiemstra, P.S. (1997) Effect of neutrophil serine proteinases and defensins on lung epithelial cells: modulation of cytotoxicity and IL-8 production. *J Leukoc Biol* 62, 217-226
89. van Wetering, S., Rahman, I., Hiemstra, P.S. and MacNee, W. (1998) Role of intracellular glutathione in neutrophil defensin-induced IL-8 synthesis and cytotoxicity in airway epithelial cells. *Eur Respir J* 12, 420S
90. Lichtenstein, A.K., Ganz, T., Selsted, M.E., and Lehrer, R.I. (1988) Synergistic cytolysis mediated by hydrogen peroxide combined with peptide defensins. *Cellular Immun* 114, 104-116
91. van den Berg, R.H., Faber-Krol, M.C., van Wetering, S., Hiemstra, P.S., and Daha, M.R. (1998) Inhibition of activation of the classical pathway of complement by human neutrophil defensins. *Blood* 92, 3898-3903
92. Idell, S., James, K.K., Levin, E.G., Schwartz, B.S., Manchanda, N., Maunder, R.J., Martin, T.R., McLarty, J., and Fair, D.S. (1989) Local abnormalities in coagulation and fibrinolytic pathways predispose to alveolar fibrin deposition in the adult respiratory distress syndrome. *J Clin Invest* 84, 695-705
93. Higazi, A.A.R., Ganz, T., Kariko, K., and Cines, D.B. (1996) Defensin modulates tissue-type plasminogen activator and plasminogen binding to fibrin and endothelial cells. *J Biol Chem* 271, 17650-17655
94. Donnelly, T.J., Meade, P., Jagels, M., Cryer, H.G., Law, M.M., Hugli, T.H., Shoemaker, W.C., and Abraham, E. (1994) Cytokine, complement, endotoxin profiles associated with the development of the adult respiratory distress syndrome after severe injury. *Crit Care Med* 22, 768-777
95. Hiratsuka, T., Nakazato, M., Date, Y., Ashitani, J.I., Minematsu, T., Chino, N., and Matsukura, S. (1998) Identification of human β-defensin-2 in respiratory tract and plasma and its increase in bacterial pneumonia. *Biochem Biophys Res Commun* 249, 943-947
96. Harder, J., Meyer-Hoffert, U., Teran, L.M, Schwichtenberg, L., Bartels, J., Maune, S., and Schroder, J.M. (2000) Mucoid *Pseudomonas aeruginosa*, TNF-α, and IL-1β, but not IL-6, induce human β-defensin-2 in respiratory epithelia. *Am J Respir Cell Mol Biol* 22, 714-721
97. Goldman, M.J., Anderson, G.M., Stolzenberg, E.D., Kari, U.P., Zasloff, M., and Wilson, J.M. (1997) Human β-defensin-1 is a salt-sensitive antibiotic in lung that is inactivated in cystic fibrosis. *Cell* 88, 553-560
98. Matsui, H., Grubb, B.R., Tarran, R, Randell, S.H., Gatzy, J.T., Davis, C.W., and Boucher, R.C. (1998) Evidence for periciliary liquid layer depletion, not abnormal ion composition, in the pathogenesis of cystic fibrosis airway disease. *Cell* 95, 1005-1015

Chapter 14

LUNG REPAIR, REMODELING, AND FIBROSIS

Sujata Guharoy and Sem H. Phan
Department of Pathology, University of Michigan, Ann Arbor, MI 48109

INTRODUCTION

Various insults or injuries to the lung commonly result in fibrosis associated with an inflammatory response. The fibrotic process could progress, resulting in a fatal outcome. Pulmonary fibrosis is a heterogeneous group of diseases whose initiating factors are quite different, but the terminal stages are more stereotypical with the progressive accumulation of connective tissue replacing normal functional parenchyma. In patients who die after lung injury, airspace granulation tissue associated with the inflammatory response persists and progresses. In survivors there is an apparent resolution of this granulation tissue. The outcome of repair and remodeling after injury depends upon cellular interactions through mediators that influence both proliferation and cell death. A number of important findings in this field have emerged from experimental animal models, and some of these directly reflect what is seen in the clinical setting. This chapter highlights the current knowledge of the molecules and mechanism that have been identified as potentially important factors that regulate and determine the outcome of pulmonary fibrosis.

Early Response to Injury

The early response to injury of the alveolar epithelium, endothelium, or lung parenchyma leads to acute inflammation and rapid recruitment of neutrophils into the lung interstitium. In animal models, which tend to be self-limiting, this acute response subsides and is replaced by a mononuclear infiltrate by the end of the first week. However, in certain forms of human pulmonary fibrosis where the disease is progressive, neutrophils tend to persist. Surviving type II alveolar epithelial cells (AEC) can repair the damaged pulmonary alveolar epithelium. They can proliferate and differentiate to replace both the type II and type I pneumocytes. The efficiency with which the AEC proliferate and differentiate to repair the damage is believed to be a critical determinant of the severity of subsequent fibrosis. Efficient repair of the denuded alveolar membrane acts to suppress fibroblast proliferation and extracellular matrix (ECM) accumulation (1).

Various reports have shown that epithelial and fibroblast cell interactions are critical in the outcome of post-injury repair. For example, lungs exposed to high concentrations of oxygen develop epithelial cell damage followed by fibroblast proliferation and collagen synthesis (2). Fibroblasts, on the other hand, produce growth factors and stimulate alveolar cell proliferation (3-5). In contrast, primary fibroblasts isolated from rats with bleomycin-induced fibrosis do not generate conditioned media capable of stimulating type II cell proliferation (6). The observation that broncho-alveolar lavage (BAL) fluid obtained from patients during lung repair induces both fibroblast and endothelial cell death suggests an additional mechanism of repair. Polunovsky et al found that endothelial cells exposed to BAL fluid undergo apoptosis, while fibroblasts do not (7). Only proliferating fibroblasts are susceptible to the bioactivities of the lavage fluid. Conversely, other studies have shown that primary fibroblasts isolated from fibrotic lungs produce factors that cause apoptosis of the AEC (8). Apoptosis-related genes Fas, Fas-ligand, p53 and bax are dysregulated in alveolar epithelial cells of mice with bleomycin-induced pulmonary fibrosis (9, 10). Thus, epithelial cell apoptosis may be induced by up-regulation of, and by the imbalance between, the apoptosis-inducible and -inhibitory genes in addition to the Fas-Fas ligand pathway (see Chapter 15). These observations indicate that the final outcome of pulmonary fibrosis depends on a critical balance between the repair of the epithelium and the prevention of fibroblast proliferation.

Repair and Remodeling

One of the major effects of repair after injury is fibroblast proliferation and enhanced synthesis, deposition, and rearrangement of ECM macromolecules such as collagens, elastin, and non-collagenous glycoproteins (11). Lung fibroblasts from rats with severe lung injury, and in fibroblasts derived from patients with idiopathic pulmonary fibrosis (IPF), exhibit significantly higher proliferation and ECM synthesis compared to uninjured lung fibroblasts (12-16). Repair resulting in fibrosis is characterized by excessive deposition of non-functional connective tissue replacing normal lung parenchyma. In studies of rapidly induced pulmonary fibrosis, as in response to acute lung injury (ALI), there is increased collagen types I, III, and V synthesis, and decreased collagen degradation (17). A transient increase in α_1 (I) and α_1(III) procollagen and fibronectin gene expression is preceded by increases in cytokine gene expression including transforming growth factor β (TGFβ), tumor necrosis factor-α (TNFα) and interleukin-1 (IL-1) (18-20). *In situ* hybridization analyses identified the interstitial cells derived from fibroblasts to exhibit increased procollagen expression in the adventitia of bronchioles, terminal bronchioles, adjacent small blood vessels, and submesothelial area (21). These cells represent newly appearing myofibroblasts characterized by α-smooth muscle actin expression and are responsible for the bulk of collagen gene expression (22). Large and small proteoglycans such as biglycan, versican, and heparan sulphate proteoglycan are also expressed at higher levels after lung injury and during the remodeling/repair phase (23, 24).

Mediators of Repair

In recent years significant progress has been made in many aspects of pulmonary fibrosis biology. Among them is the identification of several cytokines as important mediators of fibrosis. Some of these cytokines are TGFα, TGFβ, TNFα, platelet derived growth factor (PDGF), fibroblast growth factors (FGFs), IL-1, monocyte chemoattaractant protein-1 (MCP-1), and macrophage inflammatory protein-1α (MIP-1α). Cytokine regulation of fibroblast growth and gene expression has been extensively examined *in vitro*. Although many of these studies have been performed using dermal fibroblasts, synovial cells, or other cell types, the underlying similarities between scleroderma, rheumatoid arthritis, and other diseases involving pulmonary fibrosis suggest that these experimental findings may be applicable to processes in the lung.

Cytokines are produced at the sites of fibrosis by inflammatory cells such as macrophages and eosinophils (see Chapter 2). Mesenchymal cells (i.e. myofibroblasts and epithelial cells) also produce a number of relevant cytokines. *In vitro* studies of these cytokines have revealed their potential roles in regulating the fibrotic process *in vivo* (25). They serve as chemoattractants, or are mitogenic for fibroblasts, stimulate extracellular matrix and α-smooth muscle gene expression, alter the contractile phenotype of fibroblasts, and regulate a number of diverse functions of lung inflammatory and epithelial cells. Although TGFβ is the most important cytokine in terms of stimulating the fibrotic response, it is more likely that a network of cytokines bring about the entire process. Accordingly, the outcome of the disease depends upon the positive and negative forces generated by the cytokines in this network (26).

TGFβ has multiple biologic effects, including strong induction of ECM deposition by simultaneously stimulating the production of matrix proteins, inhibition of proteases, and modulating the expression of matrix receptors on the cell surface. Overproduction of TGFβ has now been linked to tissue repair as well as to the pathogenesis of numerous experimental and human fibrotic disorders in many organs including the lung (27). Increased TGFβ has been correlated with the degree of fibrosis in dystrophic muscle (28), ALI (29), and experimental and human kidney disease (27). The therapeutic efficacy of neutralizing antibodies to this cytokine has been shown in fibrotic disease models involving kidney, lung, skin, joint, arterial wall, and brain. Up-regulation of TGFβ mRNA expression in rodent bleomycin-induced pulmonary fibrosis coincides with the period of increased lung ECM production (18, 20).

Platelets contain high concentrations of TGFβ and PDGF that are released into the tissue at the site of injury. Latent TGFβ bound to the ECM can also be activated after tissue injury. TGFβ is chemotactic for neutrophils, T-cells, monocytes, and fibroblasts (34-36). Activation of these cells at the site of injury causes aggregation and degranulation with release of mediators, including these potent fibrogenic cytokines. Monocytes and macrophages produce FGFs, TNFα, and IL-1, and fibroblasts increase their synthesis of ECM (34). TGFβ also induces infiltrating cells and resident cells to produce more TGFβ. This auto-induction amplifies the biological effects of TGFβ and may play a central role in chronic progressive fibrosis (37). In bleomycin-induced pulmonary fibrosis in rats, total lung TGFβ content was several times higher than in normal rats. The increased production of TGFβ preceded the synthesis of collagens, fibronectin, and proteoglycans (23). Using a combination of *in-situ* hybridization and immunochemistry, lung epithelial cells, macrophages, eosinophils, and myofibroblasts have been identified as important sources of TGFβ in human IPF and rodent bleomycin-induced pulmonary fibrosis (34-37). The principal source of TGFβ, however, appears to be the alveolar macrophage, although eosinophils, when present, also represent an additional important source of this cytokine. In human IPF, TGFβ expression is increased in the alveolar walls at sites of ECM accumulation (37). BAL-derived cells collected from patients with autoimmune diseases and lung fibrosis contain ten times more TGFβ than similar cells from normal subjects (38). TGFβ and collagen are increased in fibroproliferative disorders such as systemic sclerosis (39), keloids (40) and hypertrophic scars resulting from burns (41).

TNF-α alters fibroblast biology in many notable ways, not all of which are consistent with a fibrogenic role. It stimulates fibroblast growth at nanogram levels (42), but it has also been shown to inhibit collagen production (43-45). The latter changes are regulated at the transcriptional level, and have been observed in many cell types from fibroproliferative lesions, such as synovial cells (46), osteosarcoma cells (47) and osteoblast enriched bone cultures (48). TNFα also inhibits fibronectin synthesis (43), but increases collagenase and glycosaminoglycan production (49). TNFα appears to be a master switch, or regulator, since it is capable of inducing the expression of a myriad of additional mediators that affect fibroblast growth and differentiation. For instance, TNFα induces production of prostaglandin

E_2 (PGE_2), a potent inhibitor of collagen synthesis. TGFβ, IL-1β and IL-6 are induced by TNFα as well.

TNFα and IL-1β together play an important role in the initiation of pulmonary fibrotic responses in the architectural remodeling phase. Both cytokines are localized to alveolar macrophages and to type II pneumocytes of lungs undergoing acute pulmonary fibrosis (50). The mRNA levels of IL-1β and TNFα are increased in IPF and in asbestosis. *In vitro* studies have shown that procollagen types I and III, and fibronectin gene expression are unregulated in human diploid fibroblasts after a short-term exposure to recombinant IL-1β and TNFα (51). The existence of complex cytokine networks regulating inflammation and fibrosis and the human lung is suggested by such studies (52). These show that complex cytokine networks exist between inflammatory cells and fibroblasts, which are capable of regulating both the associated inflammation and fibrosis. IL-1β has been reported to induce fibroblast proliferation by some studies (53-55) and to inhibit proliferation in others (56, 57). An interesting study shows that while cardiac myocytes are induced to synthesize DNA by IL-1β, under the same conditions cardiac fibroblasts are inhibited (57). Another study shows that IL-1β and TNFα inhibit fibroblast proliferation both singly and when acting in concert with one another (56). However, proliferation was noted in the presence of both IL-1β and TGFβ (53). Of relevance to lung injury and fibrosis due to the pneumoconiosis, silica particles are found to induce monocytes to produce IL-1β-like factors that induced fibroblast proliferation (55). Thus the fibroblast proliferation response is influenced differently by IL-1β depending on the etiologic factors and experimental conditions.

Similarly, collagen synthesis has been shown to be enhanced or down regulated under different situations. For example, IL-1β and TNFα together are able to up regulate procollagens type-I and type-III, and fibronectin gene expression in human diploid fibroblasts (51). However, in rat cardiac fibroblasts, collagen synthesis decreased after exposure to IL-1β, TNFα, or interferon-γ (58). IL-1β has been shown to increase collagen synthesis in liver fibrosis and in endothelial cells (59, 60). Human lung fibroblasts are induced to synthesize hyaluronan and proteoglycans by IL-1β and TNFα (56). Metalloproteinases, procollagenase (matrix metalloproteinase-1) and prostromelysin (matrix metalloproteinase-3) are all induced in dermal fibroblasts exposed to IL-1β alone, or to both IL-1β and TGFβ (53).

Recent evidence indicates that IL-1β causes significant decreases of α-smooth muscle actin expression in vascular smooth muscle cells and inhibits contraction by activating nitric oxide production (61, 62). Other studies have demonstrated that TGFβ promotes myofibroblast differentiation and contraction of fibroblast-populated collagen gels (63), while IL-1β causes dose-dependent suppression of α-smooth muscle actin expression by rat lung fibroblasts and inhibits contractility of collagen gels populated by fibroblasts (64). Furthermore, IL-1β has been implicated in causing selective apoptosis of myofibroblasts while TGF-β was found to have a protective effect on these cells against this induction of apoptosis by IL-1β (65).

Chemokines represent a large family of chemotactic cytokines postulated to play major roles in the recruitment and activation of specific leukocyte populations found in lung injury, repair, and fibrosis (see Chapter 3). Among these, MIP-1α and MCP-1, belonging to the C-C chemokine family, have been found to be of potential importance in lung inflammation and fibrosis. MCP-1 appears to be important in the recruitment of macrophages to the lung interstitium. Endothelial cells produce MCP-1 at the blood-interstitial interface with the potential of direct influence on the development of interstitial fibrosis (66). Also, alveolar macrophages secrete cytokines that can induce MCP-1 production in alveolar epithelial cells (67). MIP-1α and MCP-1 both appear to contribute to the initiation and maintenance of bleomycin-induced pulmonary fibrosis (68, 69). In addition to its chemotactic properties, MCP-1 also stimulates fibroblast collagen expression via specific receptors (CC chemokine receptor-2 or CCR2) and by up regulating endogenous TGFβ expression (70). Thus, in addition to its pro-inflammatory chemotactic activity, MCP-1 has the potential of directly up regulating fibrosis at the level of fibroblast ECM production.

Cellular Response to Injury and Repair

A variety of cell types are found in the lung, but they can be broadly categorized as inflammatory cells, alveolar epithelial cells, endothelial cells, and mesenchymal cells. They have all been studied to varying extents in the context of injury and repair processes that occur during pulmonary fibrosis (71). Among the inflammatory cells, neutrophils are recruited in large numbers soon after injury in the alveolar space (17, 72). Although they play beneficial roles in terms of controlling invading foreign organisms and scavenging debris, these cells may also contribute to the destructive process due to secretion of proteases, oxidants, and hydrolases. The initial phase of recruitment of polymorphonuclear leukocytes (PMN), including neutrophils, occurs through the interaction between PMNs and the endothelial cells (see Chapters 2-4). This interaction is mediated by differential expression of cell surface adhesion molecules such as integrins, selectins, and immunoglobulin-related intercellular cell adhesion molecules (ICAMs). During early stages, the endothelial cells express higher levels of P-selectin, followed by cytokine-induced increases of E-selectin and ICAM-1 (73). The importance of these adhesion molecules in the development of pulmonary fibrosis has been demonstrated in animal models of bleomycin-induced lung injury (74, 75). Macrophages appear soon after the neutrophils in the alveolar space. They release products that may be proinflammatory (IL-1 and other interleukins, IFN-γ, TNF-α, MIP-1β), anti-inflammatory (IL-10 and TGF-β), mitogenic for mesenchymal cells (PDGF, FGF, IGF-1), or directly fibrogenic (TGFβ) (76).

In addition to neutrophils, BAL samples from some patients with IPF are rich in activated eosinophils (77). Pulmonary eosinophilia has also been noted in certain animal models of lung injury and fibrosis, however their recruitment into the lung occurs at a later stage than the neutrophils, and parallels that for monocytes, macrophages, and lymphocytes (36, 69). Eosinophils are capable of producing fibrogenic factors that induce fibroblast proliferation and activation (36, 69, 78, 79). From this standpoint, the role of the eosinophils is more direct *vis-à-vis* fibrogenesis than that of neutrophils.

The alveolar epithelium is a primary target of lung injury, especially with respect to exogenous insults arriving via the airway. The type I pneumocyte appears to be especially susceptible (2, 80). This cell is the initial vulnerable cell type and loss of type I pneumocytes is repaired by the type II pneumocytes, which are stimulated to proliferate, differentiate, and repopulate the denuded basement membranes.

Endothelial cells located between the vascular and the interstitial compartments of the lung can play active roles as well in the inflammatory and repair/fibrotic processes. Endothelial cells participate in the recruitment, activation, and transport of cells and mediators to the interstitium. Furthermore, they can secrete fibrogenic mediators and promote fibrosis directly by activation of mesenchymal cells such as the fibroblast (81-83).

Mesenchymal cells have been studied in great depth for their role in the repair process after injury. In addition to their well known structural role, they are now known to also participate actively in both the inflammatory and repair/fibrotic phases. They are the primary source for most of the ECM in both normal repair and the generation of fibrotic tissue resulting in the distorted lung architecture seen in end-stage lung disease. Their role during remodeling includes degradation of damaged (or even normal) matrix by secretion of matrix metalloproteinases, which may contribute to the distortion in lung architecture in fibrotic lesions.

There is abundant evidence, both from studies of wound healing and tissue fibrosis, that a distinct subpopulation of mesenchymal cells is especially responsible for this active fibrotic process characterized by the excessive production of ECM. It appears that an important feature of the mesenchymal cells during repair is the emergence *de novo* of these cells, which on the basis of its α-smooth muscle actin expression and contractile phenotype have been designated as myofibroblasts (84, 85). These cells appear to be activated, express high levels of TGFβ, and have been shown to be responsible for enhanced ECM deposition during fibrosis (22, 86-89). Through the production of other cytokines, myofibroblasts are also capable of regulating the inflammatory responses of adjacent epithelial and endothelial cells (90). In normal wound healing however, resolution and termination of the active (synthetic) phase is accompanied by disappearance of these cells, presumably by apoptosis. Failure or inhibition of this apoptotic mechanism could potentially account for persistence of the myofibroblast, and hence persistent and progressive disease as seen in IPF and other fatal chronic fibrotic diseases.

In addition to myofibroblasts, an additional activated fibroblast phenotype has recently been identified in recent studies of bleomycin-induced pulmonary fibrosis in animal models (91, 92). This phenotype is characterized by the induction of telomerase activity, and appears distinct from the myofibroblast by its relative lack of α-smooth muscle actin expression. Similar to myofibroblasts, these cells appear to arise *de novo* after lung injury, and gradually disappear as the active fibrotic phase wanes in this self-limiting model of pulmonary fibrosis (92). The exact role(s) of this particular subpopulation of fibroblasts remains to be investigated, but may be associated with increased proliferative capacity of lung fibroblasts isolated from lungs undergoing fibrosis.

Thus, a complex interactive process (summarized in Figure 1) between all these various cell types in the lung parenchyma appear to be important in a myriad of ways, but all of which culminate eventually in activating the mesenchymal elements to replace the normal tissue with connective tissue. Many of the findings are gleaned from *in vitro* studies, and the corresponding *in vivo* or pathophysiological significance remain to be confirmed. Nevertheless, these sorts of *in vitro* studies have provided important clues with regard to the potential *in vivo* roles, which can form the basis for future studies.

Summary and Conclusions

The process of lung parenchymal injury, remodeling, and repair/fibrosis originates in the alveolar wall, with either initial injury or damage to the alveolar epithelial cells and/or endothelial cells. This is likely to be due to the greater vulnerability of these cells because of their particular phenotype and anatomical localization. The injury initiates an inflammatory response, which is both beneficial in initiating the repair process, and detrimental due to its destructive and profibrotic potential. Interstitial fibrosis ensues when the fibroproliferative process, triggered by the associated inflammation, is sustained by persistent secretion of a myriad of cytokines from chronic inflammatory cells as well as other cells, including endothelial cells, epithelial cells, and myofibroblasts. A complex network of factors and cellular targets determine the final outcome of the disease.

Thanks to the studies made possible using animal models, various pathways have been investigated and elucidated (93).

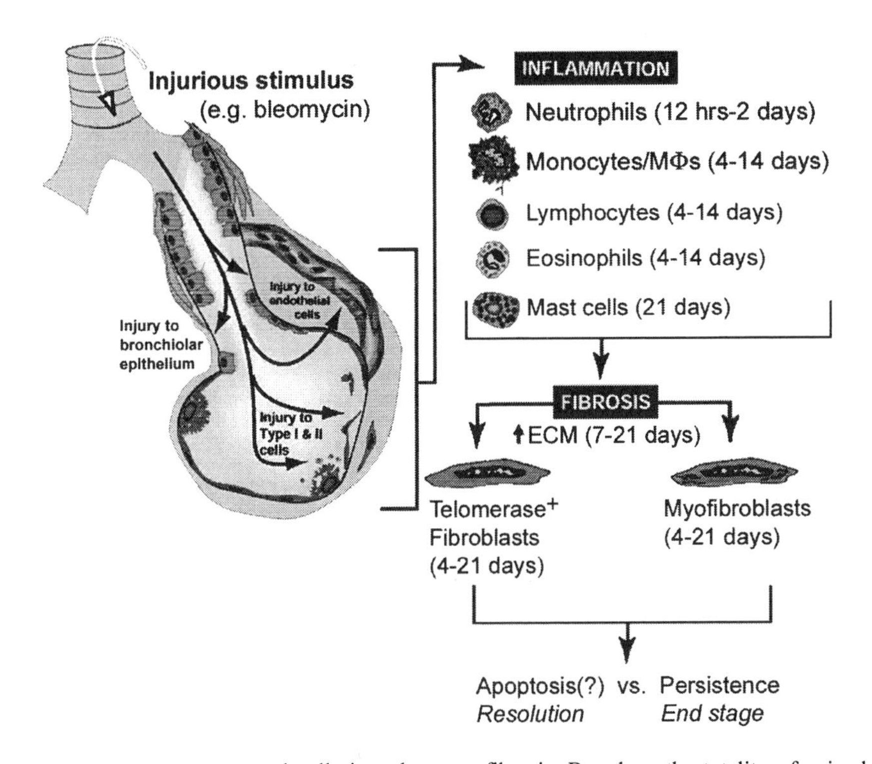

Figure 1. Key processes and cells in pulmonary fibrosis. Based on the totality of animal model and human studies, the association with inflammation is clearly established early on in lung injury, and considered among one of the most important initiators of the fibrotic process. The kinetics for inflammatory and mesenchymal cell recruitment (in parenthesis, as days after lung injury) is gleaned primarily from animal model studies. The trigger for the inflammation is presumably a response to the initial cellular/tissue injury as depicted by alveolar epithelial and endothelial injury. The inflammatory cells become an important source of mediators, including cytokines, which both amplify the inflammatory/immune response and initiate recruitment and activation of mesenchymal cells. Destructive potential of inflammatory cells could further contribute to tissue injury. Almost simultaneously the repair process under the control of cytokines/growth factors is initiated by proliferation of epithelial (e.g. type II pneumocyte), endothelial, and mesenchymal cells in an attempt to re-establish the normal alveolar architecture. However upon severe destruction with impairment of reepithelialization (and reendothelialization?), the mesenchymal elements predominate with abnormal replacement of the parenchyma with connective tissue due to the large numbers of activated fibroblasts in the form of telomerase expressing cells and myofibroblasts with their high levels of ECM and cytokine production. Cytokines (e.g. TGFβ) can function in a positive feedback loop to amplify and prolong the active fibrotic phase. Termination of this phase and successful repair with resolution occurs if these activated fibroblasts could be down regulated or removed (undergo apoptosis?). Persistent and chronic fibrosis culminating in end stage lung disease presumably can ensue if these activated fibroblast phenotypes persist.

REFERENCES

1. Terzhagi, M., Nettesheim, P., and Williams, M.L. (1978) Re-population of the denuded tracheal grafts with normal, pre-neoplastic and neoplastic epithelial cell populations. *Cancer Res* 38, 4546-4553
2. Adamson, I.Y.R., Young, L., and Bowden, H. (1988) Relationship of alveolar epithelial injury and repair to the induction of pulmonary fibrosis. *Am J Pathol* 130, 377-383.
3. Panos, R.J., Rubin, J.S., Csaky, K.G., Aaronson, S.A., and Mason, R.J. (1993) Keratinocyte growth factor and hepatocyte growth factor/ scatter factor or heparin binding growth factors for alveolar type II cells in fibroblast conditioned medium. *J Clin Invest* 92, 969-977
4. Everett , M.M., King, R.J., Jones, M.B., and Martin, H.M. (1990) Lung fibroblasts from animals breathing 100% oxygen produce growth factors for alveolar type II cells. *Am J Physiol Lung Cell Mol Physiol* 259, L247-L254
5. Tanswell, A.K. (1983) Cellular interactions in pulmonary oxygen toxicity in vitro: hyperoxic induction of fibroblast factors which alter growth and lipid metabolism in pulmonary epithelial cells. *Exp Lung Res* 5, 23-36
6. Young, L., and Adamson, I.Y.R. (1993) epithelial-fibroblast interactions in bleomycin - induced lung injury and repair. *Environ Health Perspect* 101, 56-61.
7. Polunovsky, V.A., Chen, B., Henke, C., Snover, D., Wendt, C., Ingbar, D.H., and Bitterman, P.B. (1993). Role of mesenchymal cell death in lung remodeling after injury. *J Clin Invest* 92, 388-397
8. Uhal, B.D., Joshi, I., True, A.L., Mundle, S., Raza, A., Pardo, A., and Selman, M. (1995) Fibroblasts isolated after fibrotic lung injury induce apoptosis of alveolar epithelial cells in vitro. *Am Physiol Lung Cell Mol Physiol* 269, L819-L828
9. Hagimoto, N., Kuwano, K., Nomoto, Y., Kunitake, R., and Hara, N. (1996) Apoptosis and expression of Fas/ Fas ligand mRNA in bleomycin induced pulmonary fibrosis in mice. *Am J Respir Cell Mol Biol* 16, 91-101
10. Kuwano, K., Hagimoto, N., Tanaka T., Kawasaki, M., Kunitake, R., Miyazaki, H., Kaneko, Y., Matsuba, T., Maeyama, T., and Hara, N. (2000) expression of apoptosis-regulatory genes in epithelial cells in pulmonary fibrosis in mice. *J Pathol* 190, 221-229
11. Crouch, E.C. (1990) Pathobiology of pulmonary fibrosis. *Am J Physiol* 259, L159-L184
12. Dubaybo, B.A., Rubeiz, G.J. and Fligiel, S.E.G. (1992) Dynamic changes in the functional characteristics of the interstitial fibroblast during lung repair. *Exp Lung Res* 18, 461-477
13. Pardo, A., Selman, M., Ramirez, R., Ramos, C., Montano, M., Stricklin, G. and Raghu, G. (1992) Production of collagenase and tissue inhibitor of metalloproteinases by fibroblasts derived from normal and fibrotic human lungs. *Chest* 102, 1085-89
14. Gadek, J.E., Kelmar, J.A., Fells, G.A., Weinberger, S.E., Horwitz, A.L., Reynolds, H.Y., Fulmer, J.D., and Crystal, R.G. (1979) Collagenase in the lower respiratory tract of patients with idiopathic pulmonary fibrosis. *N Eng J Med* 301, 737-742
15. Phan, S.H., Thrall, R.S., and Williams, C. (1981) Bleomycin induced pulmonary fibrosis effects of steroid on lung collagen metabolism. *Am Rev Respir Dis* 124, 484-434
16. Selman, M., Montano, M., Ramos, C., and Chapel, R..(1986) Concentration, biosynthesis and degradation of collagen in idiopathic pulmonary fibrosis. *Thorax* 41, 355-359
17. Phan , S.H. Diffuse interstitial fibrosis. In *Lung Cell Biology* Massaro D, ed. New York Marcel Dekker (1989) pp 907- 979
18. Hoyt, D.G., and Lazo, J.S. (1988) Alterations in pulmonary mRNA encoding pro-collagen, fibronectin and transforming growth factor-b precede bleomycin-induced pulmonary fibrosis in mice. *J Pharmacol Exp Ther* 246, 765-771
19. Piguet, P.F., Collartm M., Grau, G.E., Kapanci, Y., and Vassalli, P. (1987) Tumor necrosis factor/cachectin play a key role in bleomycin-induced pneumopathy and fibrosis. *J Exp Med* 170,655-663

20. Raghow, R., Irish, P., and Kang, A.H. (1989) Coordinate regulation of transforming growth factor-b expression and cell proliferation in hamster lungs undergoing bleomycin-induced pulmonary fibrosis. *J Clin Invest* 84, 1836-1842
21. Zhang, K., Gharaee-Kermani, M., McGarry, B., and Phan, S.H. (1994) *In situ* hybridization analysis of rat lung α_1(I) and α_2(I) collagen gene expression in pulmonary fibrosis induced by endotracheal bleomycin injection. *Lab Invest* 70, 192-202
22. Zhang, K., Rekhter, M.D., Gordon, D,, and Phan, S.H. (1994) Myofibroblasts and their role in lung collagen gene expression during pulmonary fibrosis. A combined immunochemical and *in situ* hybridization study. *Am J Pathol* 145, 114-125
23. Westergren-Thorsson, G., Hernnas, J., Sarnstrand, B., Oldberg, A., Heinegard, D., and Malmstrom, A. (1993) Altered expression of small proteoglycans, collagen and transforming growth factor-beta 1 in developing bleomycin-induced pulmonary fibrosis in rats. *J Clin Invest* 92, 632-637
24. Venkatesan, N., Takae, E., Roughley, P.J., and Ludwid, M. S. (2000) Alterations in large and small proteoglycans in bleomycin-induced pulmonary fibrosis in rats. *Am J Respir Crit Care Med* 161, 2066-2073
25. Remick, D.G., and DeForge, L.E. (1995) Cytokines and Pulmonary fibrosis. In Pulmonary Fibrosis. *Lung Biology in Health and Disease*, Phan, S.H. and Thrall, R.S., ed. 80, 599-626
26. Zhang, K., and Phan, S.H. (1996) Cytokines and pulmonary fibrosis. *Biol Signals* 5, 232-239
27. Border, W.A. and Noble, N.A. (1994) Transforming growth factor-beta in tissue fibrosis. *N Eng J Med* 331, 1286-1292
28. Bernasconi, P., Torchiana, E., Confalconieri, P., Brugnoni, R., Barresi, R., Mora, M., Cornelio, F., Morandi, L., and Mantegazza, R. (1995) Expression of TGF-beta 1 in dystrophic patient muscles correlates with fibrosis: pathogenic role of a fibrogenic cytokine. *J Clin Invest* 96, 1137-1144
29. Shenkar ,R., Coulson, W.F., and Abraham, E. (1994) Anti-transforming growth factor beta monoclonal antibodies prevent lung injury in hemorrhaged mice. *Am J Respir Cell Mol Biol* 11, 351-357.
30. Roberts, A.B., Sporn, M.B., Assosian, R.K., et al. (1990) Transforming growth factor type beta a multifunctional effector of both soft and hard tissue regeneration. In; Westermark B, Betsholtz C, Hokfelt B eds. Growth factors in health and disease: basic and clinical aspects. *Amsterdam: Excerpta Medica* 89-101
31. Wahl, S.M., Hunt, D.A., Wakefield, L.M., McCartney-Francis, N., Wahl, L.M., Roberts, A.B., and Sporn, M.B. (1987) Transforming growth factor type beta induces monocyte chemotaxis and growth factor production. *Proc Natl Acad Sci USA* 84, 5788-5792
32. Postlewaite, A.E., Keski-Oja, J., Moses, H.L., and Kang, A.H. (1987) Stimulation of chemotactic migration of human fibroblasts by transforming growth factor beta. *J Exp Med* 165, 251-256
33. Kim, S.-J., Angel, P., Lafyatis, R., Hattori, K., Kim, K.Y., Sporn, M.B., Karin, M., and Roberts, A.B. (1990) Autoinduction of transforming growth factor beta I is mediated by AP-1 complex. *Mol Cell Biol* 10, 1492-1497
34. Khalil, N., Bereznay, O., Sporn, M., and Greenberg, A.H. (1989) Macrophage production of transforming growth factor β and fibroblast collagen synthesis in chronic pulmonary fibrosis. *J Exp Med* 170, 727-737
35. Khalil, N., O'Connor, R.N., Unruh, H.W., Warren, P.W., Flander, K.C., Kemp, A., Bereznay, O.H., and Greenberg, A.H. (1991) Increased production and immunohistochemical localization of transforming growth factor β in idiopathic pulmonary fibrosis. *Am J Respir Cell Mol* 5, 155-162
36. Zhang, K., Flanders, K.C., and Phan, S.H. (1995) Cellular localization of transforming growth factor β expression in bleomycin-induced pulmonary fibrosis. *Am J Pathol* 147, 352-361

37. Broekelmann, T.J., Limper, A.H., Colby, T.V., and McDonald, J.A. (1991) Transforming growth factor beta is present at sites of extracellular matrix gene expression in human pulmonary fibrosis. *Proc Natl Acad Sci* 88, 6642-6646
38. Deguchi, Y. (1992) Spontaneous increase of transforming growth factor beta production by bronchoalveolar mononuclear cells of patients with systemic autoimmune diseases affecting the lung. *Ann Rheum Dis* 51, 362-365
39. Kulozik, M., Hogg, A., Lankat-Buttgereit, B., and Krieg, T. (1990) Co-localization of transforming growth factor-beta, type I and type III procollagen mRNA in tissue sections of patients with systemic sclerosis. *J Clin Invest* 86, 917-922
40. Peltonen, J., Hsiao, L.L., Jaakkola, S., Sollberg, S., Aumailley, M., Timpl, R., Chu, M.L., and Uitto, J. (1991) Activation of collagen gene expression in keloids: co-localization of type I and VI collagen and transforming growth factor-beta 1 mRNA. *J Dermatol* 97, 240-248
41. Ghahary, A., Shen, Y.J., Scott, P.G., Gong, Y., and Tredget, E.E. (1993) Enhanced expression of mRNA for transforming growth factor-beta, type I and type III procollagen in human post burn hypertrophic scar tissues. *J Lab Clin Med* 122, 465-473
42. Vilcek, J., Palombella, V.J., Henriksen-DeStefano, D., Swenson, C., Feinmann, R., Hirai, M., and Tsujimoto, M. (1986) Fibroblast growth enhancing activity of tumor necrosis factor and its relationship to other polypeptide growth factors. *J Exp Med* 63, 632-643
43. Mauviel A, Daireaux M, Redini F, Galera P, Loyau G, and Pujol JP. (1988) Tumor necrosis factor inhibits collagen and fibronectin synthesis in human dermal fibroblasts. FEBS Lett 236, 47-52
44. Mauviel, A., Heino, J., Kahari, V.M., Hartmann, D.J., Loyau, G., Pujol, J.P., and Vuorio, E. (1991) Comparative effects of interleukin-1 and tumor necrosis factor-alpha on collagen production and corresponding procollagen mRNA levels in dermal fibroblasts. J Invest Dermatol 96, 243-249
45. Solis Heruzzo, JA., Brenner, D.A., and Chojkier, M. (1988) Tumor necrosis factor alpha inhibits collagen gene transcription and collagen synthesis in cultured human fibroblasts. J Biol Chem 263, 5841-5845
46. Diareaux, M., Redini, F., Loyau, G., and Pujol, J.P. (1990) effects of associated cytokines (IL-1, TNF-alpha, IFN-gamma and TGF-beta) on collagen and glycosaminoglycan production by cultured human synovial cells. Int J Tissue React 12, 21-31
47. Nanes, M.S., Mckoy, W.M., and Marx, S.J., (1989) Inhibitory effects of tumor necrosis factor-alpha and interferon-gamma on deoxyribonucleic acid and collagen synthesis by rat osteosarcoma cells (ROS 17/2.8). Endocrinology 124, 339-345
48. Centrella, M., McCarthy, T..L, and Canalis, E., (1988) Tumor necrosis factor-alpha inhibits collagen synthesis and alkaline phosphatase activity independently of its effect on deoxyribonucleic acid synthesis in osteoblast-enriched bone cell cultures. Endocrinology 123, 1442-1448
49. Duncan, M.R., and Berman, B. (1989) Differential regulation of collagen, glycoseaminoglycan, fibronectin, and collagenase activity production in cultured human adult dermal fibroblasts by interleukin 1- alpha and beta and tumor necrosis factor alpha and beta. J Invest Dermatol 92, 699-706
50. Pan, L.H., Otani, H., Yamauchi, K., and Nagura, H. (1996) Co-expression of TNF alpha and IL-1 beta in human acute pulmonary fibrotic diseases an immunohistochemical analysis. Pathology International 46, 91-99
51. Zhang, Y., Lee, T.C., Guillemin, B., Yu, M.C., Rom, W.N. (1993) Enhanced IL-1 beta and tumor necrosis factor-alpha release and messenger RNA expression in macrophages from idiopathic pulmonary fibrosis or after asbestos exposure. J Immunol 150, 4188-4196
52. Elias, J.A., Freundlich, B., Kern, J.A., and Rosenbloom, J. (1990) Cytokine networks in the regulation of inflammation and fibrosis in the lung. Chest 97, 1439-1445
53. Unemori, E.N., Ehsani, N.. Wang, M., Lee, S., McGuire, J., and Amento, E.P. (1994) Interleukin-1 and transforming growth factor-alpha: synergistic stimulation of

metalloproteinases, PGE2, and proliferation in human fibroblasts. Exp Cell Res 210, 166-171
54. Dukovich, M., Severin, J.M., White, S,J., Yamazaki, S., and Mizel, S.B. (1986) Stimulation of fibroblast proliferation and prostaglandin production by purified recombinant murine interleukin 1. Clin Immunol Immunopathol 38, 381-389
55. Schmidt, J.A., Oliver, C.N., Lepe-Zuniga, J.L., Green, I., and Gery, I. (1984) Silica-stimulated monocytes release fibroblast proliferation factors identical to interleukin 1. A potential role for interleukin 1 in the pathogenesis of silicosis. *J Clin Invest* 73, 1462-1472
56. Tufvesson, E., and Westergren-Thorsson, G. (2000) Alteration of proteoglycan synthesis in human lung fibroblasts induced by interleukin-1beta and tumor necrosis factor-alpha. *J Cell Biochem* 77, 298-309
57. Palmer, J.N., Hartogensis, W.E., Patten, M.. Fortuin, F,D., and Long C.S. (1995) Interleukin-1 beta induces cardiac myocyte growth but inhibits cardiac fibroblast proliferation in culture. *J Clin Invest* 95,2555-2564
58. Siwik, D.A., Chang, D.L., and Colucci, W.S. (2000) Interleukin-1beta and tumor necrosis factor-alpha decrease collagen synthesis and increase matrix metalloproteinase activity in cardiac fibroblasts in vitro. *Circ Res* 86, 1259-1265
59. Okada, H., Woodcock-Mitchell , J., Mitchell, J., Sakamoto, T., Marutsuka, K., Sobel, B.E., and Fujii, S. (1998) Induction of plasminogen activator inhibitor type 1 and type 1 collagen expression in rat cardiac microvascular endothelial cells by interleukin-1 and its dependence on oxygen-centered free radicals. *Circulation* 97, 2175-2182
60. Irwin, C.R., Myrillas, T., Smyth, M., Doogan, J., Rice, C., and Schor, S.L. (1998) Regulation of fibroblast-induced collagen gel contraction by interleukin-1 beta. *J Oral Pathol Med* 27, 255-259
61. Trinkle, L.A., Beasley, D., and Moreland, R.S. (1992) Interleukin 1β alters actin expression and inhibits contraction of rat thoracic aorta. *Am J Physiol* 262, C828-833
62. Beasley, D., Cohen, R.A., and Levinsly, N.G. (1989) Interleukin 1 inhibits contraction of smooth muscle. *J Clin Invest* 83, 331-335
63. Zhang, H.-Y., Gharaee-Kermani, M., Zhang, K., and Phan, S.H. (1996) lung fibroblast α-smooth muscle actin expression and contractile phenotype in bleomycin-induced pulmonary fibrosis. *Am J Pathol* 148, 527-537
64. Zhang, H.-Y., Gharaee-Kermani, M., and Phan, S.H. (1997) regulation of lug fibroblast a-smooth muscle actin expression, contractile phenotype and apoptosis by IL-1β. *J Immunol* 158, 1392-1399
65. Zhang, H.-Y. and Phan, S.H. (1999) Inhibition of myofibroblast apoptosis by transforming growth factor β_1. *Am J Respir Cell Mol Biol* 21, 658-665
66. Smith, R.E., Strieter, R.M., Zhang, K., Phan, S.H., Standiford, T.J., Lukacs, N.W., and Kunkel, S.L. (1995) A role for C-C chemokines in fibrotic lung disease. *J Leukoc Biol* 57, 782-787
67. Strieter, R.M., Wiggins, R., Phan, S.H., Wharram, B.L., Showell, H.J., Remick, D.G., Chensue, S.W., and Kunkel, S.L. (1989) Monocyte chemotactic protein gene expression by cytokine-treated human fibroblasts and endothelial cells. *Biochem Biophys Res Commun* 162, 694-700
68. Standiford, T.J., Kunkel, S.L., Phan, S.H., Rollins, B.J., and Streiter, R.M. (1991) Alveolar macrophage-derived cytokines induce monocyte chemoattractant protein -1 expression from human pulmonary type II -like epithelial cells. *J Biol Chem* 266, 9912-9918
69. Zhang, K., Gharaee-Kermani, M., Jones, M.L., Warren, J.S., and Phan, S.H. (1994) Lung monocyte chemoattractant protein-1 gene expression in bleomycin-induced pulmonary fibrosis. *J Immunol* 153, 10, 4733-4741
70. Gharaee-Kermani, M., Denholm, E.M., and Phan, S.H. (1996) Costimulation of fibroblast collagen and transforming growth factor β_1 gene expression by monocyte chemoattractant protein-1 via specific receptors. *J Biol Chem* 271, 17779-17784

71. Strieter, R.M., Phan, S.H., and Ward, P. (1998) Inflammation, injury and repair. In: *Textbook of Respiratory Medicine*, IInd edition Volume 1: pp 469-497. Eds. Murray JF and Nadel JA
72. Brieland, J.K., and Fantone, J.C. (1995) Neutrophils and pulmonary fibrosis In *Pulmonary Fibrosis: Lung Biology in Health and Disease*, Phan, S.H., and Thrall, R.S., ed. Vol.80, pp 383-404
73. Stoolman, L.M. (1993) Adhesion molecules involved in leukocyte recruitment and lymphocyte recirculation. *Chest* 103, 79S-86S
74. Takahashi, A.A., Nose, M., Araki, K., Araki, M., Takahashi, T., Hirose, M., Kawashima, H., Miyasaka, M., and Kudoh, S. (2000) Role of E-selectin in bleomycin-induced lung fibrosis in mice. *Thorax* 55, 147-152
75. Sato, N., Suzuki, Y., Nishio, K., Naoki, K., Takeshita, K., Kudo, H., Miyao, N., Tsumura, H., Serizawa, H., Suematsu, M., and Yamaguchi, K. (2000) Roles of ICAM-1 for abnormal leukocyte recruitment in the microcirculation of bleomycin-induced fibrotic lung injury *Am J Respir Crit Care Med* 161, 1681-1688
76. Shaw, R.J. and Kelley, J. (1995) Macrophages/Monocytes. *Lung Biology in Health and Disease* Phan, S.H. and Thral, R.S., ed. Vol.80, pp 405-444
77. Hallgren, R., Bjermer, L., Lundgren, R., and Venge, P. (1989) The eosinophil component of the alveolitis in idiopathic pulmonary fibrosis. *Am Rev Respir Dis* 139, 373-377
78. Shock, A., Rabe, K.F., Dent, G., Chambers, R.C., Gray, A.J., Chung, K.F., Barnes, P.J., and Laurent, G.J. (1991) Eosinophils adhere to and stimulate replication of lung fibroblasts "in vitro". *Clin Exp Immunol* 86, 185-190
79. Ruoss, S.J. and Caughey, G.H. (1995) Mast cells, Basophils and eosinophils in the evolution of pulmonary fibrosis In *Pulmonary Fibrosis: Lung Biology in Health and Disease,* Phan, S.H. and Thrall, R.S., ed. Vol.80, pp 445-480
80. Simon, H.R. (1995) Alveolar epithelial cells in pulmonary fibrosis. In *Pulmonary Fibrosis: Lung Biology in Health and Disease,* Phan SH and Thrall RS, ed. Vol.80, pp 511-540
81. Phan, S.H., Gharaee-Kermani, M., McGarry, B., Kunkel, S.L., and Wolber, F.W. (1992) Regulation of rat pulmonary artery endothelial cell transforming growth factor beta production by IL-1beta and tumor necrosis factor-alpha. *J Immunol* 149, 103-106
82. Phan, S.H., Gharaee-Kermani, M., Wolber, F., and Ryan, U.S. (1991) Stimulation of rat endothelial cell transforming growth factor-beta production by bleomycin. *J Clin Invest* 87, 148-154
83. Phan, S.H. (1995) Endothelial cells in pulmonary fibrosis. In *Pulmonary Fibrosis: Lung Biology in Health and Disease*, Phan, S.H. and Thrall, R.S.,ed. Vol.80: pp481-510
84. Kuhn, C., and McDonald, J.A. (1991) The roles of the myofibroblast in idiopathic pulmonary fibrosis. Ultrastructural and immunohistochemical features of sites of active extracellular matrix synthesis. *Am J Pathol* 138, 1257-1265
85. Phan, S.H. (1996) Role of myofibroblast in pulmonary fibrosis. *Kidney Int Suppl* 54, S46-48
86. Darby, I., Skalli, O., and Gabbiani, G. (1990) α-smooth muscle actin is transiently expressed by myofibroblasts during experimental wound healing. *Lab Invest* 63, 21-29
87. Mitchell, J.J., Woodcock-Mitchell, J.L., Perry, L., Zhao, J., Low, R.B., Baldor, L., and Absher, P.M. (1993) *In vitro* expression of α–smooth muscle actin isoform by rat lung mesenchymal cells: regulation by culture condition and transformıng growth factor-β. *Am J Respir Cell Mol Biol* 9, 10-18
88. Vyalov, S.L., Gabbiani, G., and Kapanci, Y. (1993) Rat alveolar myofibroblasts acquire α-smooth muscle actin expression during bleomycin -induced pulmonary fibrosis. *Am J Pathol* 143, 1754-1765
89. Desmouliere, A., Geinoz, A., Gabbiani, F. and Gabbiani, G. (1993) Transforming growth factor-β_1 induces α-smooth muscle actin expression in granulation tissue myofibroblasts and in quiescent and growing cultures fibroblasts. *J Cell Biol* 122, 103-111

90. Phan, S.H., Zhang, K., Zhang, H.Y., and Gharaee-Kermani, M. (1999) The myofibroblast as an inflammatory cell in pulmonary fibrosis. *Current Topics in Pathology* 93, 173-182
91. Kim, J.K., Lim, Y., Kim, K.A., Seo, M.S., Kim, J.D., Lee, K.H., and Park, C.Y. (2000) Activation of telomerase by silica in rat lung. *Toxicol Letts* 111, 263-270, 2000
92. Nozaki, Y., Liu, T., Hatano, K., Gharaee-Kermani, M., and Phan, S.H. (2000) Induction of telomerase activity in fibroblasts from bleomycin-injured lungs. *Am J Respir Cell Mol Biol* 23, 460-465
93. Allen, J. and Cooper, D. (2000) Pulmonary fibrosis pathways are slowly coming to light. *Am J Respir Cell Mol Biol* 22, 520-523

Chapter 15

APOPTOSIS IN ACUTE LUNG INJURY

Karine Faure, Benoit Guery, and Jean Francois Pittet
Departments of Anesthesia and Surgery, University of California San Francisco, San Francisco, CA, and Laboratoire de Recherche en Pathologie Infectieuse EA 2689, University of Lille II, Lille, France

INTRODUCTION

In 1972, Kerr and Currie described a new type of cell death characterized by morphological changes distinct from the features observed during necrosis (1). The term of "apoptosis" was adopted to describe this highly conserved genetic program leading to regulated cellular self-destruction. Subsequent investigations showed that this programmed cell death is crucial during fetal development and critical for controlling harmful mechanisms triggered by environmental stresses. Recent literature has defined new roles for apoptosis in the normal and injured lung. For example, apoptosis plays an important role in postnatal lung development (2), and it is evident that resolution of lung inflammation after bacterial infection occurs by neutrophil apoptosis (3). While there is an increasing body of evidence supporting a role for apoptosis in the remodeling of the lung after acute lung injury (ALI) (3, 4), the factors that regulate normal and abnormal apoptotic cell death during ALI remain to be fully elucidated.

In this chapter, we will first summarize the process of apoptosis and the relevant cellular signals; then we will briefly describe the current methods that are used to detect the presence of apoptosis in the lung and in lung cells. Finally, we will review some of the more exciting data regarding the role of apoptosis during the acute and repair phases of ALI.

Morphology

Apoptosis is commonly divided into three phases: 1) initiation phase defined by the activation of death receptors and/or change in the permeability of the outer mitochondrial membrane in response to external or internal stimuli; 2) an effector phase characterized by the activation of caspases, a proteolytic system that is specifically activated in apoptotic cells; and 3) a degradation phase resulting in the classical morphological appearance of apoptotic cells.

Apoptosis is characterized by a typical set of morphological changes. The nucleus becomes pyknotic and basophilic, and the chromatin marginates to the periphery forming a crescent halo. The internucleosomic DNA is progressively fragmented by endonucleases. The cytoskeleton is reorganized and the cytoplasm is segmented, while the integrity of intracellular organelles is initially maintained. Adherent cells lose their attachment and changes in the plasma membrane are characterized by a loss of the membrane asymmetry and exposure of phosphatidylserine. Subsequently, nuclear fragments and cytoplasmic blebs appear on the cell surface, causing cell rupture with the formation of a number of membrane bounded apoptotic bodies. These apoptotic bodies are taken up (i.e. phagocytosis) by neighboring cells. Apart from preventing extravasation of intracellular components into the extracellular space, phagocytosis of apoptotic bodies also appears to modulate phagocyte function (5).

In contrast to apoptosis, necrosis is described as an unregulated event associated with random DNA fragmentation and swelling of cellular components leading to cell lysis and subsequent release of potentially toxic mediators into the surrounding microenvironment (Table 1). Recently, the concept of "necrapoptosis" has been introduced to describe a mechanism of cell death that begins with a common stress or death signal. Necrapoptosis progresses by shared pathways, but culminates in either cell necrosis or apoptosis depending on factors such as the availability of ATP to the cell. If cellular ATP levels fall profoundly, necrotic cell death occurs. If cellular ATP levels are sufficiently maintained, apoptotic cell death can be triggered (6).

Table 1. Differences in cell morphology between apoptosis and necrosis

Apoptosis	Necrosis
Physiological or pathological stimulus	Pathological stimulus
Energy requirement	No energy requirement
Regulated	Not regulated
Cell shrinkage	Cell swelling
Condensation of the chromatin	Flocculation of the chromatin
Internucleosomic DNA fragmentation	Random DNA breakdown
No decomposition of the organelles	Decomposition of the organelles
Plasma membrane rearrangement	Plasma membrane rupture
Apoptotic bodies	Cytolysis and excessive inflammation
Phagocytosis	Extracellular spilling of cell components

Apoptotic Cell Signaling

Most of the morphological changes characteristic of apoptotic cells are caused by a set of cysteine proteases that selectively cleave substrates at the level of an aspartate residue. Caspases (Cysteine ASP Aspartate Enzyme) are thought to be the central executioners of the apoptotic process because the elimination of caspase activity will slow down or prevent apoptosis. There are two major pathways that trigger the caspase proteolytic activity (Figure 1). One pathway, the death receptor pathway, is characterized by the activation of specialized cell surface receptors belonging to the tumor necrosis factor (TNF) receptor (R) gene superfamily (or death receptors). The second pathway, the mitochondrial pathway, is triggered either by extracellular signals such as reactive nitrogen species, reactive oxygen species and cytokines, or internal signals, such as DNA damage (e.g. after exposure to ionizing radiation), that result in the release of cytochrome c across the outer membrane of mitochondria into the cytosol. Both pathways converge at the level of the caspase-3, an effector caspase, that once activated, causes proteolysis of key proteins resulting in cell death.

To date, five death receptors have been identified (TNFR1, CD95, DR3, DR4, and DR5) that can activate death caspases within seconds causing an apoptotic demise of the cell within hours. These death receptors are defined by cysteine-rich extracellular domains and a homologous cytoplasmic sequence named death domain (DD). CD95 (or Fas, or Apo1), and TNFR1 (or p55, or CD120a) are well characterized. The death ligands, CD95L or Fas L (recognized by Fas or Apo1) and TNFα (recognized by TNFR1), bind to and cause trimerization of its specific death receptor. After binding to its ligand, the death receptor rapidly initiates cell apoptosis through the direct death effector domain (DED)-induced activation of caspase-8. In turn, caspase-8 activates caspase-3 (7). Unlike CD95L, TNFα rarely triggers apoptosis unless protein synthesis is blocked, which suggests the existence of cellular factors that can suppress the apoptotic stimulus generated by TNFα. Expression of these suppressive proteins is, in large part, controlled by the transcription factors NF-κB and AP-1 (see Chapter 1), as inhibition of either pathway sensitizes cell to apoptosis induction by TNFα (8). A similar mechanism has been described for the death receptor DR3 (7).

Apoptosis may also be activated through mitochondria-dependent mechanisms. Multiple stimuli, (e.g. Bax, oxidants, calcium overload, active caspases, and ceramide) can trigger mitochondria to release caspase activating proteins, including cytochrome c, apoptosis inducible factor (AIF), and intramitochondrial caspases. Two general mechanisms have been proposed to explain the release of caspase activating proteins from mitochondria. One mechanism involves osmotic disequilibrium leading to an expansion of the matrix space, organelle swelling, and subsequent rupture of the outer mitochondrial membrane. The second mechanism is characterized by the opening of channels in the outer mitochondrial membrane, thus releasing cytochrome c into the cytosol. In presence of ATP, cytochrome c complexes and activates Apaf-1 (cytoplasmic adaptor protein), which binds and activates procaspase-9, resulting in the formation of a complex named apoptosome. Subsequently, activated caspase-9 activates caspase-3 (9).

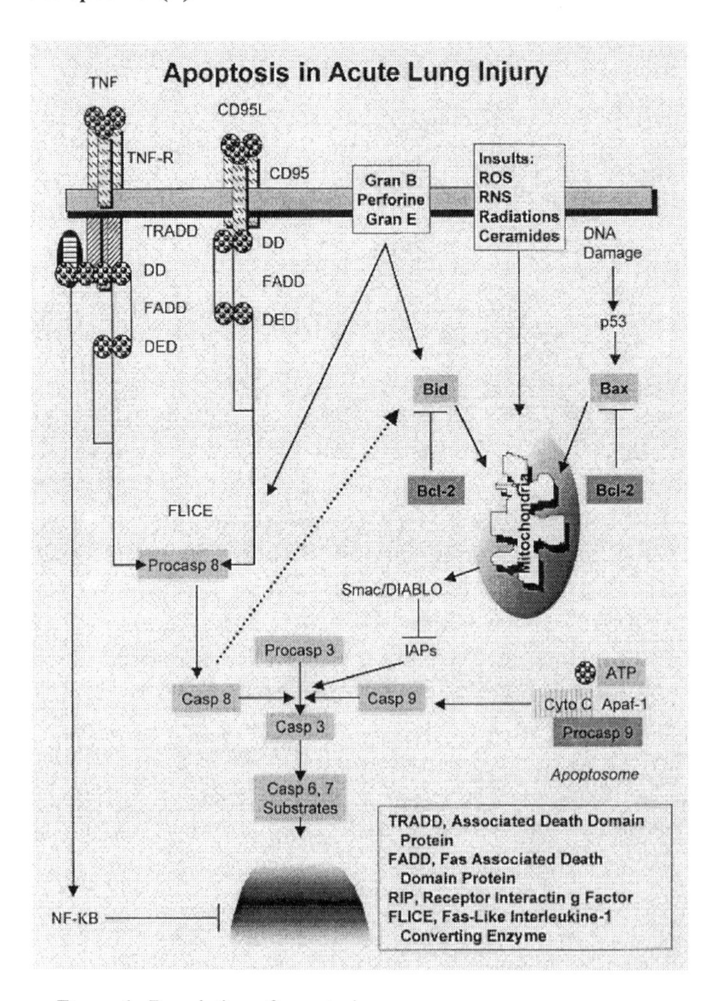

Figure 1. Regulation of apoptosis

Activation of mitochondria is modulated by the Bcl-2 protein family (pro-oncogene B Cell Lymphoma 2), a set of apoptotic regulators. To date, fifteen members have been identified in mammalian cells, and divided into three subfamilies (Table 2): the Bcl-2 subfamily possesses anti-apoptotic activity, whereas the Bax and BH3 subfamilies promote cell death. All proteins contain at least one of four conserved motifs known as Bcl-2 homology domains (BH). Pro- and anti-apoptotic proteins can hetero-dimerize to modulate their respective activities, suggesting that their relative concentration may act as a rheostat for the activation of programmed cell death (10). Unlike Bcl-2, which stays attached to the inner membrane of the mitochondria, proteins of the Bax and BH3 subfamilies can shuttle between the cytosol and mitochondria and thus modulate apoptosis. At the surface of mitochondria, they compete to regulate cytochrome c release by mechanisms that are still debated. Three basic models are proposed: 1) Bcl-2 members form channels in the outer mitochondrial membrane that facilitate protein transport; 2) Bcl-2 members interact with other proteins (VDAC) to form channels in the outer mitochondrial membrane; and/or 3) Bcl-2 members induce rupture of the outer mitochondrial membrane.

Table 2. Classification of the Bcl-2 protein family.

Proapoptotic members		Antiapoptotic members
Bax subfamily (BH1, 2, 3)	BH3 sub-family	BH1, 2
Bak	Bid	Bcl-2
Bax	Bik	Bcl-w
Bok	Bim	Bcl-Xl
	Hrk	Mcl-1
	Blk	A1
	Bad	
	BNIP3	

Another level of the regulation of the apoptotic cascade exists at the level of caspase-3 activation. Caspase 3 activity is antagonized by the IAP proteins, which in turn are antagonized by the Smac/DIABLO proteins, protein agonists for apoptosis that are released from the mitochondria into cytosol (9). Moreover, cross talk and integration between death receptor and mitochondrial pathways are provided by Bid protein. Caspase-8-mediated cleavage of Bid, a proapoptotic member of the Bcl-2 member family, promotes its translocation to mitochondria and results in the release of cytochrome c into the cytoplasm (9).

Because caspase proteins are all expressed as inert proenzymes, activation involves proteolytic processing between three domains, the N-terminal prodomain (ProD), the large subunit (p20), and the small subunit (p10) (Figure 2). The mature enzyme is a heterotetramer composed of two p20/p10 heterodimers and two catalytic sites included into the small subunit. Because distinct death signals result in the same manifestation of apoptosis, caspases have been divided in two groups describing a proteolytic cascade controlled by an upstream caspase (Table 3): effector caspases are activated by different initiator caspases, each of which is activated by a set of pro-apoptotic signals, through regulated protein-protein interactions (apoptosome for caspase-9 and DED for caspase-8), which are not completely elucidated (9, 11). Activation of effector caspases results in cell death via a multitude of subprograms, involving the cleavage of more than 100 substrate proteins (Table 4).

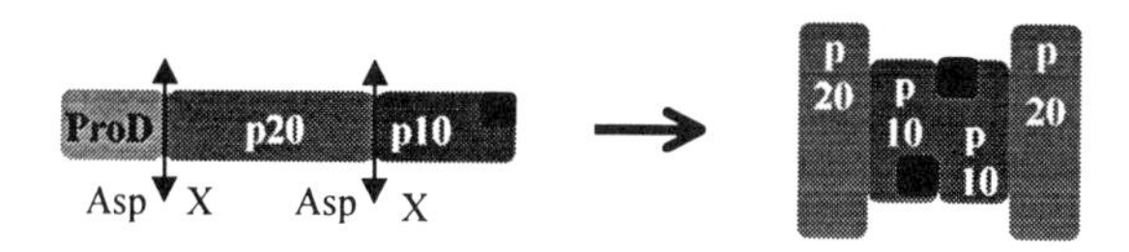

Figure 2. Structure of caspase proteins and mechanism of caspase activation. The proenzyme is cleaved at the level of aspartate residues (Asp-X) between the prodomain (ProD), the large subunit (p20) and the small subunit (p10). The active enzyme is a heterotetramer composed of two large subunits and 2 small subunits, each containing one cysteine catalytic site.

Table 3. Classification of caspase proteins.

Caspase	Proposed role in the cascade
8	Initiator via the death-receptor pathway (substrate: caspases 3 and 7)
9	Initiator via the mitochondrial pathway (substrate: caspases 3 and 6)
3, 6, 7	Effector
2	Initiator or effector?
10	Initiator or effector? Initiator via DR4 (substrate: caspase 7)
1	Major role in inflammation. Role in apoptosis?
4, 5, 11, 12, 13	Role in apoptosis?

Table 4. Substrate proteins for effector caspases.

Cytoskeletal and structural proteins (lamins, actin)
Cell cycle and replication (NDML: p53 stabilizing protein)
Transcription and translocation (IκB: inhibitory subunit of nuclear factor κB)
DNA cleavage and repair (PARP: poly ADP ribose polymerase, DNA topoisomerase)
Signal transduction (Protein Kinase, Procaspase, Bcl-2 protein family)
Cytokine precursors

Assessment of Apoptosis

It is important not only to identify and but also to quantify apoptosis in cells and tissues. The kinetics of apoptosis are highly dependent on the types of cell that becomes apoptotic and on the insults that initiate this process, and may therefore have important implications for the quantification of programmed cell death.

Although the morphologic hallmarks of cell suicide are accessible to light and electron microscopy, DNA fragmentation in apoptosis is regarded as a key parameter of this process. DNA fragmentation can be assessed by conventional agarose gel electrophoresis (CAGE), in which constant electric field at low voltage is used to resolve DNA fragments purified from apoptotic cells. The DNA ladder, however, does not reflect the full pattern of DNA fragmentation occurring during apoptosis, since CAGE cannot resolve fragments greater than approximately 30 kb (12). DNA fragmentation in cells and tissues can also be detected by using specific enzymes that add labeled nucleotides to the DNA ends. This labeling process is termed TUNEL (terminal deoxynucleotide transferase mediated UTP nick endlabeling) and is considered to be more sensitive than ISEL (*In situ* end-labeling techniques using Klenow fragment of DNA polymerase). A variety of labels have been used, but non-radioactive nucleotides (fluorescence allowing detection *in situ* by flow cytometry or by immunochemistry) appear to be superior and easier of use (13). The TUNEL technique, however, is prone to false positive or negative findings. This problem has been explained by the dependence of the staining kinetics, the reagent concentration, fixation of tissues, and by non-specific staining of active RNA synthesis and DNA damage in necrotic cells. Therefore, the TUNEL technique should be carefully standardized, using positive (treatment with DNase) and negative controls, and confirmed by other methods for detection of apoptosis (14). Dyes such as DAPI, propidium iodide or acridine orange are useful in visualizing condensed chromatin, or to analyze the DNA histogram where the accumulation of cells in subG1 phase reflects the apoptotic cell population (15).

New methods have recently been developed to detect earlier changes during the apoptotic process. These methods include detection of changes in mitochondrial potential, staining of cytochrome c in the cytosol and in the mitochondria, detection of cleaved caspases or of other proteins that are involved in the apoptotic process, and labeling of phosphatidylserine on the plasma membrane. Unfortunately, none of these techniques are universal or specific. A phospholipid-binding protein, such as annexin V, that has a high affinity for phosphatidylserine, may serve as a sensitive probe for staining cells that are undergoing apoptosis by detecting loss of membrane phospholipid asymmetry. Annexin V can be linked to a variety of labels for the purposes of flow cytometry and light microscopy in both vital and fixed materials (16). Annexin V can also allow *in vivo* detection and imaging of phosphatidylserine expression during programmed cell death (17). The annexin V binding method, however, is considered more accurate for the detection of cell lysis and should be used in combination with other techniques, such as vital dyes, in order to increase the specificity for the detection of apoptosis. Finally, this method appears to be more reliable with non-adherent cells.

Fluorochromes, which incorporate into cells depending on their mitochondrial transmembrane potential, are usually used to detect the reduction of mitochondrial potential (18). This method is commonly combined with flow cytometry analysis using annexin V. A large number of antibodies are currently available for staining proteins involved in the apoptotic machinery, including specific antibody against cytochrome c, native or mature caspase, and diverse proteins undergoing various degrees of phosphorylation. They can be used for Western blot analysis, immuno-precipitation, and immunohistochemistry. Active caspases can also be detected and quantified by measuring their enzymatic activity with specific fluorogenic substrates. In conclusion, to assess the process of apoptosis it is critical to combine at least two or three different methods in order to obtain reliable and reproducible results.

Apoptosis in the Early Phase of ALI

Fas/FasL system. A characteristic feature of ALI is a widespread destruction of the alveolar epithelium that affects the outcome of patients with ALI. Although the mechanism of cell death is uncertain, experimental studies have demonstrated that Fas, originally characterized in lymphocyte T cells as the primary effector molecule of Th1 cytotoxicity, is expressed on the apical surface of a subset of alveolar epithelial type II (ATII) cells and may induce apoptosis of these cells (19-21). Soluble Fas ligand is released in the airspaces of patients with ALI, but not in patients at risk for the syndrome (22). In addition, BAL fluid recovered from patients with ALI induced apoptosis of distal lung epithelial cells. These results suggest that activation of the Fas/FasL system contribute to the severe epithelial damage that occurs in ALI (22). Similar observation were recently made in patients with ALI secondary to sepsis (23).

Experimental activation of the Fas/FasL system in mice not only induced apoptosis and injury to the alveolar epithelium, but was also associated with the development of lung inflammation characterized by the airspace release of proinflammatory mediators (24) by a mechanism implicating activation of the NF-κB pathway (7). Interestingly, in this experimental model, damage to the alveolar epithelium occurred before neutrophil recruitment to the alveolar spaces. This observation is of crucial importance because it is still not clear whether the first event leading to lung injury, in humans, involves neutrophil migration or whether epithelial injury occurs before neutrophil infiltration. Answering this question will be a critical step to develop new specific strategies to treat ALI.

Finally, there are new experimental reports that link Fas-induced apoptosis of the alveolar epithelium to development of lung fibrosis. Repeated inhalation of anti-Fas antibody, mimicking Fas-Fas ligand cross-linking, induced excessive apoptosis of the alveolar epithelium and an inflammatory response in the airspaces of the lung, which resulted in pulmonary fibrosis in mice (25, 26). To date, there is no evidence establishing a direct link between excessive apoptosis of the lung epithelium and development of pulmonary fibrosis in humans with ALI. Nevertheless, it is reasonable to hypothesize that excessive apoptosis of the alveolar epithelium may prolong airspace inflammation and interfere with re-epithelialization, which may result in overgrowth of mesenchymal cells and lung fibrosis.

In conclusion, the Fas/Fas ligand system may serve a dual role in the pathogenesis of ALI by causing direct damage to the alveolar epithelial barrier and by recruiting neutrophils to the site of injury, thus perhaps promoting the development of lung fibrosis later during the course of the disease.

Local renin-angiotensin system (RAS). Recent evidence indicates that a local RAS is expressed in the distal lung parenchyma and plays a central role in the signaling of apoptosis in ATII cells (27-29). For example, apoptosis in response to Fas activation can be abrogated by antisense oligonucleotides against angiotensinogen, by angiotensin-converting enzyme inhibitors, or by angiotensin receptor antagonists. These results indicate that the *de novo* synthesis of angiotensin II and receptor interaction are required for the induction of apoptosis of ATII cells by Fas (29). Moreover, angiotensinogen is also secreted by human lung myofibroblasts isolated from patients suffering from interstitial pulmonary fibrosis (30) and its conversion to angiotensin II by the alveolar epithelium (29) provides a mechanism to explain alveolar epithelial cell death adjacent to underlying myofibroblasts within the fibrotic human lung (31). Taken together, these data support the hypothesis that a local RAS plays a critical role as a second messenger during the apoptotic process of the alveolar epithelium. They also suggest that therapeutic manipulation of Fas-induced apoptosis *in vivo* may be feasible in a near future with a variety of well-characterized pharmacological antagonists of the renin-angiotensin system.

Reactive oxygen and nitrogen species. The lung is exposed to a variety of reactive oxygen (ROS) and nitrogen species (RNS) during the acute phase of ALI (see Chapters 6, 7, and 9). Earlier studies showed that both ROS and RNS caused apoptosis of a rat alveolar epithelial cell line (32). The apoptosis was preceded by upregulation of c-fos and c-jun and activation of the transcription factor AP-1. In contrast, high concentration of peroxynitrite, a metabolite of NO, did not lead to AP-1 activation and failed to induce apoptosis of lung epithelial cells (32). Further studies reported that NO may actually inhibit stretch- or hyperoxia-induced AT II cell apoptosis (33, 34). To explain this discrepancy, recent studies suggest that the cellular responses to nitrating species may be different under conditions of injury and repair. Specifically, marked induction of apoptosis can occur in log-phase cultures exposed to peroxynitrite (20). Similarly, under conditions that mimic wound healing, RNS cause apoptosis selectively in cells that are migrating in to the wound (20), indicating that the response of the lung epithelium under conditions of repair may be different than the response of quiescent cells. These results indicate that the presence of RNS in the lung may interfere with epithelial repair and contribute to the continuous shedding of epithelial cells in ALI.

Infectious agents. Apoptosis of parenchymal cells may be protective by promoting pathogen removal and diminishing its infectivity. Uncontrolled cell apoptosis, however, will induce extensive epithelial injury and thereby can increase the degree of ALI. To promote infectivity, intracellular pathogens tend to inhibit apoptosis in host cells, in order to allow their own growth and survival. In contrast, extracellular pathogens promote apoptosis of the host cells (35). There is good experimental evidence that various bacteria, including *Staphylococcus aureus* (36-39), *Streptococcus (S) pneumonia* (40, 41), and *Pseudomonas aeruginosa* (42, 43) cause apoptosis of the lung epithelium *in vitro*. The mechanisms by which bacteria induce apoptosis in lung epithelial cells are not completely understood although recent data suggest that *P aeruginosa* may activate the Fas/FasL pathway in epithelial cells (44 and Chapter 10).

Despite these *in vitro* data there are currently few direct data supporting the hypothesis that excessive apoptosis of lung epithelium affects the outcome of bacterial pneumonia-induced ALI. Location and timing of apoptosis appears to determine whether pneumonia resolves, or progresses to fibrosis, in an experimental model of *Streptococcus*–induced pneumonia in rats (45). Moreover, preliminary data from our own laboratory indicate that apoptosis of the alveolar epithelium induced by a clinical isolate of *P. aeruginosa* was associated with a significant decrease in the vectorial fluid transport across this barrier in anesthetized rats. When apoptosis of the lung epithelium was prevented by pretreatment with KGF or a pharmacological inhibitor of the caspases, there was complete restoration of physiological alveolar fluid clearance (46).

In summary, there is experimental evidence that bacterial pathogens in the lung may cause excessive apoptosis of the lung epithelium, although the molecular mechanisms remain unclear and are likely to be different for each pathogen, depending on the bacterial toxicity, the response of the host, and the host/pathogen interactions. Nevertheless, the importance of apoptosis of the lung epithelium during pneumonia-induced ALI remains to be demonstrated in humans.

Neutrophil apoptosis. Several clinical studies have reported that neutrophils infiltrate the lungs in large numbers after the onset of ALI and their persistence is an important determinant of poor outcome (47, 48). Therefore, elucidation of the mechanisms that maintain pulmonary neutrophilia during ALI may be of considerable prognostic and therapeutic significance. Apoptosis and necrosis are the main mechanisms of neutrophil clearance from the alveolar spaces, although apoptosis may be of greater importance because it is a biological process subject to regulation (49, 50).

Several studies have investigated the modulation of neutrophil apoptosis in the airspaces of patients with ALI. The number of apoptotic neutrophils is significantly decreased during the early phase of ALI (51-53). In contrast, during later stages of ALI the percentage of apoptotic neutrophils in the airspaces appears to be comparable to that measured in patients without ALI or at risk for the syndrome (52). Moreover, decreased neutrophil apoptosis was associated with increased levels of G-CSF and GM-CSF in the BAL fluid of patients with ALI (51-53). Although both growth factors prolong the survival of blood neutrophils *in vitro* (54), higher BAL fluid levels of G-CSF were found in non-survivors in one study (51), whereas significantly higher BAL fluid levels of GM-CSF were reported in survivors in another study (52). The differences between these studies may be explained by the fact that GM-CSF is less specific than G-CSF to induce neutrophil proliferation (55). Moreover, the association between GM-CSF and survival could be independent of the effect of GM-CSF on neutrophil apoptosis and explained by the proliferative effect of this cytokine on alveolar macrophages and epithelial cells (52).

BAL fluid from patients with early, but not late ALI, inhibited apoptosis of normal blood neutrophils (52). These findings are important because it demonstrates that the inhibitory effect of BAL fluid changed over time in patients with ALI, suggesting that the rate at which neutrophils become apoptotic varies depending on the stage of inflammation. In summary, although neutrophil apoptosis may be a potential major mechanism for resolution of airspace inflammation in patients with ALI, it is still unclear whether inhibition of neutrophil apoptosis during the acute phase of inflammation is beneficial or harmful for the host (56).

Apoptosis in the Remodeling Phase of ALI

The role of apoptosis in the remodeling/repair phase of ALI is even less well understood compared to its role in the early phases of ALI (see chapter 14). There is limited experimental and clinical evidence that controlled apoptosis of parenchymal and inflammatory cells participates to the resolution of the inflammatory process in ALI. For example, apoptosis appears to be an important pathway for resolving KGF-induced ATII cell hyperplasia in rats, by a mechanism implicating Fas and FasL (57). Surfactant lipids released by ATII cells have been involved in the protection against intraluminal fibrogenesis by inducing fibroblast apoptosis and thus preventing collagen accumulation (58). Similarly, in the repair phase of human ALI, extensive apoptosis of ATII cells is largely responsible for the disappearance of these cells, which extensively proliferate during the early phases of ALI (3, 4, 59). As noted above, apoptosis of inflammatory cells may also be important for the resolution of airspace inflammation in resolving ALI. For example, recovery from oleic acid-induced ALI is associated with apoptosis of airspace neutrophils and their clearance by alveolar macrophages (60). Comparable findings have been reported in a experimental model of pneumonia in rats in which controlled and localized apoptosis is a prominent feature in the resolution of pneumonia (45). Taken together, these results suggest that controlled apoptosis is beneficial for the resolution of ALI by limiting the fibroproliferative response.

In contrast, abnormalities in the apoptotic process during the repair phase of ALI may promote the persistence of airspace inflammation and development of lung fibrosis. Wang et al recently reported that myo-fibroblasts isolated from patients with fibrotic lung injury, and rats treated with paraquat, release apoptotic factors that could potentially prevent a normal repopulation of the alveolar epithelium (30). This mechanism could facilitate the proliferation of underlying interstitial fibroblasts and therefore promote lung fibrosis. Apoptosis appears to play an important role as a homeostatic mechanism during the remodeling phase of ALI by removing inflammatory cells, granulation tissue, and excess ATII cells. Excessive apoptosis of lung epithelial cells and/or inefficient apoptosis of inflammatory cells, particularly neutrophils within the airspaces of the lung, may prevent the restitution of a normal alveolar architecture and therefore facilitate the development of fibrosis in patients with ALI.

Conclusions

Apoptosis of lung parenchymal and inflammatory cells is part of the host response to noxious stimuli during both the acute inflammatory and resolution phases of ALI. There are experimental and clinical data demonstrating that abnormalities of the apoptotic process (i.e., inefficient or excessive apoptosis) may increase damage to the lung parenchyma, prevent resolution of inflammation, and perhaps promote abnormal tissue repair (i.e., lung fibrosis) in ALI. Modulating the apoptotic process with Bcl-2 antisense, recombinant TRAIL, caspase inhibitors, and inhibitors of the renin-angiotensin system (61) could be considered in the future to control abnormal apoptosis in ALI. However, before these new therapeutic strategies are used in patients with ALI, there is a need for a better understanding of the significance of the apoptotic process in clinically relevant experimental models of ALI and of the importance of specific clinical disorders associated with ALI (i.e., pneumonia, trauma, sepsis, aspiration of gastric contents) in modulating the activation of apoptosis in the lung.

REFERENCES

1. Kerr, J.F., Wyllie, A.H., and Currie, A.R. (1972) Apoptosis: a basic biological phenomenon with wide-ranging implications in tissue kinetics. *Br J Cancer* 26, 239-257
2. Schittney, J., Djonov, V., Fine, A., and Burri, P.H. (1998) Programmed cell death contributes to postnatal lung development. *Am J Respir Cell Mol Biol* 18, 786-793
3. Polunovsky, V., Chen, B., Henke, C., Snover, D., Wendt, C., Ingbar, D. and Bitterman, P. (1993) Role of mesenchymal cell death in lung remodeling after injury. *J Clin Invest* 92, 388-397
4. Bardales, R.H., Xie, S.S., Schaefer, R.F., and Hsu, S.M. (1997) Apoptosis is a major pathway responsible for the resolution of type II pneumocytes in acute lung injury. *Am J Pathol* 149, 845-852
5. Savill, J., and Fadok, V. (2000) Corpse clearance defines the meaning of cell death. *Nature* 407, 784-788
6. Lemasters, J.J., Qian, T., Bradham, C.A., Brenner, D.A., Cascia, W.E., Trost, L.C., Nishimura, Y., Nieminen, A.L., and Herman, B. (1999) Mitochondrial dysfunction in the pathogenesis of necrotic and apoptotic cell death. *J Bioenerg Biomembr* 31, 305-319
7. Ashkenazi, A., and Dixit, V.M. (1998) Death receptors: signaling and modulation. *Science* 281.1305-1308
8. Beg, A.A., and Baltimore, D. (1996) An essential role for NF-κB in preventing TNF-α-induced cell death. *Science* 274,782-784

9. Hengartner, MO. (2000) The biochemistry of apoptosis. *Nature* 407, 770-776
10. Adams, J.M., and Cory, S. (1998) The Bcl-2 protein family: arbiters of cell survival. *Science* 281, 1322-1326
11. Thornberry, N.A., and Lazebnik, Y. (1998) Caspases: enemies within. *Science* 281, 1312-1316
12. Walker, P.R., Leblanc, J., Smith, B., Pandey, S., and Sikorska, M. (1999) Detection of DNA fragmentation and endonucleases in apoptosis. *Methods* 17, 329-338
13. Hall, P.A. (1999) Assessing apoptosis: a critical survey. *Endocr Relat Cancer* 6, 3-8
14. Saraste, A. (1999) Morphologic criteria and detection of apoptosis. *Herz* 24, 189-195
15. Nicoletti, I., Migliorati, G., Pagliacci, M.C., Grignani, F., and Riccardi, C. (1991) A rapid and simple method for measuring thymocyte apoptosis by propidium iodide staining and flow cytometry. *J Immunol Methods* 139, 271-279
16. van Engeland, M., Nieland, L.J., Ramaekers, F.C., Schutte, B., and Reutelingsperger, C.P. (1998) Annexin V-affinity assay: a review on an apoptosis detection system based on phosphatidylserine exposure. *Cytometry* 31, 1-9
17. Blankenberg, F.G., Katsikis, P.D., Tait, J.F., Davis, R.E., Naumovski, L., Ohtsuki, K., Kopiwoda, S., Abrams, M.J., Darkes, M., Robbins, R.C., Maecker, H.T., and Strauss, H.W. (1998) In vivo detection and imaging of phosphatidylserine expression during programmed cell death. *Proc Natl Acad Sci U S A* 95, 6349-6354
18. Zamzami, N., Marchetti, P., Castedo, M., Decaudin, D., Macho, A., Hirsch, T., Susin, S.A., Petit, P.X., Mignotte, B., and Kroemer, G. (1995) Sequential reduction of mitochondrial transmembrane potential and generation of reactive oxygen species in early programmed cell death. *J Exp Med* 182, 367-377
19. Matthay, M.A., and Wierner-Kronish, J.P. (1990) Intact epithelial barrier function is critical for the resolution of alveolar edema in humans. *Am Rev Respir Dis* 142, 1250-1257
20. Fine A, Janssen-Heininger Y, Soultanakis RP, Swisher SG, and Uhal, B.D. (2000) Apoptosis in lung pathophysiology. *Am J Physiol Lung Cell Mol Physiol* 279, L423-L427
21. Wen, L.P., Madani, K., Fahrni, J.A., Duncan, S.R., and Rosen, G.D. (1997) Dexamethasone inhibits lung epithelial cell apoptosis induced by IFN-gamma and Fas. *Am J Physiol* 273, L921-L929
22. Matute-Bello, G., Liles, W.C., Steinberg, K.P., Kiener, P.A., Mongovin, S., Chi, E.Y., Jonas, M., and Martin, T.R. (1999) Soluble fas ligand induces epithelial cell apoptosis in humans with acute lung injury (ARDS). *J Immunol* 163, 2217-2225
23. Hashimoto, S., Kobayashi, A., Kooguchi, K., Kitamura, Y., Onodera, H., and Nakajima, H. (2000) Upregulation of two death pathways of perforine/granzyme and FasL/Fas in septic acute respiratory distress syndrome. *Am J Respir Crit Care Med* 161, 237-243
24. Matute-Bello, G., Winn, R.K., Jonas, M., Chi, E.Y., Martin, T.R., and Liles, W.C. (2001) Fas (CD95) induces alveolar epithelial cell apoptosis in vivo. *Am J Pathol* 158, 153-161
25. Hagimoto, N., Kuwano, K., Miyazaki, H. Kunitake, R., Fujita, M., Kawasaki, M., Kaneko, Y., and Hara, N. (1997) Induction of apoptosis and pulmonary fibrosis in mice in response to ligation of fas antigen. *Am J Respir Cell Mol Biol* 17, 272-278
26. Kuwano, K., Miyazaki, H., Hagimoto, N., Kawasaki, M., Fujita, M., Kunitake, R., Kaneko, Y., and Hara, N. (1999) The involvement of Fas-Fas ligand pathway in fibrosing lung diseases. *Am J Respir Cell Mol Biol* 20, 53-60
27. Wang, R., Alam, G., Zagariya, A., Gidea, C., Pinillos, H., Lalude, O., Choudhary, G., Oezatalay, D., and Uhal, B.D. (2000) Apoptosis of lung epithelial cells in response to TNF-alpha requires angiotensin II generation de novo. *J Cell Physiol* 185, 253-259

28. Wang, R., Ibarra-Sunga, O., Verlinski, L., Pick, R., and Uhal, B.D. (2000) Abrogation of bleomycin-induced epithelial) apoptosis and lung fibrosis by captopril or a caspase inhibitor. *Am J Physiol Lung Cell Mol Physiol* 279, L143-L151
29. Wang, R., Zagariya, A., Ibarra-Sunga, O., Gidea, C., Ang, E., Deshmukh, S., Chaudhary, G., Baraboutis, J., Filippatos, G., and Uhal, B.D. (1999). Angiotensin II induces apoptosis in human and rat alveolar epithelial cells. *Am J Physiol Lung Cell Mol Physiol* 276, L885-L889
30. Wang, R., Ramos, C., Joshi, I., Zagariya, A., Pardo, A., Selman, M., and Uhal, B.D. (1999). Human lung fibroblast-derived inducers of alveolar epithelial apoptosis identified as angiotensin peptides. *Am J Physiol Lung Cell Mol Physiol* 277, L1158-L1164
31. Uhal, B.D., Joshi, I., Hugues, W.F., Ramos, C., Pardo, A., and Selman, M. (1998) Alveolar epithelial cell death adjacent to underlying myofibroblasts in advanced fibrotic human lung. *Am J Physiol Lung Cell Mol Physiol* 275, L1192-L1199
32. Janssen, Y.M., Matalon, S., and Mossman B.T. (1997) Differential induction of c-fos, c-jun, and apoptosis in lung epithelial cells exposed to ROS or RNS. *Am J Physiol Lung Cell Mol Physiol* 273, L789-796
33. Edwards, Y.S., Sutherland, L.M., and Murray, A.W. (2000) NO protects alveolar type II cells from stretch-induced apoptosis. A novel role for macrophages in the lung. *Am J Physiol Lung Cell Mol Physiol* 279, L1236-1242
34. Howlett, C.E., Hutchison, J.S., Veinot, J.P., Chiu, A., Merchant, P., and Fliss, H. (1999) Inhaled nitric oxide protects against hyperoxia-induced apoptosis in rat lungs. *Am J Physiol Lung Cell Mol Physiol* 277, L596-L605
35. Behnia, M., Robertson, K.A., and Martin, II W,J. (2000) Role of apoptosis in host defense and pathogenesis of disease. *Chest* 117, 1771-1777
36. Bayles, K.W., Wesson, C.A., Liou, L.E., Fox, L.K., Bohach, G.A., and Trumble, W.R. (1998) Intracellular Staphylococcus aureus escapes the endosome and induces apoptosis in epithelial cells. *Infect Immun* 66, 336-342
37. Kahl, B.C., Goulian, M., van Wamel, W., Herrmann, M., Simon, S.M., Kaplan, G., Peters, G, and Cheung, A.L. (2000) Staphylococcus aureus RN6390 replicates and induces apoptosis in a pulmonary epithelial cell line. *Infect Immun* 68, 5385-5392
38. Wesson, C.A., Deringer, J., Liou, L.E., Bayles, K.W., Bohach, G.A., and Trumble, W.R. (2000) Apoptosis induced by Staphylococcus aureus in epithelial cells utilizes a mechanism involving caspases 8 and 3. *Infect Immun* 68, 2998-3001
39. Wesson, C.A., Liou, L.E., Todd, K.M., Bohach, G.A., Trumble, W.R., and Bayles, K.W. (1998) Staphylococcus aureus Agr and Sar global regulators influence internalization and induction of apoptosis. *Infect Immun* 66, 5238-5243
40. Kuo, C.F., Wu, J.J., Tsai, P.J., Kao, F.J., Lei, H.Y., Lin, M.T., and Lin, Y.S. (1999) Streptococcal pyrogenic exotoxin B induces apoptosis and reduces phagocytic activity in U937 cells. *Infect Immun* 67, 126-130
41. Tsai, P.J., Lin, Y.S., Kuo, C.F., Lei, H.Y., and Wu, J.J. (1999) Group A Streptococcus induces apoptosis in human epithelial cells. *Infect Immun* 67, 4334-4339
42. Hauser, A.R., and Engel, J.N. (1999) Pseudomonas aeruginosa induces type-III-secretion-mediated apoptosis of macrophages and epithelial cells. *Infect Immun* 67, 530-5537
43. Rajan, S., Cacalano, G., Bryan, R., Ratner, A.J., Sontich, C.U., Van Heerckeren, A., Davis, P., and Prince, A. (2000) Pseudomonas aeruginosa induction of apoptosis in respiratory epithelial cells. *Am J Respir Cell Mol Biol* 23, 304-312
44. Grassme, H., Kirschnek, S., Riethmueller, J., Riehle, A., Von Kurthy, G., Lang, F., Weller, M., and Gulbins, E. (2000) CD95/CD95 Ligand interactions on epithelial cells in host defense to Pseudomonas aeruginosa. *Science* 290, 527-530

45. Kazzaz, J.A., Horowitz, S., Xu, J., Khullar, P., Niederman, M.S., Fein, A.M., Zakeri, Z., Lin, L., and Rhodes, G.C. (2000) Differential patterns of apoptosis in resolving and nonresolving bacterial pneumonia. *Am J Respir Crit Care Med* 161, 2043-2050
46. Faure, K., Le Berre, R., Fauvel, H., Thomas, A.M., Marchetti, P., and Guery, B. (2001) Curative injection of keratinocyte growth factor decreases Pseudomonas aeruginosa-induced apoptosis in the lung. *Am J Respir Crit Care Med* (in press).
47. Baughman, R.P., Gunther, K.L., Rashkin, M.C., Keeton, D.A., and Pattishall, E.N. (1996) Changes in the inflammatory response of the lung during acute respiratory distress syndrome: prognostic indicators. *Am J Respir Crit Care Med* 154, 76-81
48. Steinberg, K.P., Milberg, J.A., Martin, T.R., Maunder, R.J., Cockrill, B.A., and Hudson, L.D. (1994) Evolution of bronchoalveolar cell populations in the adult respiratory distress syndrome. *Am J Respir Crit Care Med* 150, 113-122
49. Haslett, C. (1999) Granulocyte apoptosis and its role in the resolution and control of lung inflammation. *Am J Respir Crit Care Med* 160, S5-S11
50. Savill, J. (1994) Apoptosis in disease. *Eur J Clin Invest* 24, 715-723
51. Aggarwal, A., Baker, C.S., Evans, T.W., and Haslam, P.L. (2000) G-CSF and IL-8 not GM-CSF correlate with severity of pulmonary neutrophilia in acute respiratory distress syndrome. *Eur Respir J* 15, 895-901
52. Matute-Bello, G., Liles, W.C., Radella, II F., Steinberg, K.P., Ruzinski, J.T., Hudson, L.D., and Martin, T.R. (2000) Modulation of neutrophil apoptosis by granulocyte colony-stimulating factor and granulocyte/macrophage colony-stimulating factor during the course of acute respiratory distress syndrome. *Crit Care Med* 28, 1-7
53. Matute-Bello, G., Liles, W.C., Radella, II F., Steinberg, K.P., Ruzinski, J.T., Jonas, M., Chi, E.Y., Hudson, L.D., and Martin, T.R. (1997) Neutrophil apoptosis in the acute respiratory distress syndrome. *Am J Respir Crit Care Med* 156, 1969-1977
54. Brach, M.A., de Vos, S., and Gruss, H.J. (1992) Prolongation of survival of human polymorphonuclear neutrophils by granulocyte-macrophage colony stimulating factor is caused by inhibition of programmed cell death. *Blood* 80, 2920-2924
55. Aglietta, M., Piacibello, W., Sanavio, F., Apra, F., Schena, M., Mossetti, C., Carnino, F., Caligaris-Cappio, F., and Gavosto, F. (1989) Kinetics of human hematopoietic cells after in vivo administration of granulocyte-macrophage colony-stimulating factor. *J Clin Invest* 83, 551-557
56. Pittet, J.F., Mackersie, R.C., Martin, T.R., and Matthay, M.A. (1997) Biological markers of acute lung injury: prognostic and pathogenetic significance. *Am J Respir Crit Care Med* 155, 1187-1205
57. Fehrenbach, H., Kasper, M., Koslowski, R., Pan, T., Schuh, D., Muller, M., and Mason, R.J. (2000) Alveolar epithelial type II cell apoptosis in vivo during resolution of keratinocyte growth factor-induced hyperplasia in the rat. *Histochem Cell Biol* 114, 49-61
58. De Lara, L.V., Becerril, C., Montano, M., Ramos, C., Maldonado, V., Melendez, J,. Phelps, D.S., Pardo, A., and Selman, M. (2000) Surfactant components modulate fibroblast apoptosis and type I collagen and collagenase-1 expression. *Am J Physiol Lung Cell Mol Physiol* 279, L950-957
59. Mantell, L.L., Kazzaz, J.A., Xu, J., Palaia, T.A., Piedboeuf, B., Hale, S., Rhodes, G.C., Niu, G., Fein, A.F., and Horowitz, S. (1997) Unscheduled apoptosis during acute inflammatory lung injury. *Cell Death Diff* 4, 600-607
60. Hussain, N., Wu, F., Zhu, L., Thrall, R.S., and Kresch, M.J. (1998) Neutrophil apoptosis during the development and resolution of oleic acid-induced acute lung injury in the rat. *Am J Respir Cell Mol Biol* 19, 867-874
61. Nicholson, D.W. (2000) From bench to clinic with apoptosis-based therapeutic agents. *Nature* 407, 810-816

Chapter 16

CORTICOSTEROID TREATMENT IN UNRESOLVING ARDS

G. Umberto Meduri
University of Tennessee Health Science Center, Division of Pulmonary and Critical Care Medicine; The Memphis Lung Research Program, and Baptist Memorial Hospitals.

INTRODUCTION

Acute respiratory distress syndrome (ARDS) is a severe form of acute lung injury, having a multi-factorial etiology characterized by rapid development of severe diffuse and non-homogenous inflammation of the pulmonary lobules leading to life-threatening hypoxemic respiratory failure. We have tested a therapeutic intervention on a previously defined pathophysiological model of ARDS (1). The model was defined by investigating during the longitudinal course of ARDS, the relationship among the three fundamental elements of a disease process: pathogenesis, structural alterations, and functional consequences (2). In these studies, we provided biological and morphological evidence indicating that ARDS patients failing to improve after one week of mechanical ventilation (unresolving ARDS) have intense and protracted pulmonary and systemic inflammatory and neo-fibrogenetic activity. In a controlled study, prolonged methylprednisolone (MP) treatment in patients with unresolving ARDS was associated with a significant reduction in (a) laboratory indices of inflammation and fibrogenesis, (b) physiological severity of ARDS, and (c) mortality. Our findings indicate that the failure of older trials investigating massive doses of MP in early ARDS was attributable to the short duration of administration.

Inflammation During ARDS

We previously reported data to support a single "hit" model for ARDS progression, where the degree and duration of the host inflammatory response (HIR) determined the *adaptive* versus *maladaptive* evolution of the reparative process and final outcome. In a series of studies, we have shown that patients with ARDS failing to improve in the first week of mechanical ventilation (unresolving ARDS) had biological and morphological evidence of intense and protracted pulmonary and systemic inflammatory and neo-fibrogenetic activity (3-7). Over time, patients with unresolving ARDS had persistent and exaggerated increases of plasma and bronchoalveolar lavage (BAL) levels of tumor necrosis factor-α (TNF-α), interleukins (IL)-1β, IL-6, IL-8, soluble intercellular adhesion molecule-1 (sICAM-1), and procollagen aminoterminal propeptide type I (PINP) and type III (PIIINP) (3-7). During the first week of ARDS, pro-inflammatory cytokine levels declined in all survivors, while levels remained persistently increased in all nonsurvivors. New unpublished data from our group indicate that cytokine levels reflected true biological activity. Furthermore, histological findings of open lung biopsies obtained in patients with unresolving ARDS (day 15 ± 7 of mechanical ventilation) provided morphological evidence of persistent activation of the HIR. Histological findings in previously spared pulmonary lobules included new injury to the endothelial and epithelial surfaces with associated intravascular coagulation and extravascular fibrin deposition (8, 9). Histological findings in previously involved pulmonary lobules included progressive fibroproliferative obliteration with transformation of the initially fibrinous exudate into myxoid connective tissue matrix and eventually into dense acellular fibrous tissue (8, 9). Histological differences between survivors and nonsurvivors placed advanced pulmonary fibrosis with acellular fibrosis and loss of alveolar architecture, at the upper boundary of disease reversibility (8).

Cellular Regulation of Inflammation

Two cellular signaling pathways have been identified as central to the evolution of the host inflammatory response: the nuclear factor-κB (NF-κB) and the glucocorticoid-mediated signal transduction cascades (10). NF-κB is a heterodimeric protein composed of the DNA-binding proteins p65 (potent transactivating capacity) and p50 (little or no transactivating capacity) constitutively present in the cytoplasm in an inactive form stabilized by the inhibitory protein, IκBα (11 and Chapter 1). Cellular activation by a multitude of adverse stimuli leads to phosphorylation and proteolytic degradation of IκBα. The liberated NF-κB then translocates into the nucleus and binds to promoter regions of target genes to initiate the transcription of multiple cytokines (TNF-α), IL-1β, IL-2, and IL-6), chemokines (IL-8, etc.), cell adhesion molecules (intercellular adhesion molecule-1, E selectin, etc.), and inflammation-associated enzymes. Products of the genes that are regulated by NF-κB can also cause the activation of NF-κB. For example, TNF-α and IL-1β both activate and are activated by NF-κB, a positive regulatory loop that may amplify and perpetuate inflammation (12).

Glucocorticoid receptor-α (GRα) and NF-κB are inducible DNA-binding transcription factors with diametrically opposed functions in the regulation of the host inflammatory response (10). Glucocorticoids (GC) are a family of hormones produced by the adrenal cortex glands, and are the most important physiological inhibitors of inflammation. GC exert most of their effects by activating ubiquitously distributed cytoplasmic heat shock protein 90-complexed glucocorticoid receptors (GR), leading to formation of the GC-GR complex (13). It is now appreciated that the GC-GR complexes modulate transcription in a hormone-dependent manner by binding to glucocorticoid response elements (GRE) in the promoters of glucocorticoid responsive genes and by interfering with the activity of other transcription factors such as NF-κB (14). GR-mediated transcriptional interference is achieved by five important mechanisms: 1.) by physically interacting with the p65 subunit and formation of an inactive (GR-NF-κB) complex (13); 2.) by inducing transcription of the IκBα gene (13, 15, 16); 3.) by blocking degradation of IκBα via enhanced synthesis of IL-10 (17); 4.) by impairing TNF-α–induced degradation of IκBα (18); and 5.) by competing for limited amounts of GR co-activators such as CREB-binding protein (CBP) and steroid receptor coactivator-1 (SRC-1) (19). Unpublished work from our group has shown that prolonged MP administration in patients with unresolving ARDS is associated with a significant enhancement of multiple aspects of GR function and with a significant reduction in NF-κB nuclear activity including transcription of pro-inflammatory cytokines.

Glucocorticoid Treatment of ARDS

Large randomized studies have previously shown that a short course (<24 hours) of high-dose methylprednisolone (MP) in early ARDS is ineffective (20-23). Table 1 shows the differences between the older trials in early ARDS and the newer one in unresolving ARDS. Because the half-life of MP is approximately 180 minutes, a sustained pharmacological effect in life-threatening, protracted lung inflammation (i.e., status asthmaticus, *Pneumocystis carinii* pneumonia, etc.) can be achieved only with prolonged administration aimed at disease resolution. In a rat model of butylated hydroxytoluene-induced acute lung injury, GC administration was shown to be effective in decreasing lung collagen and edema formation as long as treatment was prolonged; while withdrawal rapidly negated the positive effects of therapy (24-26). Two clinical studies demonstrated that premature discontinuation of prolonged GC administration in unresolving ARDS was associated with physiological deterioration that resolved with reinstitution of treatment (27, 28). Moreover, studies suggest that premature discontinuation of GC therapy may not only lead to loss of the early treatment benefits, but may in fact be harmful. Indeed, in another study the cytokine response to lipopolysaccharide challenge in humans was significantly enhanced by a prior (12 to 144 hours) short course of GCs (29). This observation may explain the differences in infection-related mortality between studies utilizing a short course (24 hours) versus a prolonged course of methylprednisolone treatment (21, 30, 31).

Table 1. Differences between old and new trials investigating methylprednisolone treatment in ARDS

	1980s trials	1990s trials
Timing of ARDS	< 2 days	7-14 days (unresolved ARDS)
Dosage	120 mg/kg/day	2 mg/kg/day
Duration	1 day	Average 30 days
Understanding of systemic Inflammation in ARDS	Massive, short-lived	Prolonged, initial intensity Affects duration
Reversibility with Glucocorticoid treatment	Lost early	Lost with end-stage fibrosis
Glucocorticoid treatment	Massive, short-course	Lower dose, prolonged until Disease resolution

We and others have reported significant improvement in lung function during prolonged methylprednisolone administration in medical (8, 28, 32, 33) and surgical (27, 34) patients with unresolving ARDS and have found that survival correlated with improvement in lung function. In our studies (8, 32), we defined improvement in lung function as a reduction in lung injury score (LIS, reference 35) of at least one point by day 10 of MP treatment. In phase II trials involving 34 patients, we reported mortalities of 17% in 29 patients who improved lung function (responders) and 100% in 5 nonresponders (8, 32). Additional findings of phase II trials are shown in Tables 2 and 3.

Table 2. Phase II studies evaluating methylprednisolone in unresolving ARDS. Findings at initiation of methylprednisolone treatment.

General Findings

- Fever was frequently present in the absence of infection (8, 40)
- Plasma and BAL inflammatory cytokine levels*
  - were similar to historical control nonsurvivors (32)
  - no spontaneous reduction over time (32)
- Lack of significant bacterial growth in BAL correlated with histological absence of pneumonia (8, 40)†
- ^{67}Gallium scintigraphy with marked and diffuse pulmonary uptake in the absence of pneumonia (41)†

Comparison between responders versus nonresponders

- Responders:

(Table 2 continued)
- Histological evidence of preserved alveolar architecture and myxoid cellular fibroproliferation (8)†
- Improvement in lung injury score was slower in those with higher plasma IL-6 levels(32)‡
- Histological evidence of more advanced fibroproliferation(8)†

- Nonresponders:
  - Histological evidence of acellular fibrosis, arteriolar subintimal fibroproliferation, and loss of alveolar architecture(8)†
  - Higher incidence of liver failure (8)
  - Higher plasma IL-6 levels on days 1-3 of ARDS (32)‡
  - Physiological evidence of accelerated fibroproliferation (8)

---

* *Inflammatory cytokines included TNF-α, IL-1β, IL-6, and IL-8.*
† *Histological findings from lung specimens obtained by open lung biopsy in 13 patients with unresolving ARDS on MV for 15 ± 7 days, and prior to initiating prolonged methylprednisolone treatment.*
‡ *Plasma IL-6 levels on days 1-3 of ARDS: nonresponders vs responders: 1903 ± 302 pg/ml vs 1050 ± 191 pg/ml (P = 0.004); at initiation of methylprednisolone therapy: delayed vs rapid responders: 1011 ± 44 pg/ml vs 526 ± 57 pg/ml (P =0.03).*

Table 3. Phase II studies evaluating methylprednisolone in unresolving ARDS. Findings during prolonged methylprednisolone treatment.

General Findings
- Infections frequently developed in the absence of fever
  - Warning signs for infection included
    - ≥ 10% increase in immature neutrophils
    - ≥ 30% increase in minute ventilation
- Ventilator- associated pneumonia was the most common infection
- Surveillance bronchoscopy with bilateral BAL was useful for detecting pneumonia
- Accurately identified and treated infections did not increase plasma and BAL cytokine levels above pre-infection levels.
  - Premature discontinuation of treatment was associated with deterioration in lung function (27, 28)

Comparison between responders versus nonresponders
- Responders:
  - Reduction in plasma and BAL inflammatory cytokine levels*
    - significant by day 5 of treatment, and
    - parallel to improvements in lung injury score and indices of ACM† permeability
      - > 1-point reduction in lung injury score by day 10 day of treatment
      - 17% mortality
- Nonresponders had
  - No reduction in plasma and BAL inflammatory cytokine levels*
    - no improvements in lung injury score and ACM† permeability
    - 100% mortality

---

* Inflammatory cytokines included TNF-α, IL-1ß, IL-6, and IL-8.
† Alveolo-capillary membrane (ACM) permeability was assessed by measuring bronchoalveolar lavage (BAL) albumin concentration over time.

We recently completed a prospective, randomized, double-blind, placebo-controlled trial designed to evaluate the efficacy and safety of prolonged MP therapy in patients with unresolving ARDS (30). The therapeutic anti-inflammatory and anti-fibrotic efficacy of the investigated pharmacological regimen was assessed with serial laboratory measurements of inflammation (TNF-α, IL-1β, IL-6, IL-8, sICAM-1), alveolo-capillary membrane permeability (BAL total proteins), and fibrosis (procollagen). The primary clinical outcome measures were: 1.) improvement in lung injury score (LIS) by day 10 of therapy, and 2.) decrease in intensive care unit (ICU) mortality. Over a 25-month period, 24 patients entered the study: 16 patients received MP and 8 received placebo. Twelve patients had undergone recent surgery (4 $\pm$ 3 days prior to development of ARDS), and 2 had immediate postoperative ARDS. Physiologic and laboratory characteristics at the onset of ARDS were similar in both groups (30).

Methylprednisolone or placebo was given daily as intravenous push every 6 h (one-fourth of the daily dose) and changed to a single oral dose when oral intake was restored. The methylprednisolone dosage regimen is shown in Table 4. If the patient was extubated prior to day 14, treatment was advanced to day 15 of drug therapy and tapered according to schedule (30). The protocol contained: 1.) a provision for blindly crossing over patients who did not improve LIS by at least 1 point after 10 days of treatment and 2.) procedures for infection surveillance, including weekly bronchoscopy with bilateral BAL (30). Because MP blunts the febrile response to an infection, this latter intervention was essential for minimizing the random variation generated by the potential morbidity and mortality of untreated nosocomial infections. The study was designed as a sequential phase III clinical trial and projected to recruit 100 patients. The decision to end the trial was made when the test statistic exceeded the upper boundary of the triangular test of Whitehead, and the null hypothesis was rejected at a significance less than 0.05 and a power greater than 0.95.

Table 4. Methylprednisolone treatment for unresolving ARDS.

Loading dose of 2 mg/kg, IV bolus, followed by:
Days 1 to 14: 2 mg/kg/day, IV, q 6 hrs
Days 15 to 21: 1 mg/kg/day, IV, q 6 hrs
Days 22 to 28*: 0.5 mg/kg/day, IV, q 6 hrs

* From days 29 to 32, methylprednisolone was given in a single oral dose of 0.25 mg/kg/day for 2 days and 0.125 mg/kg/day for 2 days (30).

At study entry (day 9 $\pm$ 1 of ARDS), the two groups had similar LIS, PaO_2:FiO_2, and multiple organ dysfunction syndrome (MODS) scores. By study day 10, all patients in the MP group improved (> 1 point reduction in LIS) versus 2 of 8 (25%) in the placebo group; four patients who failed to improve were blindly crossed over to MP (as dictated by the protocol). Changes observed by study day 10 are reported for MP versus placebo

(Figure 1): LIS (1.7 ± 0.1 vs. 3.0 ± 0.2; $p < 0.0001$), PaO_2:FiO_2 (262 ± 19 vs. 148 ± 35; $p = 0.0003$), MODS score (0.7 ± 0.2 vs 1.8 ± 0.3; $p = 0.0008$), and successful extubation (7 vs 0; $p = 0.051$). Thus, improvement in LIS was observed in two out of four of those crossed over to MP (day 18 of ARDS) after 10 days of placebo. Extubated patients were discharged from the ICU on unassisted breathing, all but one within 4 days of removal from mechanical ventilation. ICU mortality for the treatment group versus the placebo group was 0% versus 62% ($P = 0.002$); hospital-associated mortality for the two groups was 12% versus 62% ($P = 0.03$). Improvement in LIS after 10 days of treatment correlated with hospital survival ($r = 0.688 \pm 0.165$).

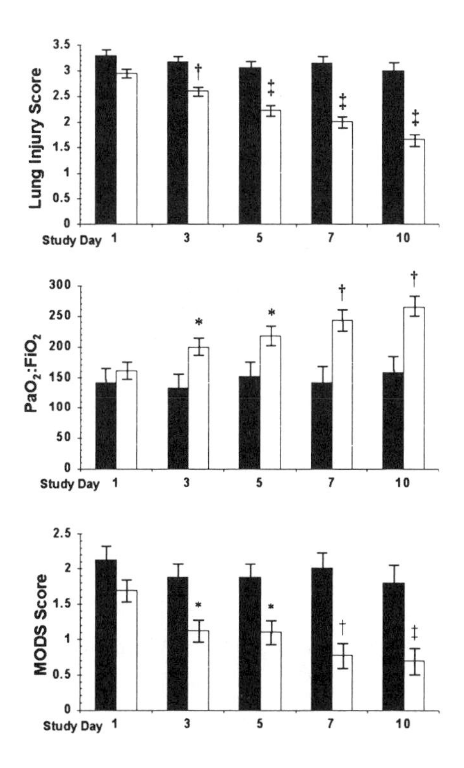

Figure 1. Mean (± SE) changes in PaO_2:FiO_2, lung injury score (LIS), and multiple organ dysfunction score (MODS) during the first 10 days of treatment in the methylprednisolone group and placebo group. There was no statistical difference between the methylprednisolone and placebo groups at time of entry into the study. The values on day 1 were obtained prior to initiating treatment. In the methylprednisolone group, a statistically significant improvement was achieved for PaO_2:FiO_2 on day 5 ($P < 0.01$), LIS on day 5 ($P < 0.0001$), and MODS score on day 7 ($P = 0.0002$). In the placebo group, no statistically significant improvement was achieved during the first 10 days of treatment. The number of patients in the methylprednisolone groups on study days 7 and 10 were 14 and 9, and in the placebo group 6 and 6, respectively. See text for definitions of LIS and MODS score. * $p < 0.01$; † $p < 0.001$; ‡ $p < 0.0001$.

The rate of complications between the two groups was similar (Figure 2). During MP treatment, pneumonia frequently developed in patients without fever (44%). Therefore, infection surveillance, including weekly bilateral bronchoscopic BAL, was useful for early detection of pneumonia and other serious infections. None of the recognized and appropriately treated infections developing during MP therapy affected resolution of ARDS or clinical outcome.

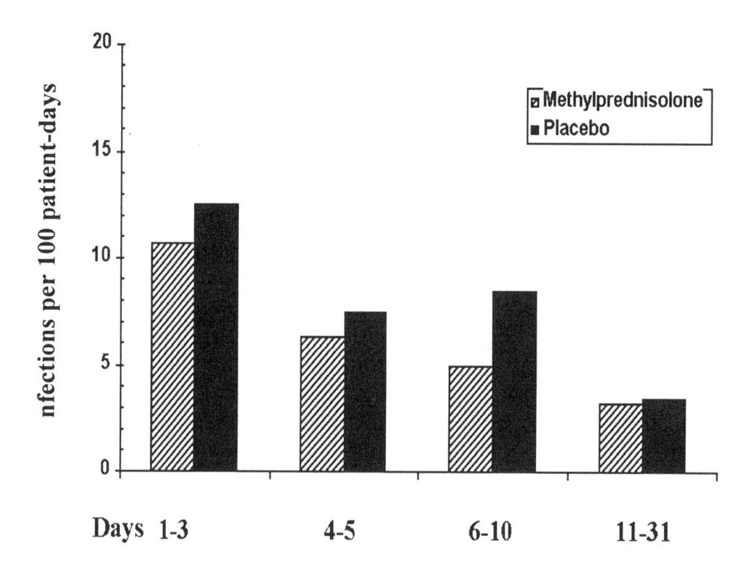

Figure 2. Incidence of infections per 100 patients-days in patients randomized to methyl-prednisolone and placebo.

Terminating this sequential trial very quickly biases the estimate of the treatment effect and raises the concern that the statistically significant difference in outcome was not due to a true effect of the tested intervention (MP treatment and infection surveillance), but to a potential lack of comparability between the two groups to the extent that the treatment effect might merely have reflected some confounding variable, such as severity of illness. Based on the Apache III, LIS, and MODS scores at entry and on the day of randomization, we have not detected any confounding variable able to explain the differences in outcome between the two groups. Moreover, longitudinal measurements of inflammation and fibrosis demonstrated a significant and sustained biological effect, making the possibility of a type I error unlikely.

A uniform connective tissue repair response has been demonstrated to occur in various organs following different types of injury (36). Pulmonary fibroproliferation in ARDS shares a common pathogenetic mechanism with other fibroproliferative diseases, where in the absence of inhibitory signals, the continued production of inflammatory mediators sustains connective tissue accumulation, which results in permanent alteration in tissue structure and function ("maladaptive response," reference 37 and Chapter 14). Within this theoretical paradigm, defined by Bitterman and Henke as the "linear response" to injury, fibrosis ensues when the inflammatory response is overexuberant and prolonged (38). Several studies have substantiated that increased plasma procollagen aminoterminal propeptide levels reflect collagen synthesis at the site of disease, and may be used as a marker of the reparative process independent of etiology (36, 39). We have found that during the first week of ARDS, nonimprovers (patients recruited for the randomized study), as opposed to improvers (patients with > 1 point reduction in LIS), had a progressive increase in plasma PINP and PIIINP levels (day 5; p = 0.004), and BAL PIIINP correlated with static pulmonary compliance (r = - 0.75; p = 0.03) and PaO_2:FiO_2 (r = 0.69; p = 0.05). MP treatment, initiated on day 9 ± 1 of ARDS, was associated with a rapid and sustained decrease in mean plasma and BAL PINP and PIIINP levels, while no decrease was observed during placebo administration. By day 3 of treatment, mean plasma PINP and PIIINP levels (ng/ml) decreased from 100 ± 9 to 45 ± 8 (p = 0.0001) and 31 ± 3 to 12 ± 3 (p = 0.0008), respectively. After 8 to 15 days of MP, mean BAL PINP and PIIINP levels (ng/ml) decreased from 63 ± 25 to 6 ± 23 (p = 0.002) and 42 ± 5 to 10 ± 3 (p = 0.003), respectively. Estimated partial correlation coefficients indicated that as plasma PINP and PIIINP levels decreased over the first 7 days of MP treatment, positive end-expiratory pressure decreased, while PaO_2:FiO_2 increased.

In conclusion, we have conducted a "holistic" level of inquiry: biology, pathology, physiology, and outcome-in both uncontrolled and controlled studies involving 92 patients with unresolving ARDS, 50 of whom received prolonged MP treatment, and provided findings to support a single "hit" model of ARDS where progression of the disease and final outcome are related to a dysregulated pulmonary and systemic HIR. These studies provide strong support for a linkage between biological and physiological response that can be affected by prolonged MP administration. We believe that failure of older trials investigating massive doses of MP in early ARDS was attributable to the short duration of administration. A randomized trial is in progress evaluating the effectiveness of prolonged MP initiated in early ARDS.

ACKNOWLEDGEMENTS

This work was supported by the Baptist Memorial Health Care Foundation and the Assisi Foundation of Memphis.

REFERENCES

1. Meduri, G.U. (1996) The role of the host defence response in the progression and outcome of ARDS: pathophysiological correlations and response to glucocorticoid treatment. *Eur Respir J* 9, 2650-2670
2. Cotran, R.S., Kumar, V., and Robbins, S.L. (1994) Cellular injury and cellular death. In R.S. Cotran, V. Kumar and S.L. Robbins, editors. *Pathologic basis of disease*, 5 ed. W.B. Saunders, Philadelphia. 1-34
3. Meduri, G.U., Headley, S., Kohler, G., Stentz, F., Tolley, E., Umberger, R., and Leeper, K. (1995) Persistent elevation of inflammatory cytokines predicts a poor outcome in ARDS. Plasma IL-1 beta and IL-6 levels are consistent and efficient predictors of outcome over time. *Chest* 107, 1062-1073
4. Meduri, G.U., Kohler, G., Headley, S., Tolley, E., Stentz, F., and Postlethwaite, A. (1995) Inflammatory cytokines in the BAL of patients with ARDS. Persistent elevation over time predicts poor outcome. *Chest* 108, 1303-1314
5. Golden, E., John, B., Stentz, F., Tolley, E.A., and Meduri, G.U. (2000) Interleukin-8 and soluble intercellular adhesion molecule-1 during acute respiratory distress syndrome and in response to prolonged methylprednisolone treatment. *Shock* 13, 42S (Abstract)
6. Headley, A.S., Meduri, G.U., Tolley, E., and Stentz, F. (2000) Infections, SIRS, and CARS during ARDS and in response to prolonged glucocorticoid treatment. *Am J Respir Crit Care Med* 161, A378 (Abstract)
7. Meduri, G.U., Tolley, E.A., Chinn, A., Stentz, F., and Postlethwaite, A. (1998) Procollagen types I and III aminoterminal propeptide levels during acute respiratory distress syndrome and in response to methylprednisolone treatment. *Am J Respir Crit Care Med* 158, 1432-1441
8. Meduri, G.U., Chinn, A.J., Leeper, K.V., Wunderink, R.G., Tolley, E., Winer-Muram, H.T., Khare, V., and Eltorky, M. (1994) Corticosteroid rescue treatment of progressive fibroproliferation in late ARDS. Patterns of response and predictors of outcome. *Chest* 105, 1516-1527
9. Meduri, G.U., Eltorky, M., and Winer-Muram, H.T. (1995) The fibroproliferative phase of late adult respiratory distress syndrome. *Semin Respir Infect* 10, 154-175
10. McKay, L.I., and Cidlowski, J.A. (1999) Molecular control of immune/inflammatory responses: interactions between nuclear factor-kappa B and steroid receptor-signaling pathways. *Endocr Rev* 20, 435-459
11. Baeuerle, P.A., and Baltimore, D. (1996) NF-kappa B: ten years after. *Cell* 87 13-20
12. Barnes, P.J., and Karin, M. (1997) Nuclear factor-kappa B: a pivotal transcription factor in chronic inflammatory diseases. *N Engl J Med* 336, 1066-1071
13. Bamberger, C.M., Schulte, H.M., and Chrousos, G.P. (1996) Molecular determinants of glucocorticoid receptor function and tissue sensitivity to glucocorticoids. *Endocr Rev* 17, 245-261
14. Didonato, J.A., Saatcioglu, F., and Karin, M. (1996) Molecular mechanisms of immunosuppression and anti-inflammatory activities by glucocorticoids. *Am J Respir Crit Care Med* 154, S11-15
15. Scheinman, R.I., Cogswell, P.C., Lofquist, A.K., and Baldwin, A.S., Jr. (1995) Role of transcriptional activation of I kappa B alpha in mediation of immunosuppression by glucocorticoids. *Science* 270, 283-286
16. Wissink, S., van Heerde, E.C., vand der Burg, B., and van der Saag, P.T. (1998) A dual mechanism mediates repression of NF-kappa B activity by glucocorticoids. *Mol Endocrinol* 12, 355-363
17. Wang, P., Wu, P., Siegel, M.I., Egan, R.W., and Billah, M.M. (1995) Interleukin (IL)-10 inhibits nuclear factor kappa B (NF kappa B) activation in human monocytes. IL-10 and IL-4 suppress cytokine synthesis by different mechanisms. *J Biol Chem* 270, 9558-9563

18. Hofman, T.G., Hehner, S.P., Bacher, S., Droge, W., and Schmitz, M.L. (1998) Various glucocorticoids differ in their ability to induce gene expression, apoptosis and to repress NF-kappaB-dependent transcription. *FEBS Lett* 441, 441-446
19. Sheppard, K.A., Phelps, K.M., Williams, A.J., Thanos, D., Glass, C.K., Rosenfeld, M.G., Gerritsen, M.E., and Collins, T. (1998) Nuclear integration of glucocorticoid receptor and nuclear factor- kappaB signaling by CREB-binding protein and steroid receptor coactivator-1. *J Biol Chem* 273, 29291-29294
20. Weigelt, J.A., Norcross, J.F., Borman, K.R., and Snyder, W.H. (1985) Early steroid therapy for respiratory failure. *Arch Surg* 120, 536-540
21. Bone, R.C., Fisher, C.J., Jr., Clemmer, T.P., Slotman, G.J., and Metz, C.A. (1987) Early methylprednisolone treatment for septic syndrome and the adult respiratory distress syndrome. *Chest* 92, 1032-1036
22. Bernard, G.R., Luce, J.M., Sprung, C.L., Rinaldo, J.E., Tate, R.M., Sibbald, W.J., Kariman, K., Higgins, S., Bradley, R., Metz, C.A., and et al. (1987) High-dose corticosteroids in patients with the adult respiratory distress syndrome. *N Engl J Med* 317, 1565-1570
23. Luce, J.M., Montgomery, A.B., Marks, J.D., Turner, J., Metz, C.A., and Murray, J.F. (1988) Ineffectiveness of high-dose methylprednisolone in preventing parenchymal lung injury and improving mortality in patients with septic shock. *Am Rev Respir Dis* 138, 62-68
24. Hesterberg, T.W., and Last, J.A. (1981) Ozone-induced acute pulmonary fibrosis in rats. Prevention of increased rates of collagen synthesis by methylprednisolone. *Am Rev Respir Dis* 123, 47-52
25. Hakkinen, P.J., Schmoyer, R.L., and Witschi, H.P. (1983) Potentiation of butylated-hydroxytoluene-induced acute lung damage by oxygen. Effects of prednisolone and indomethacin. *Am Rev Respir Dis* 128, 648-651
26. Kehrer, J.P., Klein-Szanto, A.J., Sorensen, E.M., Pearlman, R., and Rosner, M.H. (1984) Enhanced acute lung damage following corticosteroid treatment. *Am Rev Respir Dis* 130, 256-261
27. Ashbaugh, D.G., and Maier, R.V. (1985) Idiopathic pulmonary fibrosis in adult respiratory distress syndrome. Diagnosis and treatment. *Arch Surg* 120, 530-535
28. Hooper, R.G., and Kearl, R.A. (1990) Established ARDS treated with a sustained course of adrenocortical steroids. *Chest* 97, 138-143
29. Barber, A.E., Coyle, S.M., Marano, M.A., Fischer, E., Calvano, S.E., Fong, Y., Moldawer, L.L., and Lowry, S.F. (1993) Glucocorticoid therapy alters hormonal and cytokine responses to endotoxin in man. *J Immunol* 150, 1999-2006
30. Meduri, G.U., Headley, S., Carson, S., Umberger, R., Kelso, T., and Tolley, E. (1998) Prolonged methylprednisolone treatment improves lung function and outcome of unresolving ARDS. A randomized, double-blind, placebo-controlled trial. *JAMA* 280, 159-165
31. Headley, A.S., Tolley, E., and Meduri, G.U. (1997) Infections and the inflammatory response in acute respiratory distress syndrome. *Chest* 111, 1306-1321
32. Meduri, G.U., Headley, S., Tolley, E., Shelby, M., Stentz, F., and Postlethwaite, A. (1995) Plasma and BAL cytokine response to corticosteroid rescue treatment in late ARDS. *Chest* 108, 1315-1325
33. Hooper, R.G., and Kearl, R.A. (1996) Established adult respiratory distress syndrome successfully treated with corticosteroids. *South Med J* 89, 359-364
34. Biffl, W.L., Moore, F.A., Moore, E.E., Haenel, J.B., McIntyre, R.C., Jr., and Burch, J.M. (1995) Are corticosteroids salvage therapy for refractory acute respiratory distress syndrome? *Am J Surg* 170, 591-595
35. Murray, J.F., Matthay, M.A., Luce, J.M., and Flick, M.R. (1988) An expanded definition of the adult respiratory distress syndrome. *Am Rev Respir Dis* 138, 720-723
36. Horslev-Petersen, K. (1990) Circulating extracellular matrix components as markers for connective tissue response to inflammation. A clinical and experimental study with

special emphasis on serum aminoterminal type III procollagen peptide in rheumatic diseases. *Dan Med Bull* 37, 308-329

37. Kovacs, E.J., and DiPietro, L.A. (1994) Fibrogenic cytokines and connective tissue production. *FASEB J* 8, 854-861
38. Bitterman, P.B., and Henke, C.A. (1991) Fibroproliferative disorders. *Chest* 99, Suppl, 81S-84S
39. Kirk, J.M., Bateman, E.D., Haslam, P.L., Laurent, G.J., and Turner-Warwick, M. (1984) Serum type III procollagen peptide concentration in cryptogenic fibrosing alveolitis and its clinical relevance. *Thorax* 39, 726-732
40. Meduri, G.U., Belenchia, J.M., Estes, R.J., Wunderink, R.G., Eltorky, M., and Leeper, K.V., Jr. (1991) Fibroproliferative phase of ARDS. Clinical findings and effects of corticosteroids. *Chest* 100, 943-952
41. Meduri, G.U., Belenchia, J.M., Massie, J.D., Eltorky, M., and Tolley, E.A. (1996) The role of gallium-67 scintigraphy in diagnosing sources of fever in ventilated patients. *Intensive Care Med* 22, 395-403

Chapter 17

ANTI-INFLAMMATORY CYTOKINES: ROLE IN REGULATION OF ACUTE LUNG INJURY

Thomas P. Shanley
Division of Critical Care Medicine, Children's Hospital Medical Center and Children's Hospital Research Foundation, Cincinnati, OH

"To every action, there exists an equal and opposite reaction" ... Sir Isaac Newton

INTRODUCTION

The concept of balance has been appreciated in the physical sciences for centuries. In the early study of inflammation, attention was focussed primarily on molecules categorized as proinflammatory (e.g. TNF) by virtue of the role they played in mediating leukocyte recruitment, endothelial damage, and tissue injury in critical illness. It was concluded that clinical states such as acute lung injury (ALI) and sepsis were a reflection of an overwhelming proinflammatory state of the host. Over the past decade, investigators have observed that at times the host is able to control or regulate the initial proinflammatory response resulting from activation of innate immune mechanisms. This response confers containment of the inflammation and is thought to afford protection from tissue injury and perhaps, hasten resolution. In the setting of the systemic inflammatory response syndrome (SIRS), this compensatory response was assigned the acronym CARS, or compensatory anti-inflammatory response syndrome (1). Perturbation of this response by exogenous factors, such as pathogens, or endogenous factors, such as genetic regulation of anti-inflammatory cytokine expression, can have important consequences on host survival. While a number of endogenous host factors participate in this necessary regulatory response, this chapter will focus on a series of cytokines that have been demonstrated to possess anti-inflammatory properties that serve to regulate the inflammation associated with ALI.

Cytokines with Anti-inflammatory Properties

While cytokines were initially thought of as strictly proinflammatory molecules on the basis of their biological properties outlined above, more recent data has documented the role of a number of cytokines as anti-inflammatory molecules (Table 1). Interleukin-10 (IL-10) may be the most well-studied member of this group that includes: IL-4, IL-13, TGF-ß and in some circumstances, the gp130 signaling proteins, IL-6 and IL-11. These molecules are linked by their common ability to inhibit the expression of proinflammatory molecules (e.g TNFα and IL-1β) in a variety of both *in vitro* and *in vivo* experimental models.

Table 1. Anti-inflammatory Cytokines

Cytokine	Sources	Actions
IL-10	Mononuclear cells, Th2 T-cells, B-cells	"Deactivation of monocytes", Inhibition of Th1 type response, Upregulation of regulatory molecules (e.g IL-1Ra)
IL-13	Th2 T-cells	Inhibition of monocyte cytokine production, NF-κB inhibition
IL-4	Th2 T-cells, B-cells, mast cells	Inhibition of monocyte cytokine production, Drives Th2 T-cell development, Increases VCAM-1
IL-6	Mononuclear cells, Neutrophils, B-cells	Inhibits TNF and IL-1, Drives acute phase response
IL-11	Fibroblasts,	Inhibits TNF and IL-1, Drives Th2 type response
TGF-ß	Mononuclear cells, most mammalian cell types	"Deactivation of monocytes"

Interleukin-10

Interleukin-10 (IL-10) is perhaps the best studied of the anti-inflammatory cytokines. IL-10 is an 18 kDa protein that was initially identified from activated Th2 helper T-cells on the basis of its ability to inhibit interferon production from Th1 T-cell clones (2). Subsequent cloning of the human cDNA has revealed its location on chromosome 1q (3). With the subsequent recombinant expression of the mature protein, IL-10 was demonstrated to be a potent inhibitor of TNFα production from stimulated monocytes (4). Over time, a number of anti-inflammatory properties have been ascribed to IL-10 (reviewed 5, 6). For example, *in vitro*, IL-10 has been shown to inhibit proinflammatory cytokines known to contribute to the development of ALI such as TNFα, IL-1β, IL-12 and chemokines, such as IL-8, MIP-1α and MIP-2. IL-10 down-regulates the expression of a number of cell surface receptors that are crucial to the host innate immune defense system such as MHC class II molecules, CD-14, and the co-stimulatory molecule, B7 (7). IL-10 can also upregulate additional anti-inflammatory molecules such as the IL-1 receptor antagonist protein (IL-1Ra) (8) and soluble TNF receptors (9). The mechanisms by which IL-10 is expressed and subsequently mediates these various effects has been the target of much investigation over the past decade.

Gene expression of IL-10 appears to be regulated by both transcriptional and post-transcriptional mechanisms. Promoter analysis using the mouse gene has shown that expression of IL-10 is regulated by binding of the ubiquitous transcription factors, Sp1 and Sp3 (10). This unusual regulation may explain how IL-10 expression can occur when gene expression of other cytokines is low or non-detectable. An additional level of regulation occurs via a post-transcriptional mechanism in which the 3'-untranslated region of the IL-10 gene that contains multiple AUUUA motifs confers destabilization of the IL-10 mRNA which may be reversed following stimulation (11). Once expressed in its mature form, IL-10 circulates as an active homodimer binding to the IL-10 receptor complex that is ubiquitously expressed on a variety of cell types.

The IL-10 receptor complex is a structurally related member of the type II cytokine receptor family similar to the IFN-γ receptor. Two subunits have been identified, IL-10R1 that is principally responsible for IL-10 binding and IL-10R2 that is an accessory subunit, constitutively expressed in all cells and necessary for efficient IL-10 signaling (12). Members of the IFN receptor family are known to utilize the Jak-stat family of proteins for signal transduction. The Jak-stat family of proteins are tyrosine phosphorylated and translocate from the cytoplasm to the nucleus where they bind to enhancer elements such as the IFN-γ response region to activate the transcriptional machinery (13). Binding of IL-10 to its receptor on both monocytes and T-cells activates jak1 and tyk2 tyrosine kinases with which the IL-10 receptor is complexed, and results in phosphorylation and activation of stat-1α and stat3 (14, 15). Further insight into the mechanism of the recruitment and activation of stat proteins by IL-10 was provided by structural analyses showing that the ligand binding subunit of the IL-10 receptor contained tyrosine residues that were required for activation of stat3 (16). Despite this molecular insight, however, the linkage between these signal transduction pathways and the regulation of inflammatory gene expression remains incomplete.

Because of the recruitment of the Jak-stat pathway following IL-10 stimulation, investigators have focussed on these molecules, particularly stat3. Work performed in macrophages in which stat3 expression was eliminated showed that stat3 was necessary for IL-10 signaling, but not sufficient for cytokine inhibition (17). Other investigations have linked the inhibition of IFN-driven gene expression to IL-10-induced inhibition of stat1 tyrosine phosphorylation and activation (18). Finally, IL-10 can induce the suppressor of cytokine synthesis (SOCS)-3 protein, which may negatively regulate the transcription machinery driving proinflammatory cytokine expression (19). Other work has focussed on the effect of IL-10 on the MAP kinase signal transduction pathway (see Chapter 1). In these studies, IL-10 inhibited LPS-induced tyrosine phosphorylation of the Ras signaling pathway with subsequent inhibition of downstream MAP kinase activation (20).

Because the mechanism by which IL-10 regulates proinflammation appears to be multifactorial, the effects of IL-10 on additional signal transduction pathways have been examined *in vitro*. In light of the multiple inflammatory gene products under transcriptional regulation by NF-κB, inhibition of this pathway by IL-10 has been studied. Several investigations have shown that pretreatment with IL-10 abrogates NF-κB activation with a variety of mechanisms described. It has been shown that phosphorylation and subsequent degradation of the inhibitory protein, IκBα, was impaired by IL-10 in a manner associated with inhibition of monocyte IκB kinase (IKK) activity (21). An additional study suggested that IL-10 negatively regulated this pathway by stabilizing the mRNA for IκBα, thereby sequestering NF-κB in the cytoplasm on the basis of increased IκBα (22). Interestingly, additional studies have supported a contrasting mechanism by which IL-10 decreases chemokine expression via destabilization of chemokine mRNA (23, 24). The AUUUA-rich sequences of the 3'-untranslated regions of many cytokine/chemokine transcripts provide a target for this mechanism (25). Together, these multiple inhibitory mechanisms mediated by IL-10 would be anticipated to substantially impair the expression of a number of pro-inflammatory genes.

Armed with this substantial body of *in vitro* data, investigators began to examine the ability of IL-10 to regulate proinflammation associated with ALI and sepsis *in vivo*. Exogenous administration of IL-10 in a number of experimental models was associated with decreased inflammatory makers and diminished organ injury. For example, in the setting of immune complex-mediated ALI in rats, IL-10 given intratracheally resulted in decreased inflammation and lung permeability that was associated with reduced bronchoalveolar lavage (BAL) fluid levels of TNF and IL-1ß (26). In other models of silica- and bleomycin-induced lung inflammation, bacterial pneumonia, and hypersensitivity pneumonitis, IL-10 administration abrogated the degree of lung injury (27-30). However, perhaps the most elucidating data concerning the role of IL-10 in modulating inflammation has been derived from studies using the IL-10 null mutant mouse (IL-10 -/-). Notably, IL-10 -/- mice bred and developed normally. However, over time these animals developed spontaneous colitis mimicking inflammatory bowel disease (31). If IL-10 -/- mice are kept in sterile, isolated conditions and fed sterile chow, the onset of this condition can be substantially abrogated, supporting the hypothesis that IL-10 is required to regulate the intestinal flora-induced, sub-clinical inflammation of the intestine. Further immunologic characterization of IL-10 -/- mice revealed that sub-lethal doses of endotoxin resulted in 100% mortality and was associated with substantially increased levels of proinflammatory cytokines (32). These data suggested that IL-10 was a key endogenous anti-inflammatory molecule serving to regulate production of pro-inflammatory mediators.

That this regulation extended to the lung was supported by several studies using models of ALI. For example, in the immune complex-induced model of ALI, antibody neutralization of IL-10, resulted in increased inflammation that was associated with increased BAL fluid levels of TNFα and IL-1ß (33). Further work employing IL-10 -/- mice demonstrated that IL-10 regulated chemokine expression (MIP-1α and MIP-2) as both these mediators were significantly increased after intratracheal LPS in the IL-10 -/- mice, a finding that was reversed by the co-administration of IL-10 at the time of LPS challenge (23). A correlative finding in humans was demonstrated by Donnelly et al who showed that patients with higher mortality rates from ARDS had lower levels of IL-10 in their BAL fluid (34). Furthermore, the inability to sufficiently increase IL-10 expression in response to meningococcal infection was associated with increased mortality (35). Genetic mutations in the promoter region of the IL-10 gene were associated with a decreased level of IL-10 expression and support the concept that genetic regulation of the anti-inflammatory response may be a key modifier of disease outcome (36, 37).

In summary, IL-10 has been pursued as a potential therapeutic option in ALI as it possesses a number of anti-inflammatory properties. First, it inhibits cytokine synthesis in a negative, auto-regulatory manner to dampen the autocrine effect of proinflammatory cytokines. Second, it inhibits the adhesion of leukocytes to activated endothelial cells, thereby disrupting the leukocyte-endothelial cell adhesion cascade. Third, it inhibits many of the key signal transduction pathways associated with activation of the proinflammatory response. Fourth, IL-10 upregulates the expression of naturally occurring cytokine antagonists including IL-1Ra and soluble TNF receptors. Finally, IL-10 may serve to destabilize the mRNAs of cytokines possessing the AU-rich element. Thus, in light of the multiple mechanisms by which IL-10 regulates inflammation, exogenous administration of IL-10 appears promising and is currently being studied in clinical trials. Enthusiasm for this therapeutic approach is tempered by the evidence that IL-10 limits the hosts immune response directed at pathogen eradication in several *in vivo* models of infection, as well as human observations (6). Thus, titrating IL-10 to strike a balance between protection against dysregulated proinflammation and pathogen clearance remains a substantial clinical challenge.

Interleukin-4 and Interleukin-13

In a series of pioneering studies performed on T helper lymphocytes in the mouse, cDNA clones were discovered to be exclusively expressed by the Th2 subset (38). These cytokines have been broadly classified as Th2, or type 2, cytokines and include most notably, IL-4 and IL-13. These type 2 cytokines are linked by their location on the long arm of chromosome 5, by modest sequence homology at receptor ligand binding sites (39), and by the receptor complexes they employ to initiate signal transduction (40). They have also been shown to be important modifiers of the lung immune response (see Chapter 4). Together, these cytokines display a number of anti-inflammatory properties.

Although, expressed exclusively in mouse Th2 cells, studies using human cells have shown that both Th1 and Th2 cells, as well as mast cells, basophils and eosinophils are important sources of IL-4 and IL-13 expression. The cloning and eventual recombinant expression of the human homologues of IL-4 and IL-13 have shown them to exert potent immunomodulatory properties on monocytes (41). For example, IL-13 was demonstrated to inhibit inflammatory cytokine (TNFα) and chemokine (MIP-1α) expression from macrophages, while also increasing the anti-inflammatory molecule, IL-1Ra (42, 43). Further *in vivo* work examined the regulatory role of IL-4 and IL-13 in lung inflammatory models. In the immune complex model of lung injury described above, the intratracheal instillation of recombinant IL-4 and IL-13 at the initiation of injury resulted in decreased pulmonary vascular permeability, and diminished neutrophil numbers that was thought to reflect decreased lung production of TNFα; though in these studies, the effect of IL-13 was substantially more than IL-4 (44). Other models of lung inflammation described similar results for IL-13. In guinea pigs, airway instillation of IL-13 inhibited the leukocyte accumulation following TNFα- and antigen-induced inflammation (45). Furthermore, in models of endotoxemia, transgenic over-expression of IL-4 and IL-13 were shown to reduce TNFα production and improve outcome (46, 47). Interestingly, in a cecal-ligation and puncture model of sepsis, increased local organ (e.g. lung, liver), but not systemic, expression of IL-13 was observed and neutralization of endogenous IL-13 resulted in increased mortality. These data suggested that IL-13 was key to the regulation of organ-specific inflammatory responses (48).

The mechanisms by which these cytokines exert their anti-inflammatory effects remains to be fully elucidated, though a great deal is understood with regards to the signal transduction pathways initiated by them. IL-4 and IL-13 share a common receptor, the IL-4Rα. In the case of IL-4, this receptor conjugates with the common γ–chain (a component of the IL-2 receptor) to form the high affinity IL-4R complex, whereas the IL-13Rα or IL-13Rα′ receptor appears to complex with the IL-4Rα subunit to form a high affinity IL-13R complex (reviewed 49, 50). Of note, in non-hematopoietic cells IL-13Rα′ may serve as the accessory chain comprising the IL-4 responsive receptor (49). On the basis of these shared common receptor subunits, it is no surprise that several overlapping biologic effects have been described. Subsequent signaling proceeds through activation of the Jak kinases, Jak1 and Jak3, resulting in the phosphorylation of the IL-4Rα receptor with consequent recruitment and activation of Stat6 (51-53). Support for this pathway has been substantiated by studies employing the Stat6 null mutant mouse in which IL-4 and IL-13 signaling are substantially impaired (49, 54). However, the role of this pathway in mediating the observed anti-inflammatory effects is unclear. Further work in the immune complex-mediated model of lung injury not only identified IL-13 to be endogenously expressed in the lung, but also showed that exogenous IL-13 inhibited NF-κB activation in association with the preservation of the NF-κB inhibitory protein, IκB-α, suggesting an alternative mechanism of cytokine inhibition (55). Additionally, IL-4 has been demonstrated to increase the expression of the cytokine-induced SH2-containing (CIS) protein as well as SOCS1, 2 and 3 proteins that serve as negative regulators of the Jak-stat signaling pathway (56). Whether these proteins are necessary for the observed anti-inflammatory effects remains to be determined. Finally, the role of IL-4-mediated induction of phosphatases, which can deactivate kinase activities via dephosphorylation in the observed anti-inflammatory effects is also unknown (57).

The consideration of a therapeutic use for IL-4 and IL-13 in ALI is assuaged by the important observations regarding the role these cytokines play in mediating allergic lung inflammation. Although beyond the scope of this chapter, the reader is referred to recent reviews that summarize the mechanisms by which these Th2 cytokines, IL-4 and IL-13, contribute to the pathophysiology of airway hyperreactivity (58, 59). As an example, Elias and colleagues using a model of lung-specific, transgenic over-expression of IL-13 demonstrated increased inflammation, mucus production, subepithelial fibrosis, increased eotaxin production, and airway hyperreactivity all characteristic of the pathophysiology associated with asthma (60). More recently, analysis of the genetic regulation of both IL-4 and IL-13 has shown that there likely exists genotypic variances of these molecules. These variants influence the degree of IL-4 and IL-13 expression with consequent effects on both the development of a Th2 immune response as well as an atopic or allergic phenotype (61-63). Thus, until further insight is gained into how the interplay between the kinetics of expression, the presence of modifying factors, the diverse use of receptor subtypes, the nature of the inflammatory stimulus and the cell population involved influence the response to IL-4 and IL-13, therapeutic use in ALI remains a distant consideration.

gp130 Receptor Ligands: IL-6 and IL-11

IL-6 is a pleiotropic cytokine with a diverse set of functions. IL-6 can drive the acute phase response in hepatocytes, stimulate hematopoiesis, induce maturation of megakaryocytes, differentiate myeloid cells and induce neuronal differentiation (reviewed 64, 65). On the basis of further structural characterization and the signal transduction pathway it utilized, IL-6 was identified as a representative member of a family of proteins that served as receptor gp130 ligands. This family now includes: interleukin-11 (IL-11), leukemia inhibitory factor, ciliary neurotropic factor, oncostatin M and cardiotrophin-1 (66). IL-6 has been viewed historically as a proinflammatory cytokine based on the frequent observation that it is increased in a number of inflammatory disease states such as sepsis and ALI (67-69); however, while IL-6 may be a valid marker for the degree of inflammation associated with sepsis and ALI, it has remained unclear as to whether it is a mediator of that inflammation.

In studies aimed at determining the role of IL-6 in regulating lung inflammation, a consistent finding of anti-inflammatory properties was observed. Initially, IL-6 was shown to inhibit LPS-induced TNFα expression from human monocytes (70) as well as rat alveolar macrophages (71). In a variety of lung injury models it was demonstrated that exogenous administration of IL-6 abrogated the amount of lung histopathology in association with decreased cytokine production and neutrophil accumulation (70, 72-74). It was subsequently shown that endogenous IL-6 was produced in the context of lung inflammation triggered by immune complex deposition, and that antibody neutralization of IL-6 resulted in a heightened inflammatory response characterized by increased BAL fluid levels of TNFα (71). Recently, investigators have employed the IL-6 null mutant (IL-6 -/-) mouse to further define the role of IL-6 in regulating acute inflammatory responses. In the IL-6 -/- mice the levels of TNFα and MIP-2 as well as the degree of neutrophilia were all significantly increased following aerosol exposure to endotoxin (75). These data supported the hypothesis that IL-6 was a key endogenous regulator of inflammation, however, the mechanism by which inflammatory cell signaling was impaired remained undefined. Because IL-6 induces acute phase response proteins, which have in turn demonstrated anti-inflammatory effects (76), it had been speculated that this is the mechanism by which IL-6 might exert its effects. Of note, following endotoxin challenge in the IL-6 -/- mice, the acute phase response was only moderately impaired in comparison to the degree of the anti-inflammatory effect, suggesting the regulatory mechanism in this model may be independent of the acute phase response (75). Additionally, no differences in IL-10 or TGF-ß levels were observed, suggesting the effect of IL-6 was unrelated to modulation of the expression of other anti-inflammatory cytokines. Of note, induction of the acute phase response in the IL-6 -/- mouse by stimuli other than endotoxin was significantly impaired (77).

A related cytokine, interleukin-11 (IL-11) employs the same gp130 receptor and has been examined for similar anti-inflammatory properties. Induction of IL-11 expression can be observed with a number of stimuli including proinflammatory cytokines (e.g. IL-1), TGF-ß, prostaglandins, and in particular viruses (e.g. respiratory syncytial virus) (78). Gene expression of IL-11 appears to be mediated by both transcriptional and post-transcriptional mechanisms. The promoter of the IL-11 gene contains AP-1, Sp-1, and NF-κB cis-elements and mRNA expression induced by an AP-1 transcriptional complex containing junD has been observed (79). Additionally, the 3'-untranslated region contains multiple copies of an AUUUA sequence conferring constitutive instability to the transcript which can be stabilized following activation in a manner dependent on tyrosine kinase activity (79).

Because of its potent effect on megakaryocytopoiesis (80), IL-11 has entered clinical use as an inducer of platelet production in the setting of chemotherapy-induced thrombocytopenia (81). However, its role as an immunomodulating agent remains incompletely defined. Similar to IL-6, IL-11 has been shown to inhibit the inflammation observed in various models of ALI. For example, following immune-complex-induced lung injury, intrapulmonary administration of IL-11 abrogated the increases in lung vascular permeability, lung neutrophils, as well as BAL fluid TNF and C5a content (82). IL-11 has also been shown to attenuate lung inflammation resulting from endotoxin challenge (83), hyperoxia (84) and radiation-induced thoracic injury (85). As a result of these findings, investigators have attempted to elucidate the mechanisms by which both IL-6 and IL-11 exert their anti-inflammatory effects.

As mentioned previously, all members of this family share gp130 as at least one of the subunits of the receptor complex to which they bind. For IL-6 and IL-11, the signal transduction pathways are activated by engagement of the protein with homodimerization of the gp130 receptor. Following ligand engagement, multiple signaling pathways are activate including: Jak-stat tyrosine kinases (86, 87), MAP kinases (79), Src family kinases (88), and phosphatidylinositol 3-kinase (PI3K) (89). Homodimerization of gp130 activates the Jak family tyrosine kinases, Jak1, Jak2, and Tyk2, which in turn phosphorylate the cytoplasmic portion of gp130. This portion can then serve as a docking site for the Stat3 transcription factor, which becomes phosphorylated, dimerized, and translocated to the nucleus to drive gene transcription (90). In the context of Jak-stat coupled signaling, the novel SOCS family of proteins that serve as negative regulators of cytokine expression has recently been described to inhibit signaling by interfering with Jak kinase catalytic activity (91, 92). IL-6 has been shown to increase SOCS1 protein and may explain one mechanism by which cytokine production is disrupted by the gp130 family of proteins. Furthermore, the effect of IL-6 on inhibition of the Th1 type response (characterized by increased IFN-γ expression) is mediated via the SOCS1 expression (93). Alternatively, both IL-6 and IL-11 have been shown to impair NF-κB signaling both *in vitro* and *in vivo*. The mechanism by which this occurs appears to involve increased production of the inhibitors of NF-κB, IκBα and IκBβ (94). Whether additional regulatory pathways are activated by these gp130 ligands remains to be determined; however, these agents may provide a therapeutic potential in both acute and chronic inflammatory states.

Enthusiasm for the use of IL-6 and IL-11 in ALI is tempered by the elucidation of the role these proteins appear to play in lung remodeling. In an elegant series of experiments employing lung-specific over-expressing transgenic mice, Elias and colleagues have provided novel insight into the role that IL-6 and IL-11 play in the development of subepithelial fibrosis, collagen deposition, and accumulation of α-smooth muscle actin-containing structural cells, thus contributing to the development of lung fibrosis (reviewed 95). Of note, targeted over-expression of IL-6 or IL-11 in a lung specific manner was not associated with eosinophilia or excessive mucus production that is characteristic of allergic inflammation driven by type 2 cytokines. In fact, both cytokines decrease allergen-induced eosinophilia and the production of Th2 cytokines that contribute to eosinophil recruitment (96, 97). Together, these data support the concept that IL-11 is a key endogenous mediator of airway healing following the development of airway inflammation. Further understanding of the pleiotropic effects of this family of cytokines and the modifying factors such as the timing, etiology, and immune nature of insult will be required prior to their therapeutic consideration in ALI.

Summary

The proinflammatory cytokines and the biologic effects they orchestrate are a necessary component of the immune responses directed against host invasion. It is naive to conclude that all critically ill patients with ALI require inhibition of the proinflammatory response, which may lead to undesired immunosuppression. Traditionally, because of the proximal role cytokines play in the inflammatory cascade, clinical investigators have attempted to directly block their activity. Though these strategies have proved promising in preclinical trials, their ultimate clinical efficacy in human trials has been disappointing (98, 99). In fact, over-expression of anti-inflammatory cytokines such as IL-10 and IL-13 may contribute to host immunosuppression and impair pathogen eradication, creating a need for boosting the proinflammatory cytokine armamentarium in selected patients. Thus, it is necessary that the host strike a homeostatic cytokine balance in its attempt to accomplish pathogen eradication, but not at the expense of organ injury (Figure 1). Further understanding of the mechanisms of anti-inflammatory activity in the context of improved identification of both the current immunophenotype (i.e. a predominant proinflammatory versus anti-inflammatory state) of the patient with ALI as well as pathologic challenge (i.e. infectious versus non-infectious) being faced is likely to lead to more prudent and selected use of anti-inflammatory cytokines in this challenging disease state.

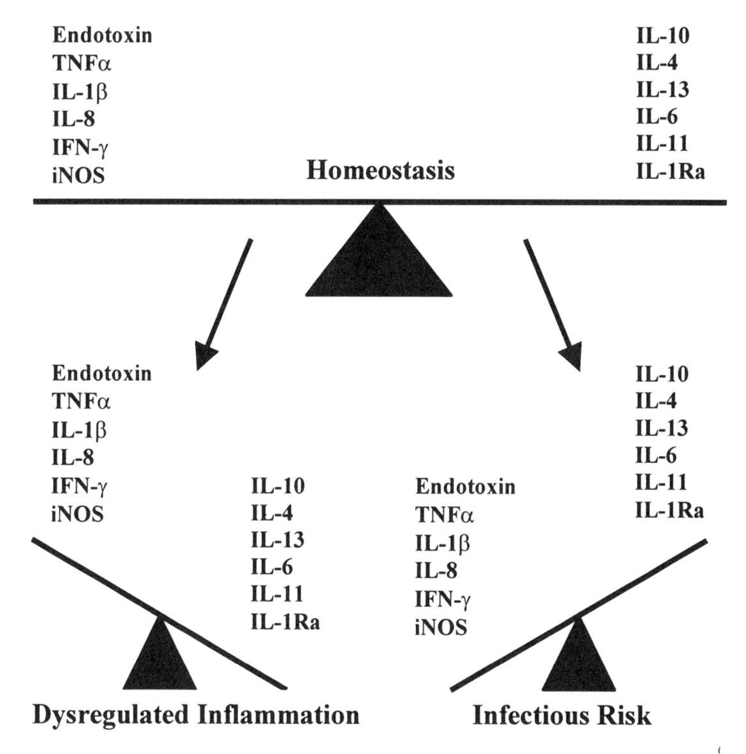

Figure 1. Balance between pro- and antiinflammatory cytokines aimed at preserving homeostasis in the context of inflammatory diseases states.

REFERENCES

1. Bone, R.C. (1996) Sir Isaac Newton, sepsis, SIRS and CARS. *Crit Care Med* 24, 1125-1128
2. Fiorentino, D.F., Bond, M.W., and Mosmann, T.R. (1989) Two types of mouse T helper cells IV. Th2 clones secrete a factor that inhibits cytokine production by Th1 clones. *J Exp Med* 170, 2081-2095
3. Vieira, P., de Waal-Malefyt, R., Dang, M.N., Johnson, K.E., Kastelein, R., Fiorentino, D.F., deVries, J.E., Roncarolo, M.G., Mosmann, T.R., and Moore KW. (1991) Isolation and expression of human cytokine synthesis inhibitory factor (CSIF.IL-10) cDNA clones: homology to Epstein Barr virus opening reading frame. *Proc Natl Acad Sci USA* 88, 1172-1176
4. de Waal Malefyt, R., Abrams, J., Bennett, B., Figdor, C.G., and de Vries, J.E. (1991) Interleukin-10 (IL-10) inhibits cytokine synthesis by human monocytes: an auto-regulatory role of IL-10 produced by monocytes. *J Exp Med* 174, 1209-1220
5. Howard, M., and O'Garra, A. (1992) Biological properties of interleukin-10. *Immunol Today* 13, 198-200
6. Moore, K.W., de Waal Malefyt, R., Coffman, R.L., and O'Garra, A. (2001) Interleukin-10 and the interleukin-10 receptor. *Annu Rev Immunol* 19, 683-765

7. Opal, S.M., and DePalo, V.A. (2000) Anti-inflammatory cytokines. *Chest* 117, 1162-1172
8. Cassatella, M.A., Meda, L., Gasperini, S., Calzetti, F., and Bonora, S. (1994) Interleukin-10 (IL-10) upregulates IL-1 receptor antagonist production from lipopolysaccharide-stimulated human polymorphonuclear leukocytes by delaying mRNA degradation. *J Exp Med* 179, 1695-1699
9. Dickensheets, H.L., Freeman, S.L., Smith, M.F. and Donnelly, R.P. (1994) Interleukin-10 upregulates tumor necrosis factor receptor type II (p75) gene expression in endotoxin-stimulated human monocytes. *Blood* 90, 4162-4171
10. Tone, M., Powell, M.J., Tone, Y., Thompson, S.A.J., and Waldmann, H. (2000) IL-10 gene expression is controlled by the transcription factors Sp1 and Sp3. *J Immunol* 165, 286-291
11. Powell, M.J., Thompson, S.A.J., Tone, Y., Waldmann, H., Tone, M. (2000) Post-transcriptional regulation of IL-10 gene expression through sequences in the 3'-untranslated region. *J Immunol* 165, 292-296
12. Kotenko, S.V., Krause, C.D., Izotova, L.S., Pollack, B.P., Wu, W., and Pestka, S. (1997) Identification and functional characterization of a second chain of the interleukin-10 receptor complex. *EMBO J* 16, 5894-5903
13. Hou, J., Schindler, U., Henzel, W.J., Ho, T.C., Brasseur, M., and McKnight, S.L. (1994) An interleukin-4-induced transcription factor: IL-4 Stat. *Science* 265, 1701-1706
14. Finbloom D.S., and Winestock, K.D. (1995) IL-10 induces the tyrosine phosphorylation of tyk2 and Jak1 and the differential assembly of Stat-1α and Stat3 complexes in human T-cells and monocytes. *J Immunol* 155, 1079-1090
15. Riley, J.K., Takeda, K., Akira, S., and Schreiber, R.D. (1990) Interleukin-10 receptor signaling through the JAK-STAT pathway. *J Immunol* 155, 1079-1090
16. Weber-Nordt R.M., Riley J.K., Greenlund A.C., Moore K.W., Darnell J.E., and Schreiber, R.D. (1996) Stat3 recruitment by two distinct ligand-induced, tyrosine-phosphorylated docking sites in the interleukin-10 receptor intracellular domain. *J Biol Chem* 271, 27954-27961
17. Takeda, K., Clausen, B.E., Kaisho, T., Tsujimura, T., Terada, N., Forster, I., and Akira, S. (1999) Enhanced Th1 activity and development of chronic enterocolitis in mice devoid of Stat3 in macrophages and neutrophils. *Immunity* 10, 39-49
18. Ito, S., Ansari, P., Sakatsume, M., Dickensheets, H., Vazquex, N., Donnelley, R.P., Larner, A.C., and Finbloom, D.S. (1999) Interleukin-10 inhibits expression of both interferon α- and interferon γ-induced genes by suppressing tyrosine phosphorylation of STAT1. *Blood* 93, 1456-1463
19. Cassatella, M.A., Gasperini, S., Bovolenta, C., Calzetti, F., Vollebregt, M., Scapini, P., Marchi, M., Suzuki, R., Suzuki, A., and Yoshimura, A. (1999) Interleukin-10 (IL-10) selectively enhances CIS3/SOCS3 mRNA expression in human neutrophils: evidence for an IL-10-induced pathway that is independent of STAT protein activation. *J Interferon Cytokine Res* 19, 679-685
20. Geng, Y., Gulbins, E., Altman, A., and Lotz, M. (1994) Monocyte deactivation by interleukin-10 via inhibition of tyrosine kinase activity and the Ras signaling pathway. *Proc Natl Acad Sci USA* 91, 8602-8606
21. Schottelius, A.J., Mayo, M.W., Sartor, R.B., and Baldwin, A.S. Jr. (1999) Interleukin-10 signaling blocks inhibitor of kappaB kinase activity and nuclear factor kappaB DNA binding *J Biol Chem* 274, 9558-9563
22. Shames, B.D., Selzman, C.H., Meldrum, D.R., Pulido, E.J., Barton, H.A., Meng, X., Harken, A.H., and McIntyre, R.C. Jr. (1998) Interleukin-10 stabilizes inhibitory IκB-α in human monocytes. *Shock* 10, 389-394
23. Shanley, T.P., Vasi, N., and Denenberg, A. (2001) Regulation of chemokine expression by IL-10 in lung inflammation. *Cytokine* 12, 1054-1064

24. Kasama, T., Strieter, R.M., Lukacs, N.W., Burdick, M.D., and Kunkel, S.L. (1994) Regulation of neutrophil-derived chemokine expression by IL-10. *J Immunol* 152, 3559-3569
25. Kishore, R., Tebo, J.M., Kolosov, M., Hamilton, T.A. (1999) Clustered AU-rich elements are the target of IL-10-mediated mRNA destabilization in mouse macrophages. *J Immunol* 162, 2457-2461
26. Mulligan, M.S., Jones, M.L., Vaporciyan, A.A., Howard, M.C., and Ward, P.A. (1993) Protective effects of IL-4 and IL-10 against immune complex-induced lung injury. *J Immunol* 151, 5666-5674
27. Huaux, F., Louahed, J., Hudsmith, B., Meredith, C., Delos, M., Renauld, J.-C., and Lison, D. (1998) Role of interleukin-10 in the lung response to silica in mice. *Am J Resp Cell Mol Biol* 18, 51-59
28. Arai, T., Abe, K., Matsuoka, H., Yoshida, M., Mori, M., Goya, S., Kida, H., Nishino, K., Osaki, T., Tachibana, I., Kaneda, Y., and Hayashi, S. (2000) Introduction of the interleukin-10 gene into mice inhibited bleomycin-induced lung injury in vivo. *Am J Physiol Lung Cell Mol Physiol* 278, L914-L922
29. Morrison, D.F., Foss, D.L., and Murtaugh, M.P. (2000) Interleukin-10 gene therapy-mediated amelioration of bacterial pneumonia. *Inf Immun* 30, 4752-4758
30. Gudmundsson, G., Bosch, A., Davidson, B.L., Berg, D.J., and Hunninghake, G.W. (1998) Interleukin-10 modulates the severity of hypersensitivity pneumonitis in mice. *Am J Resp Cell Mol Biol* 19, 812-818
31. Kuhn, R., Lohler, J., Rennick, D., Rajewsky, K., and Muller, W. (1993) Interleukin-10 deficient mice develop chronic enterocolitis. *Cell* 75, 263-274
32. Berg, D.J., Kuhn, R., Rajewsky, K., Muller, W., Menon, S., Davidson, N.J., Grunig, G., and Rennick, D. (1995) Interleukin-10 is a central regulator of the response to LPS in murine models of endotoxic shock and the Schwartzman reaction but not of endotoxin tolerance. *J Clin Invest* 96, 2339-2347
33. Shanley, T.P., Jones, M.L., Schmal, H., Friedl, H.P., and Ward, P.A. (1995) Regulatory effects of intrinsic interleukin-10 in IgG immune complex-induced lung injury. *J Immunol* 154, 3454-3460
34. Donnelly, S.C., Strieter, R.M., Reid, P.T., Kunkel, S.L., Burdick, M.D., Armstrong, I., Mackenzie, A., and Haslett, C. (1996) The association between mortality rates and decreased concentrations of interleukin-10 and interleukin-1 receptor antagonist in the lung fluids of patients with the adult respiratory distress syndrome. *Ann Intern Med* 125, 191-196
35. Westendorp, R.G., Langermans, J.A., de Bel, C.E., Meinders, A.E., Vandenbroucke, J.P., van Furth, R., and van Dissel, J.T. (1995) Release of tumor necrosis factor: an innate host characteristic that may contribute to the outcome of meningococcal disease. *J Infect Dis* 171, 1057-1060
36. Eskdale, J., and Gallagher, G. (1995) A polymorphic dinucleotide repeat in the human IL-10 promoter. *Immunogen* 42, 444-445
37. Eskdale, J., Gallagher, G., Verweij, C.L., Keijsers, V., Westendorp, R.G.J., and Huizinga, T.W.J. (1998) Interleukin 10 secretion in relation to human IL-10 locus haplotypes. *Proc Natl Acad Sci USA* 95, 9465-9470
38. Cherwinski, H.M., Schumacher, J.H., Brown, K.D., and Mosmann, T.R. (1987) Two types of mouse helper T cell clones III. Further differences in lymphokine synthesis between Th1 and Th2 clones revealed by RNA hybridization, functionally monospecific bioassays and monoclonal antibodies. *J Exp Med* 166, 1229-1244
39. Smirnov, D.V., Smirnova, M.G., Korobko, V.G. and Frolova, E.I. (1995) Tandem arrangement of human genes for interleukin-4 and interleukin-13: resemblance in their organization. *Gene* 155, 277-281

40. Zurawski, S.M., Vega, F., Huyghe, B., and Zurawski, G. (1993) Receptors for interleukin-13 and interleukin-4 are complex and share a novel component that functions in signal transduction. *EMBO J* 12, 2663-2670
41. de Vries, J.E. (1996) Molecular and biological characteristics of interleukin-13. *Chem Immunol* 66, 101-111
42. Yanagawa, H., Sone, H., Haku, T., Mizuno, K., Yano, S., Ohmoto, Y., and Ogura, T. (1995) Contrasting effect of IL-13 on interleukin-1 receptor antagonist and proinflammatory cytokine production by human alveolar macrophages. *Am J Respir Cell Mol Biol* 12, 71-76
43. Berkman, N., John, M., Roesems, G., Jose, P.J., Barnes, P.J., and Chung, K.F. (1996) Interleukin-13 inhibits macrophage inflammatory protein-1α production from human alveolar macrophages and monocytes. *Am J Respir Cell Mol Biol* 15, 382-389
44. Mulligan, M.S., Warner, R.L., Foreback, J.L., Shanley, T.P., and Ward, P.A. (1997) Protective effects of IL-4, IL-10, IL-12 and IL-13 in IgG immune complex-induced lung injury: role of endogenous IL-12. *J Immunol* 159, 3483-3489
45. Watson, M.L., White, A.-M., Campbell, E.M., Smith, A.W., Uddin, J., Yoshimura, T., and Westwick, J. (1999) Anti-inflammatory actions of interleukin-13. *Am J Respir Cell Mol Biol* 20, 1007-1012
46. Baumhofer, J.M., Beinhauer, B.G., Wang, J.E., Brandmeier, H., Geissler, K., Losert, U., Phillip, R., Aversa, G., and Rogy, M.A. (1998) Gene transfer with IL-4 and IL-13 improves survival in lethal endotoxemia in the mouse and ameliorates peritoneal macrophages immune competence. *Eur J Immunol* 28, 610-615
47. Muchamuel, T., Menon, S., Pisacane, P., Howard, M.C., and Cockayne, D.A. (1997) IL-13 protects mice from lipopolysaccharide-induced lethal endotoxemia: correlation with down-modulation of TNF-α, IFN-γ and IL-12 production. *J Immunol* 158, 2898-2903
48. Matsukawa, A., Hogaboam, C.M., Lukacs, N.W., Lincoln, P.M., Evanoff, H.L., Strieter, R.M., and Kunkel, S.L. (2000) Expression and contribution of endogenous IL-13 in an experimental model of sepsis. *J Immunol* 164, 2738-2744
49. Nelms, K., Keegan, A.D., Zamorano, J., Ryan, J.J., and Paul, W.E. (1999) The IL-4 receptor: signaling mechanisms and biologic functions. *Annu Rev Immunol* 17, 701-738
50. Hilton, D.J., Zhang, J.-G., Metcalf, D., Alexander, W.S., Nicola, N.A., and Willson, T.A. (1996) Cloning and characterization of a binding subunit of the interleukin 13 receptor that is also a component of the interleukin 4 receptor. *Proc Natl Acad Sci USA* 93, 497-501
51. Misazaki, T., Kawahara, A., Fujii, H., Nakagawa, Y., Minami, Y., Liu, Z.-J., Oishi, I., Silvennoinen, O., Witthuhn, B.A., Ihle, J.N., and Taniguchi, T. (1994) Functional activation of Jak1 and Jak3 by selective association with IL-2 receptor subunits. *Science* 266, 1045-1047
52. Taketa, K., Tanaka, T., Shi, W., Matsumoto, M., Kashiwamura, S.-I., Nakanishi, K., Yoshida, N., Kishmoto, T., and Akira, S. (1996) Essential role of STAT6 on IL-4 signaling. *Nature* 380, 627-630
53. Kaplan, M.H., Schindler, U., Smiley, S.T. and Grusby, M.J. (1996) STAT6 is required for mediating responses to IL-4 and for the development of Th2 cells. *Immunity* 4, 313-319
54. Takeda, K., Kamanaka, M., Tanaka, T., Kishimoto, T., and Akira, S. (1996) Impaired IL-13-mediated functions of macrophages in STAT6-deficient mice. *J Immunol* 157, 3220-3232
55. Lentsch, A.B., Czermak, B.J., Jordan, J.A. and Ward, P.A. (1999) Regulation of acute lung inflammatory injury by endogenous IL-13. *J Immunol* 162, 1071-1076

56. Starr, R., Willson, T.A., Viney, E.M., Murray, L.J., Rayner, J.R., Jenkins, B.J., Gonda, T.J., Alexander, W.S., Metcalf, D., Nicola, N.A., and Hilton, D.J. (1997) A family of cytokine-inducible inhibitors of signaling. *Nature* 387, 917-921
57. Jiang, H., Harris, M.B., and Rothman, P. (2000) IL-4/IL-13 signaling beyond JAK/STAT. *J Allergy Clin Immunol* 105, 1063-1070
58. Wills-Karp, M. (1999) Immunologic basis of antigen-induced airway hyper-responsiveness. *Ann Rev Immunol* 17, 255-281
59. Wills-Karp, M., Luyimbazi, J., Xu, X., Schofield, B., Neben, T.Y., Karp, C.L., and Donaldson, D.D. (1998) Interleukin-13: central mediator of allergic asthma. *Science* 282, 2258-2261
60. Zhu, Z., Homer, R.J., Wang, Z., Chen, Q., Geba, G.P., Wang, J., Zhang, Y., and Elias, J.A. (1999) Pulmonary expression of interleukin-13 causes inflammation, mucus hypersecretion, subepithelial fibrosis, physiologic abnormalities and eotaxin production. *J Clin Invest* 103, 779-788
61. Wierenga, E.A., and Messer, G. (2000) Regulation of interleukin-4 gene transcription: alterations in atopic disease? *Am J Respir Crit Care Med* 162, S81-S85
62. Heinzmann, A., Mao, X.Q., Akaiwa, M., Kreomer, R.T., Gao, P.S., Ohshima, K., Umeshita, R., Abe, Y., Braun, S., Yamashita, T., Roberts, M.H., Sugimoto, R., Arima, K., Arinobu, Y., Yu, B., Kruse, S., Enomoto, T., Dake, Y., Kawai, M., Shimazu, S., Sasaki, S., Adra, C.N., Kitaichi, M., Inoue, H., Yamauchi, K., Tomichi, N., Kurimoto, F., Hamasaki, N., Hopkin, J.M., Izuhara, K., Shirakawa, T., Deichmann, K.A. (2000) Genetic variants of IL-13 signaling and human asthma and atopy. *Hum Mol Genet* 9, 549-559
63. McKenzie, A.N. (2000) Regulation of T helper type 2 cell immunity by interleukin-4 and interleukin-13. *Pharmacol Ther* 88, 143-151
64. Gadient, R.A., and Patterson, P.H. (1999) Leukemia inhibitory factor, interleukin-6, and other cytokines using the gp130 transducing receptor: roles in inflammation and injury. *Stem Cells* 17, 127-137
65. Kishimoto, T., Akira, S., and Taga, T. (1992) Interleukin-6 and its receptor: a paradigm for cytokines. *Science* 258, 593-597
66. Chen-Kiang, S., Hsu, W., Natkunam, Y., and Zhang, X. (1993) Nuclear signaling by interleukin-6. *Curr Opin Immunol* 5, 124-128
67. Donnelly, T.J., Meade, P., Jagels, M., Cryer, G., Lae, M.M., Hugli, T.E., Shoemaker, W.C., and Abraham, E. (1994) Cytokine, complement and endotoxin profiles associated with the adult respiratory distress syndrome after severe injury. *Crit Care Med* 22, 768-776
68. Damas, P., Ledoux, D., Nys, M., DeGroote, D., Franchimont, P., and Lamy, M. (1991) Cytokine serum levels during severe sepsis in human: IL-6 as a marker of severity. *Ann Surg* 215, 356-366
69. Roumen, T., Hendriks, J., van der Ven-Jongekrijg, G.A., Nieuwenhuijzen, R., Sauerwein, W., van der Meer, J.W.M. and Goris, R.J.A. (1993) Cytokine patterns in patients after major vascular surgery, hemorrhagic shock, and severe blunt trauma. *Ann Surg* 218, 769-776
70. Aderka, D., Le, J., and Vilcek, J. (1989) IL-6 inhibits lipopolysaccharide-induced tumor necrosis factor production in cultured human monocytes, U937 cells, and in mice. *J Immunol* 143. 3517-3523
71. Shanley, T.P., Foreback, J.L., Remick, D.G., Ulich, T.R., Kunkel, S.L., and Ward, P.A. (1997) Regulatory effects of interleukin-6 in immunoglobulin G immune complex-induced lung injury. *Am J Pathol* 151, 193-203
72. Ulich, T.R., Yin, S., Guo, K., Yi, E.S., Remick, D.G., and del Castillo, J. (1991) Intratracheal injection of endotoxin and cytokines. II. Interleukin-6 and transforming growth factor-ß inhibit acute inflammation. *Am J Pathol* 138, 1095-1101

73. Ulich, T.R., Guo, K., Remick, D.G., del Castillo, J. and Yin, S. (1990) Endotoxin-induced cytokine expression in vivo. III. IL-6 mRNA and serum expression and the in vivo hematologic effects of IL-6. *J Immunol* 146, 2316-2326
74. Denis, M. (1992) Interleukin-6 in mouse hypersensitivity pneumonitis: changes in lung free cells following depletion of endogenous IL-6 or direct administration of IL-6. *J Leukocyte Biol* 52, 197-201
75. Xing, Z., Gauldie, J., Cox, G., Baumann, H., Jordana, M., Lei, X.F., and Achong, M.K. (1998) IL-6 is an antiinflammatory cytokine required for controlling local or systemic acute inflammatory responses. *J Clin Invest* 101, 311-320
76. Alcorn, J.M., Fierer, J., and Chojkier, M. (1992) The acute phase response protects mice from D-galactosamine sensitization to endotoxin and tumor necrosis factor-α. *Hepatology* 15, 122-129
77. Kopf, M., Baumann, H., Freer, G., Freudenberg, M., Lamers, M., Kishimoto, T., Zinkernagel, R., Bluethmann, H., and Kohler, G. (1994) Impaired immune and acute-phase responses in interleukin-6 deficient mice. *Nature* 368, 339-342
78. Einarsson, O., Geba, G.P., Zhou, Z., Landry, M.L., Panettieri, R.A. Jr., Tristam, D., Welliver, R., Metinko, A., and Elias, J.A. (1993) Interleukin-11 in respiratory inflammation. *Ann NY Acad Sci* 89-101
79. Yin, T.G., and Yang, Y.-C. (1994) Involvement of MAP kinase and pp90^{rsk} activation in signaling pathways shared by interleukin-11, interleukin-6, leukemia inhibitory factor and oncostatin M in mouse 3T3-L1 cells. *J Biol Chem* 269, 3731-3738
80. Teramura, M., Kobayashi, S., Hoshino, S., Oshimi, K., and Mizoguchi, H. (1992) Interleukin-11 enhances human megakaryocytopoiesis in vitro. *Blood* 79, 327-331
81. Tepler, I., Elias, L., Smith, J.W., Hussein, M., Rosen, G., Chang, A.Y., Moore, J.O., Gordon, M.S., Kuca, B., Beach, K.J., Loewy, J.W., Garnick, M.B., and Kaye, J.A. (1996) A randomized-placebo-controlled trial of recombinant human interleukin-11 in cancer patients with severe thrombocytopenia due to chemotherapy. *Blood* 87, 3607-3614
82. Lentsch, A.B., Crouch, L.D., Jordan, J.A., Czermak, B.J., Yun, E.C., Guo, R., Sarma, V., Diehl, K.M., and Ward, P.A. (1999) Regulatory effects of interleukin-11 during acute lung inflammatory injury. *J Leukoc Biol* 66, 151-157
83. Sheridan, B.C., Dinarello, C.A., Meldrum, D.R., Fullerton, D.A., Selzman, C.H., and McIntyre, R.C. Jr. (1999) Interleukin-11 attenuates pulmonary inflammation and vasomotor dysfunction in endotoxin-induced lung injury. *Am J Physiol* 277, L861-L867
84. Waxman, A.B., Einarsson, O., Seres, T., Knickelbein, R.G., Warshaw, J.B., Johnston, R., Homer, R.J., and Elias, J.A. (1998) Targeted lung expression of interleukin-11 enhances murine tolerance of 100% oxygen and diminishes hyperoxia-induced DNA fragmentation. *J Clin Invest* 101, 1970-1982
85. Redlich, C.A., Gao, X., Rockwell, S., Kelly, M., and Elias, J.A. (1996) IL-11 enhances survival and decreases TNF production after radiation-induced thoracic injury. *J Immunol* 157, 1705-1710
86. Taga, T., and Kishimoto, T. (1997) Gp130 and the interleukin-6 family of cytokines. *Annu Rev Immunol* 15, 797-819
87. Yin, T., Yasukawa, K., Taga, T., Kishimoto, T., and Yang, Y.C. (1994) Identification of a 130 kilodalton tyrosine phosphorylated protein induced by interleukin-11, interleukin-6, leukemia inhibitory factor and oncostatin M as JAK2 kinase, which associates with gp130 signal transducer. *Exp Hematol* 22, 467-472
88. Tsyganov, A. and Bolen, J. (1993) The Src family of tyrosine protein kinases in hematopoietic signal transduction. *Stem Cells* 11, 371-380
89. Parker, P.J. and Waterfield, M.D. (1992) Phosphatidylinositol 3-kinase: a novel effector. *Cell Growth Differ* 3, 747-752

90. Pflanz, S., Kurth, I., Grotzinger, J., Heinrich, P.C., and Muller-Newen, G. (2000) Two different epitopes of the signal transducer gp130 sequentially cooperate on IL-6-induced receptor activation. *J Immunol* 165, 7042-7049
91. Alexander, W.S., Starr, R., Metcalf, D., Nicholson, S.E., Farley, A., Elefanty, A.G., Brysha, M., Kile, B.T., Richardson, R., Baca, M., Zhang, J.G., Willson, T.A., Viney, E.M., Sprigg, N.S., Rakar, J., Mifsud, S., DiRago, L., Cary, D., Nicola, N.A., and Hilton, D.J. (1999) Suppressors of cytokine signaling (SOCS): negative regulators of signal transduction. *J Leukoc Biol* 66, 588-92.
92. Endo, T.A., Masura, M., Yokouchi, M., Suzuki, R., Sakamoto, H., Mitsui, K., Matsumoto, A., Tanimura, S., Ohtsubo, M., and Misawa, H. (1997) A new protein containing an SH2 domain that inhibits JAK kinases. *Nature* 387, 921-924
93. Diehl, S., Anguita, J., Hoffmeyer, A., Zapton, T., Ihle, J.N., Fikrig, E., and Rincon, M. (2000) Inhibition of Th1 differentiation by IL-6 is mediated by SOCS1. *Immunity* 13, 805-815
94. Trepicchio, W.L., Wang, L., Bozza, M., and Dorner, A.J. (1997) IL-11 regulates macrophage effector function through inhibition of nuclear factor-κB. *J Immunol* 159, 5661-5670
95. Elias, J.A., Zhu, Z., Chupp, G., and Homer, R.J. (1999) Airway remodeling in asthma. *J Clin Invest* 104, 1001-1006
96. Wang, J., Homer, R.J., Chen, O., and Elias, J.A. (2000) Endogenous and exogenous IL-6 inhibit aeroallergen-induced Th2 inflammation. *J Immunol* 165, 4051-4062
97. Wang, J., Homer, R.J., Hong, L., Cohn, L., Lee, C.G., Jung, S., and Elias, J.A. (2000) IL-11 selectively inhibits aeroallergen-induced pulmonary eosinophilia and Th2 cytokine production. *J Immunol* 165, 2222-2231
98. Bone, R.C. (1996) Why sepsis trials failed. *JAMA* 276, 565-566
99. Lemeshow, S., Teres, D., and Moseley, S. (1996) Statistical issues in clinical sepsis trials. In: Sepsis and multiple organ failure. Baltimore, MD: Williams and Wilkins. pp. 614-626

Chapter 18

THE HEAT SHOCK RESPONSE AND HEAT SHOCK PROTEIN 70: CYTOPROTECTION IN ACUTE LUNG INJURY

Hector R. Wong
Division of Critical Care Medicine, Children's Hospital Medical Center and Children's Hospital Research Foundation, Cincinnati, OH

INTRODUCTION

In 1962 Ferruccio Ritossa described a distinct, hyperthermia-dependent puffing pattern in the giant chromosomes from the salivary glands of *Drosophila* (1). Subsequent investigations linked this puffing pattern with the expression of a specific group of proteins that were fittingly named "heat shock proteins," and the expression of these proteins in response to hyperthermia was termed the "heat shock response." During the ensuing four decades the field of heat shock response biology has grown immensely and it has become evident that all organisms have the capacity to induce the heat shock response, which is now recognized as having fundamental importance in both cellular biology and medicine. One of the more important functional consequences of the heat shock response is its ability to confer cytoprotection against a variety of cellular and tissue injuries relevant to clinical medicine, including ischemia-reperfusion injury, inflammatory injury, and oxidant-mediated injury.

This chapter will focus on the heat shock response and heat shock protein 70 (HSP70), and their potential roles in ameliorating acute lung injury (ALI). For information regarding heat shock proteins, other than HSP70, the reader is referred to several recent reviews (2-8). For information regarding heme oxygenase (also known as HSP32) see Chapter 8.

Inducers of the Heat Shock Response

Hyperthermia is the best known, and first described, inducer of the heat shock response (9). In experimental mammalian systems, the threshold temperature required to induce the heat shock response is beyond the range of normal fever. Thus, the heat shock response is typically seen with exposure to temperatures of approximately 42° C. Whether or not typical fever ranges seen in humans (38.5 to 40.5° C) are capable of independently inducing the heat shock response in the clinical setting remains to be fully determined, and remains a fertile area for clinical investigation.

It is now well established that a variety of chemical compounds and clinical conditions are also capable of inducing the heat shock response (i.e. induction of heat shock protein gene expression) in a manner similar to hyperthermia. Examples of chemical inducers of the heat shock response include sodium arsenite, prostaglandins, dexamethasone, non-steroidal anti-inflammatory drugs, and curcumin (Table 1) .

Table 1. Chemical Inducers of the Heat Shock Response

Compound	Refs.	Comments
Sodium Arsenite	10,11	Used extensively *in vitro* and *in vivo*.
Prostaglandin-A_1	12-14	Other prostaglandins also active.
Dexamethasone	15,16	Variable effect.
Bimoclomol	17	Hydroxylamine derivative, non-toxic.
Herbimycin A	18	Tyrosine kinase inhibitor.
Geldanamycin	19	Tyrosine kinase inhibitor and HSP 90 inhibitor.
Aspirin	20	Lowers temperature threshold for HSP induction.
NSAID*	21	Lowers temperature threshold for HSP induction.
Serine Protease Inhibitors	22	Concomitant inhibition of NF-κB.
PDTC*	23	Antioxidant; inhibitor of NF-κB.
DETC*	†	Similar to PDTC.
Glutamine	24	May be specific for intestinal epithelium.
Heavy Metal Ions	25	Cadmium, Zinc.
Phosphatase Inhibitors	26,27	Tyrosine and Ser/Thr phosphatases.
Curcumin	28	Major constituent of turmeric; antiinflammatory.
Geranylgeranlyacetone	29	Anti-ulcer drug.

**NSAID, Non-Steroidal Antiinflammatory Drugs; PDTC, Pyrrolidine Dithiocarbamate; DETC, Diethyldithiocarbamate.*
†H. Wong, unpublished data.

It is important to note that many of these compounds do not consistently induce the heat shock response in all experimental models. In addition, some compounds, such as aspirin, do not independently induce the heat shock response, but rather, can lower the temperature threshold for induction of the heat shock response (20). Thus, the ability of these compounds to induce the heat shock response is, in large part, dependent on species, cell type, and experimental model.

From experimental models it is well established that a variety of clinically relevant conditions, including ischemia-reperfusion injury (30), endotoxemia (31), oxidant stress (32, 33), and pancreatitis (34) can also induce the heat shock response. These effects, however, are not universal;

they are highly dependent on species and experimental model. Finally, it is now established that humans express the heat shock response in the setting of disease states directly or indirectly associated with ALI (Table 2).

Table 2. Induction of the Heat Shock Response in Humans

Condition	Ref.	Comments
Lung cancer	35	Various histologic types of tumors.
ALI	36	Decreased inducibility in peripheral monocytes.
ALI	37	Increased expression in alveolar macrophages.
Interstitial lung disease	38	Increased expression in alveolar macrophages.
Severe sepsis	39	Impaired inducibility in peripheral lymphocytes.
Severe sepsis	40	*Ex vivo* inhibition of HSP70 expression by endotoxin.
Asthma	41	Increased in airway cells.
Peripheral vascular disease	42	Increased circulating levels of HSP70.
Diaphragmatic hernia	43	Neonatal patients with pulmonary hypertension.

Since it is well known that the heat shock response confers broad cytoprotection against a variety of cytotoxic stimuli, several implications stem from the observation that the heat shock response is induced by many stimuli (other than hyperthermia) and during many disease states. First, is the concept that a fundamental function of fever may be to induce the heat shock response, and thus induce a universal cytoprotective mechanism. Second, documentation that the heat shock response is induced during various animal models of organ injury, and in various human disease states, implies that the heat shock response may be a fundamental cellular defense mechanism in the clinical setting. Finally, there is potential to develop novel pharmacologic agents, or gene-based therapies, to safely induce the heat shock response as a means to confer whole organ protection in clinical conditions such as ALI.

Regulation of HSP70 Expression

The molecular mass of heat shock proteins ranges from 8 to 110 kDa (reviewed in references 2-8). The HSP70 family is highly conserved throughout vertebrate and invertebrate evolution; homologues are well characterized in bacteria and yeast. HSP70 contains ATP-binding domains and protein binding domains, which function in intracellular protein folding, maturation, and transport (molecular chaperones). Two primary isoforms of HSP70 exist: HSP73 and HSP72. HSP73 is expressed constitutively in all cells and is thought to function as a molecular chaperone under basal conditions. HSP73 is also known as "constitutive HSP70" and "HSP70 cognate." HSP72 is the inducible isoform of HSP70 and is consistently expressed at high levels in response to hyperthermia (heat shock), or variably expressed in response to other forms of cellular stress (ischemia-reperfusion, oxidant stress, endotoxemia). HSP72 is thought to be responsible for many of the cytoprotective properties of the heat shock response.

HSP72 expression is regulated primarily by increases in gene transcription (reviewed in references 2-8). Increased transcription of HSP72 is mediated by a well characterized transcription factor called heat shock factor-1 (HSF-1). In the nonstressed cell, HSF-1 exists as a large cytoplasmic pool of inactive monomers. In response to cellular stress, HSF-1 monomers rapidly form active trimers that translocate to the nucleus and bind the DNA consensus sequence in the HSP72 promoter known as the heat shock element (HSE). HSF-1 binding to the HSE directs rapid and high level expression of HSP72. The HSE is defined by tandem repeats of the pentamer nGAAn ("n" denoting less conserved sequences), arranged in alternating orientations (9). It appears that at least two consecutive pentamers are necessary to constitute a functional HSE, and the pentamers can be arranged either "head to head" (e.g. 5'-nGAAnnTTCn-3') or "tail to tail" (e.g. 5'-nTTCnnGAAn-3'). Importantly, genes encoding for HSP72 do not contain introns.

In summary, abundant levels of HSF-1 monomers are present in the nonstressed cell and can be rapidly activated to direct high level transcription of the HSP72 gene. As an intronless gene, HSP72 mRNA processing is minimized, thereby providing an additional mechanism for rapid expression of HSP72. Thus, the regulatory mechanisms that govern HSP72 expression are optimized for high level expression during cellular stress. HSP72 is highly conserved and is one of the most abundantly expressed heat shock proteins in cells subjected to stress. Many of the cytoprotective properties of the heat shock response are attributed to HSP72 expression. This assertion is well supported by experiments in which HSP72 is overexpressed by genetic manipulation (44-49). In addition, animals and cells having genetic ablation of HSF-1, and thus not capable of inducing expression of HSP72, have impaired tolerance to stresses such as hyperthermia and sepsis (50, 51).

In Vitro Cytoprotection

Several *in vitro* models demonstrate that the heat shock response and HSP70 are capable of conferring cytoprotection against a broad range of lung cell injuries. Induction of the heat shock response, by either hyperthermia or sodium arsenite, protected pulmonary artery endothelial cells against endotoxin-mediated apoptosis (48). The mechanism of protection in this model may involve an antioxidant effect, since induction of the heat shock response attenuated endotoxin-mediated superoxide anion production. Further evidence suggesting that the heat shock response can serve an antioxidant function stems from the observation that induction of the heat shock response in lung cells protected against oxidant stress secondary to hydrogen peroxide (52), nitric oxide (53), or paraquat (54).

A specific role for HSP70 in mediating protection against lung cellular injury is suggested by experiments in which cell lines were generated having increased expression of HSP70 by standard DNA recombinant techniques. For example, specific overexpression of HSP70 protected cultured lung cells against nitric oxide-mediated (53), endotoxin-mediated (48) and hyperoxia-mediated cell death (55). In the model involving hyperoxia-mediated cell death, cytoprotection involved attenuation of lipid peroxidation and intracellular ATP depletion. Recent data suggest that HSF-1 is also an important component of the heat shock response that confers resistance to oxidant stress. Embryonic fibroblasts from HSF-1 null mice are highly sensitive to hyperoxia when compared to embryonic fibroblasts from wild-type mice (*H. Wong, unpublished data*). Collectively, these data indicate that HSP70 and HSF-1 are important, endogenous, cytoprotective molecules in lung tissues.

In Vivo Cytoprotection

Villar and colleagues were the first to demonstrate a cytoprotective effect of the heat shock response in animal models of ALI (56). The heat shock response was induced by subjecting rats to whole body hyperthermia (41° to 42° C for 15 min) 18 hours before ALI. Whole body hyperthermia caused substantial expression of HSP70 in the lung. ALI secondary to installation of intratracheal phospholipase-A_1 led to a mortality rate of 22% in the normothermic group, compared to 0% mortality in the hyperthermic group. Heated animals also demonstrated decreased lung injury compared to unheated animals, as measured by histologic criteria, wet/dry weight ratios, and lung lavage cells counts. Using the same model, these investigators subsequently demonstrated that induction of the heat shock response conferred protection against sepsis-induced (cecal ligation and perforation) ALI (57). The protective effect of the heat shock response against sepsis-induced ALI was subsequently corroborated by another group of investigators using intravenous endotoxin as a model of sepsis (58).

Javadpour and colleagues recently investigated the effects of the heat shock response on ALI secondary to ischemia-reperfusion (59). In this study, rats were exposed to hyperthermia 18 h before ischemia-reperfusion. The infrarenal aorta was subsequently clamped for 30 minutes, followed by reperfusion for 2 hours. This cycle of ischemia-reperfusion led to substantial lung injury as measured by increased lung edema, increased alveolar protein content, and increased neutrophil infiltration into the lungs of unheated animals. In contrast, animals preconditioned with hyperthermia had significantly reduced levels of lung injury as measured by these same parameters.

Other investigations have tested the hypothesis that prior induction of the heat shock response can protect lung grafts from reperfusion injury after transplantation. Waddel and colleagues reported that preconditioning rabbits with hyperthermia, 8 hours before lung harvest, did not protect lung function from reperfusion injury after 18 hours of preservation (60). Lung function was tested in this study using an *ex vivo* preparation, rather than reimplantation to a recipient animal. Hiratsuka and colleagues tested this hypothesis in a rat model of lung harvest and reimplantation into recipient animals (61). Rats were subjected to hyperthermia 6 or 12 hours before lung harvest. Lung grafts were stored at 40° C for 18 hours, then reimplanted into recipient animals. Transplanted lungs from donors that were preconditioned with 6 hours of hyperthermia had improved function, as measured by arterial oxygenation, neutrophil infiltration, and wet/dry ratios, compared to transplanted lungs from donors that were not preconditioned. These same investigators subsequently presented compelling evidence that HSP70 is involved in the mechanism by which induction of the heat shock response protects lung grafts in their transplantation model (62). Donor animals were injected with an adenovirus-HSP70 vector, which led to increased expression of HSP70 in the lungs. Control animals were injected with an adenovirus-β-galactosidase vector. Lungs were harvested, stored for 18 hours, then reimplanted into recipient animals. Transplanted lungs from donors that received the adenovirus-HSP70 vector had improved function, compared to transplanted lungs from donors that received the control vector.

Collectively, these studies demonstrate that induction of the heat shock response and specific expression of HSP70 can protect against various forms of ALI. The challenges facing this field include a greater understanding of the mechanisms of protection and further animal experimentation to determine the feasibility of translating these findings to the bedside.

Potential Mechanisms of Protection

Studies involving specific overexpression of HSP70 imply that this particular component of the heat shock response plays a central role in the mechanism by which the heat shock response protects against ALI. HSP70 is well characterized as a molecular chaperone, and as such it can effectively stabilize, refold, and transport damaged intracellular proteins. Thus it is believed, although not directly proven, that the cytoprotective properties of the heat shock response during ALI involve, in part, the molecular chaperone properties of HSP70. Further elucidation and characterization of chaperone-related mechanisms of HSP70 protection in ALI is another fruitful area of investigation in the field of heat shock protein biology.

Another potential mechanism by which the heat shock response protects against ALI may involve the interaction between the heat shock response and cellular proinflammatory responses. It is now well established that cellular proinflammatory responses, when dysregulated, contribute substantially to the pathophysiology of ALI. Accordingly, various investigators have examined how the heat shock response affects cellular proinflammatory responses. Induction of the heat shock response, by either hyperthermia or chemical inducers, inhibited a variety of cellular proinflammatory responses. For example, in pulmonary artery smooth muscle cells and in alveolar epithelial cells, induction of the heat shock response inhibited cytokine-mediated expression of inducible nitric oxide synthase (10, 54). This inhibitory effect appears to be relatively specific to proinflammatory responses because in cultured alveolar epithelial cells the heat shock response did not affect surfactant protein gene expression (54). Apart from cytokine-mediated inducible nitric oxide synthase gene expression, the heat shock response also inhibited expression of a variety of proinflammatory genes that play an important role in the pathophysiology of ALI; these include interleukin-8 (12), RANTES (63), tumor necrosis factor-α (64), and interleukin-1β (65).

More recent work has elucidated the molecular mechanisms by which the heat shock response inhibits cellular proinflammatory responses. Cahil and colleagues demonstrated that HSF-1 acts as a repressor of the prointerluekin-1 promoter (66). Feinstein and colleagues demonstrated that the specific overexpression of HSP70 inhibited inducible nitric oxide synthase gene expression in astroglial cells (67). Both of these mechanisms may be relatively specific to cell type and model. A broader mechanism by which the heat shock response inhibits cellular proinflammatory responses appears to involve the NF-κB pathway.

NF-κB is a pluripotent transcription factor that regulates the expression of many proinflammatory genes and is thought to play an important role in the pathophysiology of ALI (see Chapter 20 and references 68, 69). Under basal conditions NF-κB is retained in the cytoplasm by the NF-κB inhibitory protein, IκBα. Upon stimulation by a proinflammatory signal, IκBα is rapidly phosphorylated and ubiquitinated, which in turn leads to rapid degradation of IκBα by the 26S proteasome. Rapid degradation of IκBα unmasks NF-κB nuclear translocation sequences, thus allowing NF-κB to enter the nucleus and direct expression of target proinflammatory genes. In both *in vitro* and *in vivo* models, induction of the heat shock response inhibited nuclear translocation and activation of NF-κB (references 70-72 and Figure 1). Heat shock response-mediated inhibition of NF-κB activation involves stabilization of IκBα, such that IκBα degradation following stimulation with a proinflammatory signal is inhibited by the heat shock response (71, 73).

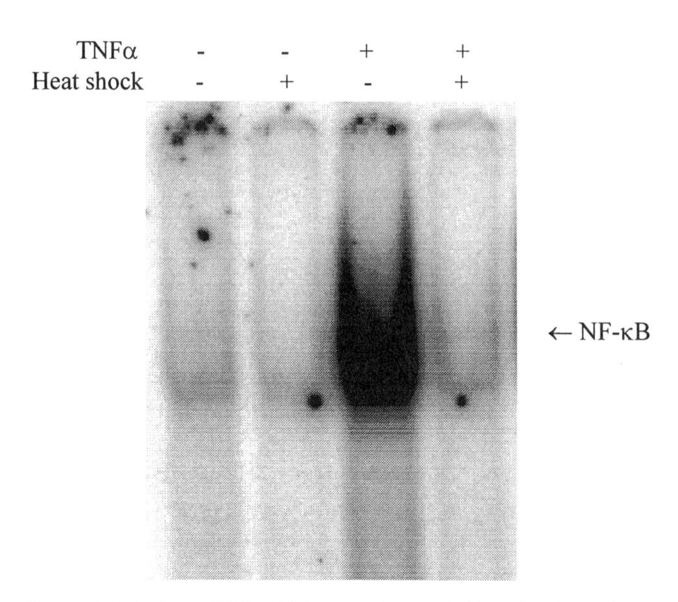

Figure 1. Electromobility shift assay demonstrating that induction of the heat shock response inhibits activation of NF-κB in cultured human respiratory epithelium. NF-κB activation was induced by treating cells with tumor necrosis factor-α (TNFα, 2 ng/ml) for 30 minutes. The heat shock response was induced by incubating cells for 1 hour, at 43° C, 1 hour before TNFα stimulation.

Thus far, the most proximal identified mechanism by which the heat shock response inhibits activation of the NF-κB pathway involves IκB kinase (IKK). IKK is a large protein complex consisting of two catalytic subunits (IKKα and IKKβ) and a regulatory/structural subunit (IKKγ). IKK is the protein kinase that regulates phosphorylation of IκBα at specific serine residues in response to cytokine and endotoxin stimulation (74). Activation of IKK appears to be a key signaling event leading to activation of NF-κB. Recent studies demonstrated that the heat shock response inhibited activation of IKK, thereby elucidating a proximal signaling mechanism by which the heat shock response inhibits NF-κB activation and subsequent expression of proinflammatory genes (references 75, 76 and Figure 2). In addition, heat shock-mediated inhibition of IKK appears to be relatively specific because the heat shock response did not inhibit endotoxin-mediated activation of c-jun N terminal kinase and subsequent phosphorylation of c-jun (*H. Wong, unpublished data* and reference 73) .

In addition to the effects of the heat shock response on IKK activity, the heat shock response also causes independent, *de novo* expression of the IκBα gene (12, 71, 72, 77). Hyperthermia (heat shock) or chemical inducers of the heat shock response cause increased expression of IκBα mRNA and protein. The mechanisms of this effect are currently under investigation, but appear to involve a combination of increased IκBα transcription, as well as increased stability of IκBα mRNA in response to heat shock (*H. Wong, unpublished data*). These observations imply that IκBα may be a novel heat shock protein. In addition, since it is known that increased intracellular levels of IκBα inhibit NF-κB activation, these observations provide another potential mechanism by which the heat shock response can inhibit activation of NF-κB and subsequent expression of NF-κB-dependent genes.

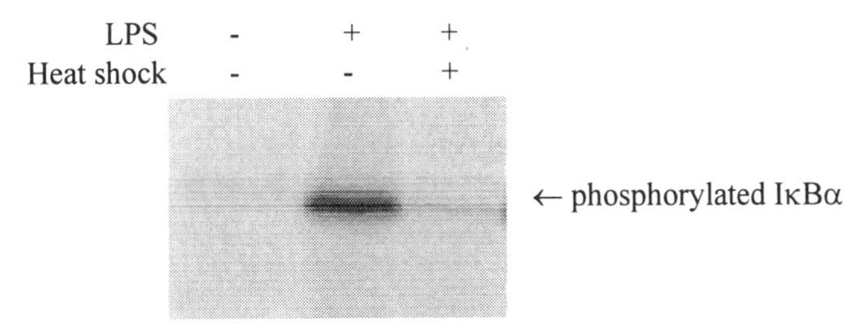

Figure 2. In vitro kinase assay demonstrating that induction of the heat shock response inhibits activation of IKK in murine macrophages. Cellular proteins from treated cells were subjected to immuno-precipitation with an anti-IKKγ antibody. Immunoprecipitation products were then incubated with recombinant IκBα and [^{32}P]-dATP, and the resulting products were resolved on Tris-glycine gradient gels. IKK activation was induced by treating cells with lipopolysaccharide (LPS, 100 ng/ml) for 15 minutes. The heat shock response was induced by incubating cells for 45 minutes, at 43° C, 1 hour before stimulation with LPS.

Summary and Conclusions

The field of heat shock protein biology began almost forty years ago as a curious observation in *Drosophila.* Since that time it has become apparent that the heat shock response is a highly conserved cellular defense mechanism that has the potential to confer broad organ and tissue protection during various forms critical illness, including ALI. The biologic relevance of the heat shock response during ALI is suggested by the expression of heat shock proteins in animal models of ALI, and in patients with ALI or having clinical conditions associated with ALI. Multiple *in vitro* and *in vivo* studies demonstrate that induction of the heat shock response confers considerable protection against ALI and intracellular events associated with ALI. The mechanisms of protection appear to involve HSP70 specifically and the ability of the heat shock response to attenuate cellular proinflammatory responses. The latter effect seems to primarily involve inhibition of the NF-κB pathway, which is being increasingly recognized as a key signaling molecule in the pathophysiology of ALI. Future work in this field will be directed towards a greater understanding of the mechanisms of protection, the regulation of the heat shock response during ALI, and the potential and expected negative consequences of the heat shock response during ALI. Elucidation of these questions could lead to the development of safe pharmacologic or gene-based therapies directed towards induction of the heat shock response as an adjunctive strategy to attenuate ALI.

ACKNOWLEGEMENTS

The author's published and unpublished work presented in this chapter was supported, in part, by grants from the National Institutes of Health (K08HL03725 and RO1GM61723) and the Children's Hospital Research Foundation.

REFERENCES

1. Ritossa, F.M. (1962) A new puffing pattern induced by a temperature shock and DNP in Drosophila. *Experientia* 18, 571-573
2. Welch, W.J. (1992) Mammalian stress response: cell physiology structure/function of stress proteins, and implications for medicine and disease. *Physiol Rev* 72, 1063-1081
3. Morimoto, R.I., and Santoro, M.G. (1998) Stress-inducible responses and heat shock proteins: new pharmacologic targets for cytoprotection. *Nat Biotechnol* 16, 833-838
4. Ciocca, D.R., Oestereich, S., Chamness, G.C., McGuire, W.L., and Fuqua, S.A.W. (1993) Biological and clinical implications of heat shock protein 27000 (Hsp27): a review. *J Natl Cancer Inst* 85, 1558-1570
5. DeMaio, A. (1999) Heat shock proteins: facts, thoughts, and dreams. *Shock* 11, 1-12
6. Leppa, S., and Sistonen, L. (1997) Heat shock response pathophysiological implications. *Ann Med* 29, 73-78
7. Wong, H.R., and Wispé, J.R. (1997) The stress response and the lung. *Am J Physiol* 273, L1-L9
8. Wong, H.R. (1998) Potential protective role of the heat shock response against sepsis. *New Horiz* 6, 194-200
9. Lanks, K.W. (1986) Modulators of the eukaryotic heat shock response. *Exp Cell Res* 165, 1-10
10. Wong, H.R., Finder, J.D., Wasserloos, K., and Pitt, B.R. (1995) Expression of inducible nitric oxide synthase in cultured rat pulmonary artery smooth muscle cells is inhibited by the heat shock response. *Am J Physiol* 269, L843-L848
11. Eaves-Pyles, T., Wong, H.R., and Alexander, J.W. (2000) Sodium arsenite induces heat shock protein-70 in the gut and decreases bacterial translocation in a burned mouse model with gut-derived sepsis. *Shock* 13, 314-319
12. Thomas, S.C., Ryan, M.A., Shanley, T.P., and Wong, H.R. (1998) Induction of the stress response with prostaglandin-A, increases I-κBα gene expression. *FASEB J* 12, 1371-1378
13. Rossi, A., Elia, G., and Santoro, M.G. (1997) Inhibition of nuclear factor κB by prostaglandin A_1: an effect associated with heat shock transcription factor activation. *Proc Natl Acad Sci USA* 94, 746-750
14. Santoro, M.G., Garaci, E., and Amici, C. (1989) Prostaglandins with antiproliferative activity induce the synthesis of a heat shock protein in human cells. *Proc Natl Acad Sci USA* 86, 8407-8411

15. Urayama, S., Musch, M.W., Retsky, J., Madonna, M.B., Straus, D., and Chang, E.B. (1998) Dexamethasone protection of rat intestinal epithelial cells against oxidant injury is mediated by induction of heat shock protein 72. *J Clin Invest* 102, 1860-1865
16. Sun, L., Chang, J., Kirchhoff, S.R., and Knowlton, A.A. (2000) Activation of HSF and selective increase in heat shock proteins by acute dexamethasone treatment. *Am J Physiol* 278, H1091-H1097
17. Vigh, L., Literati, P.N., Horvath, I., Torok, Z., Balogh, G., Glatz, A., Kovacs, E., Boros, I., Ferdinancy, P., Farkas, B., Jaszlits, L., Jednakovits, A., Korvanyi, L., and Maresca, B. (1997) Bimoclomol: a nontoxic, hydroxylamine derivative with stress protein-inducing activity and cytoprotective effects. *Nat Med* 3, 1150-1154
18. Morris, S.D., Cumming, D.V.E., Latchman, D.S., and Yellon, D.M. (1996) Specific induction of the 70-kD heat stress proteins by the tyrosine kinase inhibitor herbimycin-A protects rat neonatal cardiomyocytes. *J Clin Invest* 97, 706-712
19. Conde, A.G., Lau, S.S., Dillmann, W.H., and Mestril, R. (1997) Induction of heat shock proteins by tyrosine kinase inhibitors in rat cardiomyocytes and myogenic cells confers protection against simulated ischemia. *J Mol Cell Cardiol* 29, 1927-2938
20. Jurivich, D.A., Sistonen, L., Kroes, R.A., and Morimoto, R.I. (1992) Effect of sodium salicylate on the human heat shock response. *Science* 255, 1243-1245
21. Lee, B.S., Chen, J., Angelidis, C., Jurivich, D.A., and Morimoto, R.I. (1995) Pharmacological modulation of heat shock factor 1 by antiinflammatory drugs results in protection against stress-induced cellular damage. *Proc Natl Acad Sci USA* 1995, 7207-7211
22. Rossi, A., Elia, G., and Santoro, M.G. (1998) Activation of the heat shock factor 1 by serine protease inhibitors. *J Biol Chem* 273, 16446-16452
23. DeMeester, S. L., Buchman, T. G., Qiu, Y., Dunnigan, K., Hotchkiss, R. S., Karl, I.E., and Cobb, J.P. (1998) Pyrrolidine dithiocarbamate activates the heat shock response and thereby induces apoptosis in primed endothelial cells. *Shock* 10, 1-6
24. Wischmeyer, P.E., Musch, M.W., Madonna, M.B., Thisted, R., and Chang, E.B. (1997) Glutamine protects intestinal epithelial cells: role of inducible HSP70. *Am J Physiol* 272, G879-G884
25. Wagner, M., Hermanns, I., Bittinger, F., and Kirkpatrick, C.J. (1999) Induction of stress proteins in human endothelial cells by heavy metal ions and heat shock. *Am J Physiol* 277, L1026-L1033
26. Mivechi, N.F., Murai, T., and Hahn, G.M. (1994) Inhibitors of tyrosine and Ser/Thr phosphatases modulate the heat shock response. *J Cell Biochem* 54, 186-197
27. Chang, N.T., Huang, L.E., and Liu, A.Y. (1993) Okadaic acid markedly potentiates the heat-induced hsp 70 promoter activity. *J Biol Chem* 268, 1436-1439
28. Kato, K., Ito, H., Kamei, K., and Iwamoto, I. (1998) Stimulation of the stress-induced expression of stress proteins by curcumin in cultured cells and in rat tissues in vivo. *Cell Stress Chaperones* 3, 152-160
29. Hirakawa, T., Rokutan, K., Nikawa, T., and Kishi, K. (1996) Geranylgeranylacetone induces heat shock proteins in cultured guinea pig gastric mucosal cells and rat gastric mucosa. *Gastroenterology* 111, 345-357
30. Schoeniger, L.O., Curtis, W., Esnaola, N.F., Beck, S.C., Gardner, T., and Buchman, T. (1994) Myocardial heat shock gene expression in pigs is dependent on superoxide anion generated at reperfusion. *Shock* 1, 31-35
31. Fincato, G., Polentarutti, N., Sica, A., Mantovani, A., and Colotta, F. (1991) Expression of a heat-inducible gene of hsp70 family in human myelomonocytic cells: regulation by bacterial products and cytokines. *Blood* 77, 579-586
32. Aucoin, M.M., Barhoumi, R., Kochevar, D.T., Granger, H.J., and Burghardt, R.C. (1995) Oxidative injury of coronary venular endothelial cells depletes intracellular glutathione and induces HSP 70 mRNA. *Am J Physiol* 268, H1651-H1658
33. Kukreja, R.C., Kontos, M.C., Loesser, K.E., Batra, S.K., Qian, Y.Z., Gbur, C.J., Naseem, S.A., Jesse, R.L., and Hess, M.L. (1994) Oxidant stress increases heat shock protein 70 mRNA in isolated perfused rat heart. *Am J Physiol* 267, H2213-H2219

34. Folch, E., Closa, D., Neco, P., Sole, S., Planas, A., Gelpi, E., and Rosello-Catafau, J. (2000) Pancreatitis induces HSP72 in the lung: role of neutrophils and xanthine oxidase. *Biochem Biophys Res Commun* 273, 1078-1083
35. Bonay, M., Soler, P., Riquet, M., Battesti, J.P., Hance, A.J., and Tazi, A. (1994) Expression of heat shock proteins in human lung and lung cancers. *Am J Respir Cell Mol Biol* 10, 453-461
36. Durand, P., Bachelet, M., Brunet, F., Richard, M.J., Dhainaut, J.F., Dall'Ava, J., and Polla, B.S. (2000) Inducibility of the 70 kD heat shock protein in peripheral blood monocytes is decreased in human acute respiratory distress syndrome and recovers over time. *Am J Respir Crit Care Med* 161, 286-292
37. Kindas-Mugge, I., Pohl, W.R., Zavadova, E., Kohn, H.D., Fitzal, S., Kummer, F., and Micksche, M. (1996) Alveolar macrophages of patients with adult respiratory distress syndrome express high levels of heat shock protein 72 mRNA. *Shock* 5, 184-189
38. Polla, B.S., Kantengwa, S., Gleich, G.J., Kondo, M., Reimert, C.M., and Junod, A.F. (1993) Spontaneous heat shock protein synthesis by alveolar macrophages in interstitial lung disease associated with phagocytosis of eosinophils. *Eur Respir J* 6, 483-488
39. Schroeder, S., Lindemann, C., Hoeft, A., Putensen, C., Decker, D., Ruecker, A.A., and Stuber, F. (1999) Impaired inducibility of heat shock protein 70 in peripheral blood lymphocytes of patients with severe sepsis. *Crit Care Med* 27, 1080-1084
40. Schroeder, S., Bischoff, J., Lehmann, L.E., Hering, R., Spiegel, T.V., Putensen, C., Hoeft, A., and Stuber, E. (1999) Endotoxin inhibits heat shock protein 70 (HSP70) expression in peripheral blood mononuclear cells of patients with severe sepsis. *Intensive Care Med* 25, 52-57
41. Vignola, A.M., Chanez, P., Polla, B.S., Vic, P., Godard, P., and Bousquet, J. (1995) Increased expression of heat shock protein 70 on airway cells in asthma and chronic bronchitis. *Am J Respir Cell Mol Biol* 13, 683-691
42. Wright, B. H., Corton, J. M., El-Nahas, A.M., Wood, R.F., and Pockley, A.G. (2000) Elevated levels of circulating heat shock protein 70 (Hsp70) in peripheral and renal vascular disease. *Heart Vessels* 15, 18-22
43. Shehata, S.M., Sharma, H.S., Mooi, W.J., and Tibboel, D. (1999) Expression patterns of heat shock proteins in lungs of neonates with congenital diaphragmatic hernia. *Arch Surg* 134, 1248-1253
44. Li, G.C., Li, L., Liu, Y.-K., Mak, J.Y., Chen, L., and Lee, W.M.F. (1991) Thermal response of rat fibroblasts stably transfected with the human 70-kDa heat shock protein-encoding gene. *Proc Natl Acad Sci* 88, 1681-1685
45. Chi, S.H., and Mestril, R. (1996) Stable expression of a human hsp70 gene in a rat myogenic cell line confers protection against endotoxin. *Am J Physiol* 270, C1017-1021
46. Marber, M.S., Mestril, R., Chi, S.H., Sayen, M.R., Yellon, D.M., and Dillman, W.H. (1995) Overexpression of the rat inducible 70-kD heat stress protein in a transgenic mouse increases the resistance of the heart to ischemic injury. *J Clin Invest* 95, 1446-1456
47. Plumier, J.-C.L., Ross, B.M., Currie, R.W., Angelidis, C.E., Kazlaris, H., Kollias, G., and Pagoulatos, G.N. (1995) Transgenic mice expressing the human heat shock protein 70 have improved post-ischemic myocardial recovery. *J Clin Invest* 95, 1854-1860
48. Wong, H.R., Mannix, R.J., Rusnak, J.M., Boota, A., Zar, H., Watkins, S. C., Lazo, J.S., and Pitt, B.R. (1996) The heat shock response attenuates lipopolysaccharide-mediated apoptosis in cultured sheep pulmonary artery endothelial cells. *Am J Respir Cell Mol Biol* 15, 745-751
49. Wong, H.R., Menendez, I.Y., Ryan, M.A., Denenberg, A., and Wispé, J. R. (1998) Increased expression of heat shock protein-70 protects human respiratory epithelium against hyperoxia. *Am J Physiol* 275, L836-L841
50. McMillan, D.R., Xiao, X., Shao, L., Graves, K., and Benjamin, I.J. (1998) Targeted disruption of heat shock transcription factor 1 abolishes thermotolerance and protection against heat-inducible apoptosis. *J Biol Chem* 273, 7523-7528

51. Xiao, X.Z., Zuo, X.X., Davis, A.A., McMillan, D.R., Curry, B.B., Richardson, J.A., and Benjamin, I.J. (1999) HSF1 is required for extra-embryonic development, postnatal growth and protection during inflammatory responses in mice. *EMBO J* 18, 5943-5952
52. Wang, J.R., Xiao, X.Z., Huang, S.N., Luo, F.J., You, J.L., Luo, H., and Luo, Z.Y. (1996) Heat shock pretreatment prevents hydrogen peroxide injury of pulmonary endothelial cells and macrophages in culture. *Shock* 6, 134-141
53. Wong, H.R., Ryan, M., Menendez, I.Y., Denenberg, A., and Wispé, J.R. (1997) Heat shock protein induction protects human respiratory epithelium against nitric oxide-mediated cytotoxicity. *Shock* 8, 213-218
54. Wong, H.R., Ryan, M., Gebb, S., and Wispé, J.R. (1997) Selective and transient *in vitro* effects of heat shock on alveolar type II cell gene expression. *Am J Physiol* 272, L132-L138
55. Wong, H.R., Menendez, I.Y., Ryan, M.A., Denenberg, A., and Wispé, J.R. (1998) Increased expression of heat shock protein-70 protects human respiratory epithelium against hyperoxia. *Am J Physiol* 275, L836-L841
56. Villar, J., Edelson, J.D., Post, M., Mullen, J.B., and Slutsky, A.S. (1993) Induction of heat stress proteins is associated with decreased mortality in an animal model of acute lung injury. *Am Rev Respir Dis* 147, 177-181
57. Villar, J., Riberio, S.P., Mullen, J.B.M., Kuliszewski, M., Post, M., and Slutsky, A.S. (1994) Induction of the heat shock response reduces mortality rate and organ damage in a sepsis-induced acute lung injury model. *Crit Care Med* 22, 914-922
58. Koh, Y., Lim, C.M., Kim, M.J., Shim, T.S., Lee, S.D., Kim, W.S., Kim, D.S., and Kim, W.D. (1999) Heat shock response decreases endotoxin-induced acute lung injury in rats. *Respirology* 4, 325-330
59. Javadpour, M., Kelly, C.J., Chen, G., Stokes, K., Leahy, A., and Bouchier-Hayes, D.J. (1998) Thermotolerance induces heat shock protein 72 expression and protects against ischaemia-reperfusion-induced lung injury. *Br J Surg* 85, 943-946
60. Waddell, T.K., Hirai, T., Piovesan, J., Oka, T., Puskas, J.D., Patterson, G.A., and Slutsky, A.S. (1994) The effect of heat shock on immediate post-preservation lung function. *Clin Invest Med* 17, 405-413
61. Hiratsuka, M., Yano, M., Mora, B.N., Nagahiro, I., Cooper, J.D., and Patterson, G.A. (1998) Heat shock pretreatment protects pulmonary isografts from subsequent ischemia-reperfusion injury. *J Heart Lung Transplant* 17, 1238-1246
62. Hiratsuka, M., Mora, B.N., Yano, M., Mohanakumar, T., and Patterson, G.A. (1999) Gene transfer of heat shock protein 70 protects lung grafts from ischemia-reperfusion injury. *Ann Thoracic Surg* 67, 1421-1427
63. Ayad, O., Stark, J.M., Fiedler, M.A., Menendez, I.Y., Ryan, M.A., and Wong, H.R. (1998) The heat shock response inhibits RANTES gene expression in cultured human lung epithelium. *J Immunol* 161, 2594-2599
64. Snyder, Y.L., Guthrie, L., Evans, G.F., and Zuckerman, S. (1992) Transcriptional inhibition of endotoxin-induced monokine synthesis following heat shock in murine peritoneal macrophages. *J Leuko Biol* 51, 181-187
65. Schmidt, J.A., and Abdulla, E. (1988) Down-regulation of IL-1β biosynthesis by inducers of the heat-shock response. *J Immunol* 141, 2027-2034
66. Cahil, C.M., Waterman, W.R., Xie, Y., Auron, P.E., and Calderwood, S.K. (1996) Transcriptional repression of the prointerleukin-1β gene by heat shock factor 1. *J Biol Chem* 271, 24874-24879
67. Feinstein, D.L., Galea, E., Aquino, D.A., Li, G.C., Xu, H., and Reis, D.J. (1996) Heat shock protein 70 suppresses astroglial-inducible nitric oxide synthase expression by decreasing NF-κB activation. *J Biol Chem* 271, 17224-12732
68. Baldwin, A.S. (1996) The NF-κB and IκB proteins: new discoveries and insights. *Annu Rev Immunol* 14, 649-681
69. Ghosh, S., May, M.J., and Koop, E.B. (1998) NF-κB and Rel proteins: evolutionarily conserved mediators of immune responses. *Annu Rev Immunol* 16, 225-260

70. Wong, H.R., Ryan, M., and Wispé, J.R. (1997) The heat shock response inhibits inducible nitric oxide synthase gene expression by blocking Iκ-B degradation and NF-κB nuclear translocation. *Biochem Biophys Res Comm* 231, 257-263
71. Wong, H.R., Ryan, M., and Wispé, J.R. (1997) Stress response decreases NF-κB nuclear translocation and increases I-κBα and expression in A549 cells. *J Clin Invest* 99, 2423-2428
72. Pritts, T.A., Want, Q., Sun, X., Moon, M.R., Fischer, D.R., Fischer, J.E., Wong, H.R., and Hasselgren, P.O. (2000) Induction of the stress response in vivo decreases NF-κB activity in jejunal mucosa of endotoxemic mice. *Arch Surg* 135, 860
73. Shanley, T.P., Ryan, M.A., Eaves-Pyles, T., and Wong, H.R. (2000) Heat shock inhibits phosphorylation of I-kappaBalpha. *Shock* 14, 447-450
74. Karin, M., and Ben-Neriah, Y. (2000) Phosphorylation meets ubiquitination: the control of NF-κB activity. *Annu Rev Immunol* 18, 621-663
75. Curry, H.A., Clemens, R.A., Shah, S., Bradbury, C.M., Botero, A., Goswami, P., and Gius, D. (1999) Heat shock inhibits radiation-induced activation of NF-κB via inhibition of I-κB kinase. *J Biol Chem* 274, 23061-23067
76. Yoo, C.G., Lee, S., Lee, C.T., Kim, Y.W., Han, S.K., and Shim, Y.S. (2000) Anti-inflammatory effect of heat shock protein induction is related to stabilization of IκBα through preventing IκB kinase activation in respiratory epithelial cells. *J Immunol* 164, 5416-5423
77. Wong, H.R., Ryan, M.A., Menendez, I.Y., and Wispé, J.R. (1999) Heat shock activates the I-κBα promoter and increases I-κBα mRNA expression in BEAS-2B cells. *Cell Stress Chaperones* 4, 1-7

Chapter 19

BIOTRAUMA: SIGNAL TRANSDUCTION AND GENE EXPRESSION IN THE LUNG

Claudia C. dos Santos, Mingyao Liu, and Arthur S. Slutsky
Inter-Departmental Division of Critical Care, University of Toronto, Thoracic Surgery Research Laboratory, University Health Network, Toronto General Hospital and Department of Critical Care Medicine, St. Michael's Hospital, Toronto, Ontario, Canada

INTRODUCTION

Mechanical injury due to artificial ventilation has been thought to play a role in lung injury for over 250 years (1). However, only recently has it become apparent that this injury is likely not confined to the lung alone, but it may contribute to the development of multi-system organ failure (MSOF) in patients with acute respiratory distress syndrome (ARDS) (2). The major cause of death in patients with ARDS is multiple organ failure, not hypoxia (3). There is little doubt that mechanical ventilation can physically disrupt the lung (4, 5). However, in addition to structural damage, it has been postulated that mechanical ventilation can also induce changes in the activation and recruitment of inflammatory cells, and stimulate the production of a number of inflammatory mediators (2). One theory, is that by altering both the pattern and magnitude of stretch, mechanical ventilation may lead to alterations in gene expression and/or cellular metabolism, ultimately leading to the development of an overwhelming generalized inflammatory response that may eventually lead to MSOF and/or death. This mechanism of injury has been termed *biotrauma* (2, 6).

The clinical importance of ventilator induced lung injury (VILI) was not fully appreciated until recently, when the results of the ARDS Network trial demonstrated a significant 22% relative risk reduction in 180 day mortality rates favoring a lung protective strategy group using low tidal volume (6 ml/kg of predicted body weight) versus a conventional ventilatory strategy group (12 ml/kg) (7). Based on these data, the attributable mortality of VILI may be in the range of 9% (absolute risk reduction in mortality between the high and low volume ventilatory strategies). This sobering thought underscores both the vital importance of VILI in determining patient outcome and the importance of current and future research in identifying the clinically appropriate choice of ventilatory strategy in the management of ventilated patients. More importantly, although advances in lung protective ventilatory strategies have and will continue to be made, they clearly have their limitations. This places the onus on intensivists and researchers to conceive of novel approaches to the prevention and treatment of VILI. In this chapter, we propose that understanding how cells respond to cyclic-stretch and injury might provide us with clues as to how to alter the biology of the cell and ultimately improve morbidity and mortality in ARDS patients. To this end, this chapter will integrate information from a variety of different sources and attempt to delineate the putative path(s) by which mechanical forces, akin to the ones generated in VILI, might lead to a local inflammatory stimulus in the lung and the generation of a systemic inflammatory response which may affect the entire organism.

MECHANISMS FOR SENSING MECHANICAL FORCES

We will not review in detail the effects on direct cellular and tissue injury of physical forces resulting from regional lung over-inflation or regional lung under-inflation, which are certainly involved in the pathophysiology of VILI (for a review please see references 8 and 12). Rather, we will focus on the molecular responses initiated by the physical forces generated during mechanical ventilation that may ultimately lead to the generation of a systemic inflammatory response.

Precisely how mechanical forces are sensed by cells (mechanosensors) and converted into biochemical signals for intracellular signal transduction is still unknown. Furthermore, because of the complexity of the lung, the variety of cell types, and the variety of physical forces to which cells are exposed, mechanisms by which mechanical stimulation is perceived and the cellular responses induced, may vary widely (for a review on mechanical force-induced signal transduction in the lung see reference see references 13 and 14). Mechanotransduction is the conversion of mechanical stimuli, such as cell deformation, into biochemical and biomolecular alteration. Mechanical forces important to pulmonary structure and function are produced by gradients in gravity, motion, osmotic forces, and interactions between cells and/or cell matrix (for a review on the effects of mechanical forces on lung function see reference 15).

In the following sections, we will discuss selected mechanisms by which it has been proposed that cellular deformation is converted into cellular responses, and how these may be responsible for the generation of VILI (for a previous review on this topic see reference 16).

Stretch Sensitive Channels

Stretch activated ion-channels are known to play important roles in both pulmonary physiology and development (13, 14, 17, 18). However, the specific role they play in VILI is yet to be determined. Recently Parker et al studied the initial signaling events causing the increase in vascular permeability (hallmarks of ARDS and VILI) after high airway pressure injury. Microvascular permeability were completely abolished by gadolinium (which blocks stretch activated non-selective cation channels) and hence they concluded that stretch-activated cation channels may initiate the increase in permeability induced by mechanical ventilation through increases in intracellular calcium concentration (19 and Figure 1).

Liu et al have shown that mechanical stretch induced a rapid Ca^{2+} influx into organotypic cultured fetal rat lung cells, which can be blocked by gadolinium (20). They further demonstrated that the activation of stretch-activated ion channels are involved in stretch-induced protein kinase C (PKC) activation and cell proliferation (20 and Figure 1), and in stretch-induced glycosaminoglycan secretion from these cells (21). However, mechanical stretch-induced release of macrophage inflammatory protein-2 (MIP-2) from these cells can not be blocked by gadolinium (22). Therefore, the role of stretch-activated ion channel in VILI merits further investigation.

To examine the role of ion channels in macrophage function, Martin et al undertook patch clamp studies of human culture-derived macrophages grown under serum-free conditions (23). Macrophages have been implicated as one of the major effector cells in VILI. The major ionic current in these cells is carried by an outwardly rectifying K^+ channel. These channels opened infrequently in resting cells but were activated immediately by 1) adhesion of mobile cells onto a substrate, 2) stretch applied to isolated membrane patches in Ca^{2+}-free buffers, 3) intracellular Ca^{2+} (EC_{50} of 0.4 μM), and 4) interleukin-2. Furthermore, barium and 4-aminopyridine, blockers of this channel, altered the organization and structure of the cytoskeletal proteins actin, tubulin and vimentin; and were associated with reversible alterations in cell morphology. These experiments suggest the presence of an outwardly rectifying K^+ channel that appears to be involved in cytokine and adherence-mediated macrophage activation, and in the maintenance of cytoskeletal integrity and cell shape (23).

A stretch-activated cation channel may also regulate Na^+-K^+-ATPase activity, a factor that may play an important role in lung edema clearance. In recent experiments Waters et al demonstrated that at 30 min and after 60 min of cyclical stretch (30 cycles/min), Na^+-K^+-ATPase activity was significantly increased in murine lung epithelial cells (24). When cells were treated with amiloride (blocks amiloride sensitive Na+ entry into cells) or with gadolinium, there was no stimulation of Na^+-K^+-

ATPase. Interestingly, changes in Na^+-K^+-ATPase activity were paralleled by increases in Na^+-K^+-ATPase protein in the basolateral membrane of epithelial cells, suggesting that increased recruitment of these channel subunits from intracellular pools to the basolateral membrane may be induced by cyclic stretch (24). Moreover, ENaC (epithelial amiloride sensitive sodium channel) has been shown to be a mammalian homologue of mechanosensitive channels in *C. elegans* (25, 26). In *C. elegans*, definition of transmembrane topology has identified regions that may interact with the extracellular matrix or cytoskeleton to mediate mechanical signaling (27).

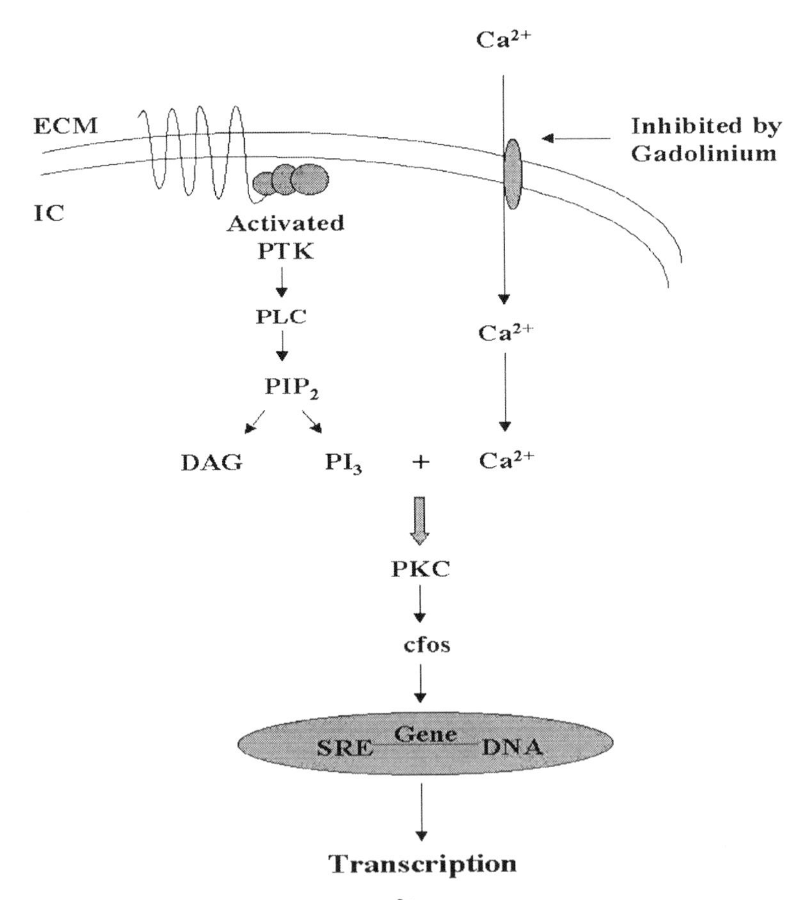

Figure 1. Mechanical stretch induces Ca^{2+} influx via a mechanosensitive cation channel that is inhibited by gadolinium (19). Mechanical stretch activates protein tyrosine kinases (PTK) and consequently activates phospholipase C-γ (PLC-γ) via tyrosine phosphorylation. PLC-γ mediates hydrolysis of phosphoinositol 4,5-diphosphate (PIP2) to generate inositol 1,4,5-triphosphate (IP3) and diacylglycerol (DAG). IP3 mobilizes intracellular Ca^{2+}. DAG in the presence of Ca^{2+} activates protein kinase C (PKC) and downstream events (20, 22). PKC may activate transcriptional factors (c-fos) that bind to "stretch response element" (SRE). Increased gene expression and production of cytokines and other pro- and anti- inflammatory molecules regulate the pathogenesis of VILI. ECM: extracellular matrix; IC: intracellular

Plasma Membrane Integrity

The role of structural damage to cells as the inciting inflammatory event is currently being investigated. To discuss the mechanisms of membrane stress failure is beyond the scope of this chapter, but for an excellent review on the topic please see reference (28). To date, there is no substantial evidence suggesting that cytopathological changes mediate gene transcription that contributes to the inflammatory response characteristic of VILI. Nevertheless, maintenance of plasma membrane integrity undoubtedly plays an integral role in intracellular, as well as extracellular signaling pathways relevant to VILI. There is evidence to illustrate the potential significance of plasma membrane stress failure as an inciting mechanism by which cells sense and respond to mechanical injury.

A number of studies have demonstrated that changes in calcium homeostasis can affect not only the pathways in which Ca^{2+} itself serves as a signaling component, but also the signaling system turned on by other sepsis-induced agonists, which may be affected by Ca^{2+} signaling. In fact, the increase in apparent basal $[Ca^{2+}]i$ in sepsis can hypersensitize PKC (for a review on alterations in calcium signaling and cellular response in septic injury see reference 29). In recent experiments, Hinman et al demonstrated that when alveolar type II cells (ATII) are injured, elevations in intracellular free Ca^{2+} concentration ($[Ca^{2+}]i$) begin at the edge of the wound and propagate outwards as a wave for at least 300 microns. The $[Ca^{2+}]i$ wave is due both to influx of extracellular calcium, and release of intracellular calcium stores. In these experiments, $[Ca^{2+}]i$ elevations propagated over the break in ATII monolayers and caused elevations of $[Ca^{2+}]i$ in uninjured cells (30). The spectrum of ultrastructural injury to the lung after mechanical ventilation includes endothelial and epithelial plasma membrane blebs, transcellular and intercellular gaps, and overt breaks in the basement membrane (11, 12). We speculate that this type of injury, above and beyond the structural damage, may cause alterations in $[Ca^{2+}]i$ which may contribute to the activation of signal transduction pathways in part responsible for the inflammatory response characterizing VILI.

Evidence to suggest that ultrastructural damage may lead to nuclear transcription comes from Grembowicz et al. This group demonstrated that plasma membrane disruption (PMD) induced an increase in Fos protein synthesis (Figure 2). This increase was shown to be highly specific and occurred primarily in cells lining the membrane disruption tracts (where breaks in the cell membrane occurred) (31). The expression of the fos gene can be induced by calcium (32), and *c-fos* gene is known to contain a calcium response element in its cis-regulatory region (33). The importance of *c-fos* resides not only in the fact that it is an early response gene with a stretch responsive promoter element, but also in the central role of Fos protein in mediating other gene transcription. c-fos m-RNA has been shown to be increased by injurious ventilatory strategies in an isolated rat lung model (34). Moreover, rapid upregulation of c-fos expression in response to the application of mechanical force has been documented in a number of cell types, including muscle (35), cardiac (36), osteocytes (37), osteoblasts (38) and endothelial cells (39). In all of these cases, the upregulation of c-fos seems to be an early mechanically stimulated transcriptional response.

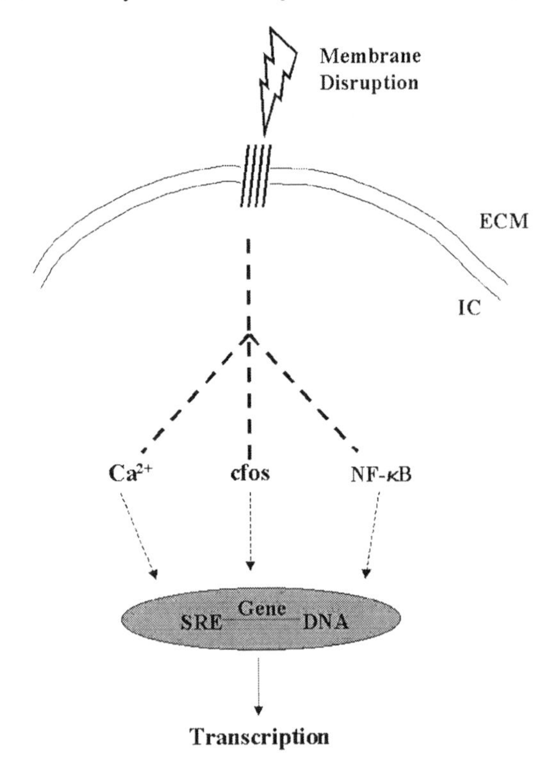

Figure 2. Structural damage to cells causes elevations in intracellular free Ca^{2+} concentrations (31). Changes in calcium homeostasis can affect signaling pathways and in sepsis has been shown to hypersensitize PKC activation. Traumatic breaks in the plasma membrane can induce activation and translocation of c-fos and NF-κB into the nucleus (31). Activated NF-κB and c-fos can bind to "stretch response elements" (SRE) to activate transcription. Increased gene expression and production of cytokines and other pro- and anti- inflammatory molecules regulate the pathogenesis of VILI.

Whether the structural disruption caused by mechanical ventilation can induce changes in Fos protein, as well as PMD needs to be elucidated. Even more interesting, is the fact that PMD has also been shown to induce translocation of nuclear factor kappa B (NF-κB) into the nucleus of mechanically injured human endothelial smooth muscle cells (31). As discussed below, NF-κB translocation and activation is an essential step for expression of several pro-inflammatory cytokines and chemokines.

In their recent review, Vlahakis and Hubmayr outlined the fact that plasma membrane breaks not only serve as a mechanism for transducing intracellular processes but might also serve as a means of releasing cytoprotective growth factors important for cell proliferation and tissue remodeling (28). An example of this process is basic fibroblast growth factor (bFGF), which has been found to be released from wounded endothelial cells, both *in vitro* and *in vivo*. Presumably, release of bFGF occurs through the induced membrane breaks because bFGF lacks the required signal sequence for exocitic-driven release from cells (40, 41). As the authors pointed out, this finding suggests that injured cells might possess autocrine mechanisms that are utilized as well for cytoprotection. Mechanical forces, such as the ones generated during mechanical ventilation, may potentially overwhelm these cytoprotective mechanisms, not solely due to their nature, but also secondary to their intensity and duration.

Direct Conformational Change in Membrane Associated Molecules

The tensegrity model provides a theoretical framework to explain how mechanical forces may be translated into biochemical responses. As well, it proposes a mechanism for integrating these signals with those generated by growth factors and molecular components of the extracellular matrix (for an integrative review see reference 42). In this model, cell surface adhesion proteins (e.g. integrins, cell-cell adhesion molecules), the cytoskeleton, and associated nuclear scaffolds, function as a structurally unified system that provides the architectural infrastructure which enables cells to respond to mechanical forces transmitted over its surface (Figure 3). Specialized anchoring complexes or focal adhesion complexes (FAC) are central to this theory, presumably enabling mechanical coupling between components of the cell's architecture and molecules that transduce signals into the cell.

Evidence suggesting that cell matrix and adhesion molecules may be important in mediating the inflammatory response characteristic of VILI comes from Tschumperlin et al (43), who subjected pulmonary epithelial cells to magnetic twisting cytometry (MTC). This technique uses ferro-magnetic beads coated with a ligand and bound to the cell surface through integrins, a family of transmembrane heterodimers that form part of the cellular cytoskeleton. The beads are then magnetized. Subsequent application of an external magnetic field applies a torque (or twisting stress). Bead rotation is opposed by reaction forces generated within the cytoskeleton to which the bead is bound. MTC measures the applied twisting stress and the resulting angular rotation of the magnetic bead and expresses the ratio of cell stiffness. This stiffness is a measure of the ability of the cell to resist distortion to shape. Using a similar technique Wang et al (44) previously demonstrated that first, endothelial cells stiffen with increasing magnitude of imposed twisting stress and that this twisting stress is transmitted across the cell membrane to the cytoskeleton via the agency of integrins, and second, that the resistance of the bead to rotation is conferred by the recruitment of adhesion proteins such as talin, vinculin, and actin - cytoskeletal network proteins.

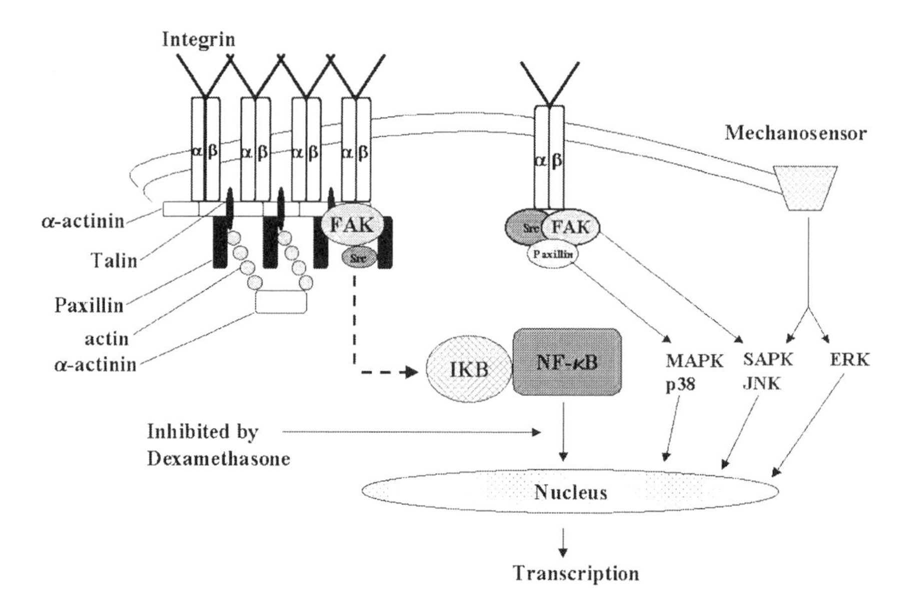

Figure 3. The structural organization and interconnectedness of the cytoskeleton provides a physical basis for translating mechanical forces into biochemical responses (50). Mechanosensitive receptors and/or integrins are activated by mechanical forces. Integrins are known to maintain close relationships with focal adhesive kinase (FAK), kinases of the Src family (Src), and paxillin. Through a series of as of yet unidentified pathways members of the mitogen activating protein kinases (p38, SAPK/JNK, and ERK) are activated (56). Subsequently, these proteins may activate NF-κB and through other mechanisms gene transcription. Members of the integrin family can mediate NF-κB regulation through activation of I-κB kinase (IKK). This in turn would mediate activation NF-κB (see Chapter 1). Dexamethasone can inhibit activation of NF-κB (77). Activation of members of the MAPK family and NF-κB can induce transcription of early response genes and bind to "stretch response element" (SRE) to activate transcription. Increased gene expression and production of cytokines and other pro- and anti- inflammatory molecules regulate the pathogenesis of VILI.

Using this technique, Tschumperlin, et al investigated the role of forces applied directly to epithelial cell adhesion molecules (43). Early growth response protein (Egr-1) and transforming growth factor β (TGF-β) transcription increased in response to magnetic twisting applied to beads coated with collagen. The collagen presumably functions as a ligand for transmembrane integrins. Data were normalized to cultures of identical beads that were not subjected to force. The epithelial response peaked at much different field strengths, corresponding to differing amounts of mechanical stress (43). These findings indicate that both the magnitude and specific cell-matrix connections are important determinants of the airway epithelial response to mechanical force. In more recent experiments, Peake et al described the role of cell-matrix interactions and calcium-mediated signaling pathways in gene up-regulation in response to dynamic mechanical loading of human osteoblasts. Using the expression of c-fos as a marker of early mechanoresponsive transcription, the authors provided important insights into the possible requirement for key matrix components that enable c-fos gene expression. Primarily they underscored the essential role of β1-integrin-mediated interactions and show that this is certainly not interchangeable with certain non-integrin-mediated mechanisms (45).

Additional evidence in support of the role of cytoskeletal proteins in signal transduction in VILI has been presented by Hubmayr and colleagues. This group demonstrated that human airway smooth muscle cells (HASM) stiffness increased when cells were exposed to contractile agonists previously documented to have an effect on actin-myosin interactions. Contractile agonists utilized in this study mediated their action by increasing intracellular calcium concentration and thereby activating the contractile apparatus (46).

Bhullar et al have shown that preincubation of endothelial cells (EC) with a monoclonal anti-integrin antibody attenuates shear stress induction of IκB kinase (IKK, see Chapter 1 and Figure 3). Inhibition of the tyrosine kinase activity caused a similar down-regulatory effect suggesting that an integrin-mediated signaling pathway regulates NF-κB through IKK, at least in EC cells (47). If this is true in pulmonary cells involved in the ALI response, it might have important implications for potential targeting in the context of novel molecular therapy.

INTRACELLULAR SIGNALING PATHWAYS RELATED TO VILI

The signal transduction pathways that may be involved in VILI are not known but, are likely varied, reflecting the complexity of this syndrome and organ-system. Stimuli generated by VILI may act through intracellular signaling pathways similar as those used by cytokines (48) and chemokines (49): 1) the cAMP/protein kinase A pathway, 2) the mitogen-activated protein kinases (MAPK) pathway; 3) the janus kinase/signal transducers and activators of transcription (JAK/STAT) pathway, and 4) the phosphatidyl-inositol/calcium/protein kinase C pathway. Alternatively, the stress/stretch signal may be directly submitted to the nucleus via the cytoskeleton (50). By promoting an increase in intracellular Ca^{2+} concentration, stretch may also stimulate the calcium/calmodulin-dependent phosphatase calcineurin, a novel signaling pathway demonstrated to be important in cardiac hypertrophic responses (36). Here we shall briefly discuss data suggesting that mechanical stimuli (akin to the stimuli generated in VILI) can activate different intracellular signaling pathways. Activation of these pathways may mediate the generation of nuclear signaling which ultimately causes changes in gene expression.

G-protein and Serine/Threonine Kinases

Interaction of a cytokine/chemokine with its receptor leads to exchange of GTP for GDP and the dissociation of the α subunit from the $\beta\gamma$ subunit of G proteins (see Chapter 3). These dissociated units can subsequently activate several major intracellular signaling pathways. The activity of protein kinases and the phosphorylation status of proteins is one of the main determinants of cellular enzyme activity and intracellular signaling mechanisms. Activity of protein kinase A (PKA), a c-AMP dependent kinase, has been shown to be increased in mechanically ventilated animals (51). Mechanical stretch-induced activation of PKC plays an important role in mediating stretch-induced fetal rat lung cell proliferation (52, 53).

MAPK is a family of serine/threonine kinases that mediate proliferative and mitogenic activities as well as cell-stress-induced responses. At least three distinct families of MAPKs exist in mammalian cells: the p42/44 extracellular signal-regulated kinase (ERK), c-Jun NH2-terminal kinases (JNK, also called stress-activated protein kinase [SAPK]), and the p38 MAPK (14, 54, and Chapter 1). Recently, it has been shown the ERK is rapidly activated by mechanical strain in a human pulmonary epithelial cell line (54). Mechanical stretch (5% strain 6 cycles/min for 2h) activated JNK in another human lung epithelial cell line (55). Moreover, higher amplitude of stretch (15%) was shown to enhance phosphorylation of JNK and p38 MAPK - which is associated with increased IL-8 mRNA and protein levels (55). Mechanical stretch activated JNK, ERK and p38 in pulmonary endothelial cells (56 and Figure 3).

As discussed previously, the barrier function of pulmonary epithelial cells depends, in part, on the activity of ENaC. This activity has been shown to be inhibited by reactive oxygen and nitrogen species (57). The major enhancer for the stimulated alpha-ENaC expression in lung epithelial cells is the glucocorticoid response element (GRE). The presence of an intact GRE is necessary and sufficient for oxidants to repress ENaC expression. Exogenous H_2O_2-mediated repression of ENaC GRE activity is partially blocked by either a specific inhibitor for ERK pathway activation, U0126, or dominant negative ERK, suggesting that, in part, activated ERK may mediate the repressive effects of H_2O_2 on alpha-ENaC expression (58). Based on genetic analysis and molecular studies, ENaC has been suggested to be mechanosensitive. Mechanical stretch and cellular deformation-induced ERK activation, may regulate ENaC expression through similar mechanisms as that of oxidative stress (60, 61).

Protein Tyrosine Kinases

The receptor associated cytoplasmic protein tyrosine kinase, termed janus kinases (JAKs) and the latent monomeric signal transducers and activators of transcription, STAT proteins, have been demonstrated to have a role in generating intracellular signaling pathways in response to cyclic stretch (62, 63). Importantly, the down-stream product of STATS is c-fos mRNA (49). Thus the JAK/STAT complex provides an alternative pathway from receptor to nucleus that is independent of G proteins.

Recently, Bhattacharya et al found that lung expansion activated rat pulmonary endothelial tyrosine kinases that may promote vascular remodeling in response to mechanical stresses. After induction with an inspiratory pressure of 22 cm H_20 Western blots of cell lysates showed markedly enhanced phosphorylation of focal adhesion kinase (FAK), paxillin, and Shc adaptor protein, which the authors speculated may function to stabilize endothelial cells to the cell matrix (64). Mechanical stretch-induced activation of pp60src, an cytosolic protein tyrosine kinase, is an up-stream event of a defined signal transduction pathway, which leads to increased proliferation of fetal rat lung cells (65). Parker et al used a phosphotyrosine phosphatase inhibitor (phenylarsine) and a tyrosine kinase inhibitor (genistein) to assess the role of these enzymes in pressure-induced lung injury. Inhibition of phosphotyrosine phosphatase was shown to lead to an increase in susceptibility of rat lungs to high peak inflation pressure (PIP) injury, while inhibition of tyrosine kinase attenuated the injury relative to the high PIP control lungs (66). Moreover, Mac Gillivray et al recently demonstrated that IL-1β receptor associated kinase (IRAK) co-localizes with focal adhesion complexes and is required for IL-1β dependent ERK activation (67). This suggests that the integrity of the actin filament arrays and the recruitment of IRAK into focal adhesion complexes are involved in

IL-1β-mediated signaling. It is unknown whether IL-1β mediated cell signaling functions in a similar way in VILI.

Transcriptional Activators

How different signal transduction pathways mediate the inflammatory response characteristic of VILI is not yet known. One possible mechanism is by activation of transcription factors. Transcription factors are DNA-binding proteins that regulate gene expression. Physical forces can exert their effects by influencing expression of immediate early response genes, some of which encode transcription factors, such as: c-Fos, c-Jun, c-Myc, and Egr-1. It has been demonstrated that most early response genes contain stretch response elements in their cis-acting regulatory sequences. Putative transcriptional factor binding sites that contain shear stress-responsive elements have also been demonstrated in cis-acting regulatory regions of various genes including PDGF-B, tissue plasminogen activator (tPA - thrombosis), intercellular adhesion molecule-1 (ICAM-1), and TGFβ (14). In addition, genes that encode for nitric oxide synthase and cyclooxygenase-2 have been shown to be regulated by shear stress (68, 69) and cyclic strain (70, 71). In this section we will primarily focus on the transcription factor NF-κB.

Current evidence suggests that the activation and control of NF-κB plays a critical role in the generation and propagation of the cytokine response in VILI. NF-κB is known to be a key transcription factor for maximal expression of many cytokines that are involved in the pathogenesis of inflammatory diseases, such as ARDS and sepsis syndrome (for a review of NF-κB in cytokine gene regulation see reference 72 and Chapter 1). NF-κB can be activated in cells by a variety of stimuli, including bacterial endotoxin, TNF-α, IL-1β, mitogens, and viral proteins. NF-κB itself contains a DNA "shear-stress" response element in its promoter region, and NF-κB protein binds to the IL-6, IL-8, IL-1β and TNFα promoter sequences (72). A number of studies in both *in vitro* and *ex vivo* whole lung preparations have demonstrated that NF-κB is up-regulated in response to stretch (73-75). Moreover, the signaling cascade mediated by NF-κB results in a chemical (mediator) response (76) in alveolar macrophages in response to cyclic stretch.

A recent study by Held et al has further enhanced our understanding of mechanotransduction in VILI by demonstrating that although signal transduction mechanisms responsible for biotrauma act, at least in part, via NF-κB (similar to lipopolysaccharide [LPS] signaling), the upstream mechanism appears to be quite distinct (77). In the case of LPS, the initial recognition of endotoxin by monocytes is dependent on the expression of

CD14 and subsequent presentation of the endotoxin molecule to a member of the Toll receptor family (TLR-4) – a highly evolutionary conserved family of transmembrane receptor proteins (78). The intracellular signaling through the TLR-4 receptor shows remarkable similarity to IL-1β signaling pathways. This cellular receptor (TLR-4/IL-1β) has been shown to activate p42/44 MAPK, JNK, p38, p65, and NF-κB (79) in response to LPS stimulation. Held et al used C3H/HeJ mice that are defective in the LPS induced NF-κB mediated signaling due to a mutation in the TLR-4 encoding gene (80), to demonstrate that overventilation is still able to activate NF-κB (77). Therefore, the upstream event of ventilation-induced signaling is different from that of LPS.

The findings reported by Held et al are of particular clinical relevance for two main reasons: 1) NF-κB activation could be abrogated by dexamethasone (77) - its steroid responsiveness implies that it will be possible to treat and/or prevent biotrauma pharmacologically, and 2) if we think of the LPS signaling pathway as a major pathway for innate immunity, then it might be possible in the future to target specific signal transduction pathways proximal to NF-κB that are important to VILI without disrupting LPS-mediated signaling (176). These results therefore have important potential bedside implications.

MECHANICAL VENTILATION INCREASES THE PRODUCTION OF MEDIATORS OF INFLAMMATION AND CELL DEATH

Cyclic motion produced by certain modes of mechanical ventilation has been shown to lead to the induction, synthesis, and release of cytokines and inflammatory mediators from the lung (34, 81-83). A number of investigators have generated evidence demonstrating the dramatic improvements in gas-exchange, histological indices of lung injury, and survival in animals ventilated using a lung protective strategy versus a conventional strategy - or injurious strategy (84-86). Moreover, Ranieri et al provided convincing evidence in support for the role of ventilation in altering cytokine levels in humans (87). It is beyond the scope of this review to list and discuss the role of each mediator known to be activated by mechanical ventilation (for excellent reviews on inflammatory mediators in VILI please see references 2, 8, 10, 12, and 88). Here we want to highlight those cytokines and mediators that we believe offer a special insight into how individual cellular responses to mechanical injury might contribute to the cascade of inflammation underlying MOSF in ventilated ARDS patients.

Cytokines

The early response cytokines TNFα and IL-1β have long been felt to be the main orchestrators of the pro-inflammatory cascade in patients with ARDS/acute lung injury (ALI) and systemic inflammatory syndrome (34, 89-91 and Chapter 2). TNFα in particular has received much attention, primarily because of its known functions, which include the release of IL-1β, IL-8, IL-6, activation of NF-κB, and inducing programmed cell death (apoptosis) (92). In surfactant-depleted rabbits, increased levels of IL-1β, IL-8, IL-6 and TNFα, have been found in the bronchoalveolar lavage (BAL) fluid of animals subjected to a conventional mechanical ventilation strategy (peak inspiratory pressure 25-28 cm H_2O) as compared to high frequency ventilation (81, 82). Takata et al found that with conventional mechanical ventilation the cells retrieved by lung lavage exhibited increased expression of TNFα mRNA - this preceded the development of physiological or histological abnormalities (81). Similarly, Imai et al found increased lavage concentrations of platelet activating factor and thromboxane B2 in rabbits ventilated with conventional ventilation (82). Tremblay et al examined the effects of four different ventilatory strategies on lung inflammatory mediators in the presence or absence of a pre-existing inflammatory stimulus. In these experiments, the injurious ventilatory strategy led to increased detection of both TNFα and c-fos mRNA (34). Von Bethmann et al found similar results by exposing an isolated perfused and ventilated mouse lungs to overdistension, which resulted in increased expression of TNFα and IL-6 m-RNA and gene product (83).

Recently, however, the role of TNF-α in VILI has been disputed (93). In an *in vivo* animal model of surfactant deficiency Verbrugge et al did not document ventilator-induced systemic prostacyclin and TNFα release, or pulmonary TNFα release (94). Moreover, in a study of pro-inflammatory activity of BAL fluid from patients with ARDS most of the pro-inflammatory activity was demonstrated to be attributable to the biologically active IL-1β, and not to TNF-α (93). In previous studies, Narimanbekov and Rozycki demonstrated that pretreatment of surfactant-depleted rabbits with an antagonist of IL-1β prior to ventilation, reduced PMNs, elastases and albumin in BAL fluid as well as histopathological evidence of lung injury (95). It is of note however that no improvement in lung function was detected, suggesting that although IL-1β contributes to VILI, other factors are also involved.

Other important cytokines in VILI include IL-8, IL-6 and IL-10 - release of all of which may be induced by both TNF-α and IL-β. IL-8, a CXC chemokine, functions as an activating cytokine for neutrophils and a neutrophil chemotactic factor. High BAL levels of IL-8 and neutrophil infiltration is characteristic of acute lung injury (ALI), ARDS and VILI (2,

8). To more precisely define mechanisms of IL-8 gene expression triggered by cellular detachment and deformation, Shibata et al exposed human bronchial epithelial cells (BECs) to neutrophil elastase (NE), a potent protease capable of deforming and detaching BECs. Exposure of BECs to NE and other proteolytic agents caused increase in the expression of IL-8 mRNA (96). This induction could be inhibited by pre-treatment with taxol (a microtubule stabilizing agent). Induction of IL-8 mRNA seemed to occur secondary to deformation of cytoskeletal structures, rather than through a specific receptor. Fluorescent microscopy observations demonstrated that both NE and colchicine (a microtubule disrupting agent) treatment disrupted the fine structure of microtubules, and taxol treatment reorganized them into bundles (96). Accordingly, up-regulation of IL-8 mRNA transcripts may be related to microtubular disruption. The expression of other genes appears to also be related to the configuration of microtubules. In murine macrophages taxol is known to induce TNFα gene expression (97). Colchicine is reported to induce IL-1β gene expression in human monocytes (98). The expression of IL-6 does not seem to be up-regulated by detachment and deformation (96, 98, 99).

In contrast, in human vascular endothelial cells cultured on a flexible silicone membrane and exposed to cyclical stretch, Cytochalasin D (disrupts the actin cytoskeleton) abolished the stretch-induced gene expression of IL-8. Neither inhibition of stretch-activated ion channels nor disruption of microtubules affected the induction of this chemokine by cyclic stretch. Using enzyme inhibitors it has been showed that phospholipase C, PK-C, and tyrosine kinase were involved in the stretch-induced gene expression of IL-8, whereas cAMP- or cGMP-dependent protein kinase was not (100).

IL-6 levels have been used in previous studies as a measure of the degree of systemic inflammation. Elevation in IL-6 levels seems to directly correlate with pulmonary and end-organ damage (7, 87, 101). IL-6 is induced often together with the pro-inflammatory cytokines TNFα and IL-1β in many alarm conditions (102), and circulating IL-6 plays an important role in the induction of acute phase reactions (103, 104). However, whether this endogenous IL-6 plays any additional pro- or anti-inflammatory roles in local or systemic responses remains unclear. Xing et al studied the role of IL-6 in acute inflammatory responses in animal models of endotoxic lung or endotoxemia by using IL-6+/+ and IL-6-/- mice. This group found that endogenous IL-6 plays a crucial anti-inflammatory role in both local and systemic acute inflammatory responses by controlling the level of pro-inflammatory, but not anti-inflammatory, cytokines, and that these anti-inflammatory activities by IL-6 cannot be compensated for by IL-10 or other IL-6 family members (105). Moreover, IL-6 has also been implicated in regulation of antigen presentation (102).

IL-10 was originally described as a product of B-cells but has since been shown to be produced by a number of cells, including alveolar macrophages and bronchial epithelial cells (106). Mechanical ventilation (31) and cytokines, such as TNFα, have been shown to induce IL-10 production (107). Blood levels of IL-10 seem to be directly related to the severity of inflammation and the development of organ failure in septic shock (108). Donnelly et al demonstrated that increased levels of IL-10 in the BAL fluid of patients with ARDS correlate with patient survival (109). IL-10 has been shown to inhibit the production of IL-1β, TNFα, and IL-6. IL-10 also down regulates class II MHC and B7, which are important molecules in antigen presentation (102). Moreover, IL-10 prevents the endotoxin-induced procoagulant effect on human monocytes and blocks the release of oxygen free radicals and nitric oxide from activated macrophages (110). This inhibition seems to be, at least in part, mediated by modulating PKC (111); this is important in as much as it may provide a link to mechanosensitive mechanisms of cellular signal transduction. Furthermore, IL-10 gene transfer improves the survival rate of mice inoculated with *E. coli*, presumably via inhibition of TNFα and IL-6 production (112). These data strongly suggests that IL-10 may have a protective effect in inflammatory diseases (see Chapter 17). IL-10 has been shown to be activated in the setting of VILI (34), presumably its primary role in VILI is secondary to its anti-inflammatory activity, but it is unclear which mechanotransduction pathways are involved in its activation.

Pro-coagulants

The inflammatory and pro-coagulant host responses are closely related (113). Inflammatory cytokines such as TNF-α, IL-1β and IL-6 are capable of activating coagulation and inhibiting fibrinolysis, whereas the pro-coagulant thrombin is capable of simulating inflammatory pathways (103). Little is known about the effects of mechanical ventilation on the pro-coagulant pathway (113). However, Chan et al have shown that mechanical stretch increased antithrombin activity due to an increase in the concentration of active chondroitin sulfate from fetal rat lung cells. Stretch also down-regulated secretion of tissue factor procoagulant activity, which may lead to decreased thrombin generation on the surface of lung cells (114). They suggested that mechanical stretch derived from fetal breathing movements may play a role in regulating coagulation in the fetal lung. Ventilator-derived physical force may affect the coagulation with similar mechanisms.

Evidence from studies in cardiac hypertrophy suggests that that mechanical, cyclical, and pressure induced signals, stimulate endothelin synthesis (115). Recently, vascular endothelial cells have been shown to respond to cyclic and pressure stimuli (116). Catarruzza et al exposed vascular endothelial cells to cyclical stretch (20% elongation) and demonstrated an eight-fold increase in the expression of pre-pro endothelin (precursor to endothelin) and endothelin A and B receptors. However, endothelin regulation in different cell types appears to be quite heterogeneous. In fact, smooth muscle cells exposed to the same forces as the ones described above, responded by decreasing endothelin synthesis by 80%. Moreover, cyclic-stretch in these cells, caused a decrease in endothelin A receptor mRNA, but an increased expression of endothelin B receptor mRNA. Increased expression of endothelin receptor B signaled (ET-1 mediated) cells to undergo apoptosis (116). The evidence that endothelin synthesis in pulmonary endothelial cells may be regulated by mechanical ventilatory stimuli is less promising. Behnia et al could not document an increase in either lung lavage, serum, or lung tissue ET-1 mRNA expression after exposing rats to as much as 7 h of positive pressure ventilation (117). In more recent experiments, Simma et al did not find any difference in ET-1 or ET-3 release in rabbits ventilated with a lung protective strategy versus a conventional ventilatory strategy (118).

This marked interest in the relationship between the procoagulant and pro-inflammatory pathways is partly due to the recently published results from the PROWESS trial. In a study of patients with severe sepsis, administration of human recombinant activated protein C (APC) resulted in a 19.4 % reduction in the relative risk of death from any cause at 28 days (119). What is of interest is that APC has also been shown to inhibit LPS-induced activation of NF-κB *in vitro* (120). The ability of APC to inhibit LPS-induced translocation of NF-κB is likely to be a significant event given the critical role of the latter in the regulation of cytokine expression (see above) (120). If these findings were to be true in pulmonary endothelial cells, this may have significant implications. For example, APC has been strongly implicated in the regulation of inflammatory reactions involving cytokines or activated leukocytes (113, 121). The precise mechanism of amplification or down-regulation of the inflammatory cascade by APC is unclear, but it has been shown to prevent endotoxin-induced pulmonary vascular injury by inhibiting leukocyte activation and cytokine production (122). APC seems to inhibit the coupling of endothelin-1 (ET-1) to CD14, which stimulates production of cytokines (121). Moreover, APC inhibits E-selectin-mediated cell adhesion (123). Because the expression of E-selectin on endothelial cells is dependent on TNFα (124), and interactions between leukocytes and endothelial cells is critical in mediating endothelial cell injury, endothelin regulated, APC-induced suppression of TNFα synthesis may attenuate acute lung injury/VILI.

How this will impact on VILI is unclear, given that to date there is no evidence suggesting that APC may be regulated by mechanical forces. Future research is required to determine if APC can also suppress activation of NF-κB induced by mechanical forces.

Surfactant-Associated Proteins

Following the loss of the macromolecular barrier in acute lung injury, protein-rich edema fluid floods the alveoli and impairs surfactant function, with ensuing atelectasis and a concomitant deterioration in gas exchange and lung compliance. Although surfactant proteins are normally only found in appreciable amounts in the lung (only trace amounts in the serum), leakage of surfactant protein-A (SP-A) (125, 126), surfactant protein-B (SP-B) (126) and surfactant protein-D (SP-D) (127) into the circulation has been reported in a number of respiratory disorders. The route by which these proteins enter the circulation is unknown; however, there is strong evidence that bidirectional plasma protein flux occurs in the lungs, the magnitude of which depends of disease severity (128). Impairment of the surfactant system, as a result of mechanical ventilation, has also been shown to contribute to lung parenchyma stretch and VILI (129-131).

Animal studies have demonstrated that mechanical ventilation of injured lungs can contribute to the changes in surfactant system (129, 130). Moreover, in a recent study, Veldhuizen et al have also demonstrated that mechanical ventilation of normal lungs can also alter surfactant (128). In addition to alveolar-capillary membrane failure, Rose et al have recently presented evidence indicating that, alveolar epithelial type II cells (ATII) respond with exocytosis of surfactant containing lamellar bodies to stimulation with mechanical stretch and secretagogues (132). Furthermore, the state of actin polymerization is intimately linked to the exocytosis process underlying surfactant secretion in these cells. Microfilament system-related compartmentalization effects and/or the impact of the state of actin assembly on signaling events may be considered as underlying events in ATII cells. Moreover, the same group has shown that in ATII cells mechanical stretch-induced surfactant secretion likely occurs via the prostaglandin PGI2-cAMP axis (133).

In the context of biotrauma, the intrinsic properties of each surfactant-associated protein might also be relevant to the pathophysiology of ARDS/ALI/VILI. SP-A is primarily important in the formation of tubular myelin and anti-bacterial host defenses (134-136 and Chapter 13). SP-B enhances spreading and stabilizes surfactant phospholipids at the air liquid interface (137 and Chapter 12). Hereditary SP-B deficiency causes ARDS in newborns and mice (138). Moreover, Arias-Diaz et al recently showed that

SP-A inhibits LPS-induced TNFα response in both interstitial and alveolar human macrophages, as well as an IL-1 response in interstitial macrophages. The authors postulated that the SP-A effect on TNFα production could be mediated by suppression in the LPS-induced increase in intracellular cGMP (139). These data lend further evidence that SP-A regulates alveolar host defenses and inflammation by suggesting a fundamental role for this apoprotein in limiting excessive pro-inflammatory cytokine release.

Mechanical Stretch-Induced Apoptosis

In a recent editorial Chandel (140) discussed data of Edwards et al indicating that exposure of alveolar type II epithelial cells (ATII) to 30% strain at 60 cycles/min caused apoptosis of ATII cells (141). Furthermore, ATII cells co-cultured with macrophages were protected against cyclic-stretch-induced apoptosis. The ability of macrophages to protect ATII cells against cyclic stretch-induced apoptosis was completely abolished by nitric oxide (NO) inhibitors. In further experiments, NO donors protected ATII cells against stretch-induced apoptosis. Davis et al showed that intratracheal administration of bleomycin led to caspase-mediated apoptosis (142). The effects of bleomycin were associated with translocation of p53 from the cytosol to the nucleus only in alveolar macrophages that had been exposed to the drug *in vivo*, suggesting that the lung microenvironment regulated p53 activation. Experiments with a thiol antioxidant (N-acetylcysteine) *in vivo* and NO donors *in vitro* confirmed that reactive oxygen species were required for p53 activation. A specific role for NO was demonstrated in experiments with inducible nitric oxide synthase (iNOS) -/- macrophages, which failed to demonstrate nuclear p53 localization after *in vivo* bleomycin exposure (142). The importance of these findings relate to the fact that NO synthesis is regulated by mechanical forces.

Artlich et al recently demonstrated that high frequency oscillating ventilation (HFOV) increases pulmonary NO production in healthy rabbits (143). This group also proposed that the increase in mean airway NO concentrations may have biological effects in the respiratory tract and may account for some of the benefits of HFOV treatment. Bannenberg et al corroborated these findings by demonstrating that positive pressure ventilation induces NO production. Moreover, this group showed that NO synthesis could be inhibited with gadolinium (144). These findings imply that increased NO synthesis induced by mechanical forces may be mediated by calcium dependent stretch responsive channels.

There are two main questions regarding stretch-induced apoptosis. At what level of strain is apoptosis triggered? What are the signaling mechanisms regulating stretch-induced apoptosis? Alveolar epithelial cells

(AEC) are exposed to 1-5% strain during normal breathing rather than 30%. Both a threshold and duration of strain may be required to trigger apoptosis, and AEC may trigger adaptive responses until the strain levels reach threshold - at which point cells undergo apoptosis (140). Ashino et al recently demonstrated that intracellular calcium oscillations and exocytosis of lamellar bodies in ATII cells could occur secondary to lung expansion (145). Loss of mitochondrial integrity associated with cytochrome c and activation of caspases has been implicated in the activation of signal transduction pathways leading to apoptosis. NO has been shown to inhibit both cytochrome c and caspases in response to various cell-death stimuli such as TNFα (146). Mechanical stretch-induced apoptosis of ATII s (141) may be mediated through similar mechanism.

Moreover, mitochondria are associated with microtubules and motor proteins such as kinesin, which participates in the shuttle movement of mitochondria between the cell surface and periphery. Disruption of kinesin results in abnormal perinuclear clustering of kinesin (147). Therefore, low tidal volume ventilation might trigger apoptosis by the activating cytoskeletal signaling mechanisms involving calcium and kinases. In contrast, apoptosis triggered by high tidal volume ventilation might be secondary to disruption of focal adhesions and microfilaments causing loss of mitochondrial integrity (140). Thus, the magnitude of cytoskeletal network change could determine whether ATII cells produce surfactant or die by apoptosis.

CELLULAR ACTIVATION

To understand the effects of the complex interplay between different inflammatory mediators in an organ system it is essential to dissect the functional interactions involving the major cells known to play a role in the generation and propagation of VILI.

Polymorphonuclear Leukocytes

Polymorphonuclear (PMN) leukocytes have been unequivocally implicated in the pathogenesis of ALI/ARDS (for a review see reference 148 and Chapters 2-4). The current literature also suggests that injurious ventilatory strategies used in normal lungs may increase cytokine production and neutrophil recruitment in lung lavage specimens (149). The current evidence seems to point to the role of PMNs as major effector cells in the generation of the tissue injury characteristic of VILI (2). One of the first studies proposing that mechanical ventilation could lead to an inflammatory response used a model of neutrophil depletion by nitrogen mustard. Kawano et al demonstrated that the neutrophil-depleted animals had markedly improved oxygenation and decreased pathologic evidence of injury following lung lavage/mechanical ventilation than a control group without neutrophil depletion (84).

More recently, Zhang et al combined these two observations and examined the hypothesis that the BAL from patients with ARDS could lead to activation of PMN by mechanical ventilation (150). In these studies, PMNs isolated from normal human volunteers were incubated with BAL fluid from ARDS patients ventilated with either conventional mechanical ventilation or a protective ventilation strategy. Treated neutrophils were assessed post-incubation for evidence of PMN activation. This group found that in the conventional ventilation strategy group, all markers of neutrophil activation were increased (150). These findings support the current evidence suggesting that mechanical ventilation can lead to release of mediators that prime neutrophils, and may provide a mechanism by which PMNs mediate tissue injury in VILI. Thus, the role of PMNs in the pathogenesis of VILI may be regulated by their interaction with other cells, including epithelial cells and possibly vascular endothelial cells.

Endothelial Cells

Pulmonary endothelial cells form a continuous monolayer on the luminal surface of the lung vasculature. Much is known about the response of vascular endothelial cells to shear stress (for a comprehensive review see references 151 and 152); however, there is relatively little that is known about the response of pulmonary endothelial cells to mechanical stress generated by artificial ventilation. A major concept in vascular biology is that endothelial cells can become activated (153). Moreover, what has become evident is that only activated endothelium participates in the inflammatory response (148). In the human lung, endothelial cell stimulation with LPS, IL-1β, or TNFα induces secretion of IL-8 and epithelial neutrophil activating peptide (ENA-78), mediators of neutrophil sequestration and degranulation (154). Cyclic strain has also been shown to cause induction of immediate response genes such as activator protein-1 (AP-1) and NF-κB (74), the significance of which is unclear. Despite advances in endothelial biology there is a paucity of data to explain the role of these cells in VILI. Currently, the search for the primary regulator of VILI has generated two important candidate cell types, alveolar macrophages and epithelial cells.

Alveolar Macrophages

Over the past few years, alveolar macrophages (AM) have gained much attention as a source of cytokines, and it has been postulated that they function as important effector cells in the orchestration of VILI (155). Pugin et al reported that AM were responsible for the highest degree of response to mechanical stress as assessed by measuring IL-8 concentrations (156). AMs were also shown to play a role in lung remodeling as they respond to mechanical stress by increasing *de novo* synthesis of matrix-metalloproteinase (MMP)-9, a type IV collagenase (156). Monocyte-derived macrophages (MDM) produced similar responses as AM and were used in experiments as surrogate cells for AM. NF-κB was shown to undergo nuclear translocation in MDM submitted to cyclic pressure stretching (156). Lancet et al have pointed to the essential role of AM in the activation of NF-κB (73). In AM-depleted rat lungs, the elevation of TNFα and MIP-2 was suppressed and the up-regulation of ICAM-1 was abrogated. Instillation of TNFα restored NF-κB activation and inflammatory mediator responses. It has been inferred from these experiments that NF-κB translocation occurs in response to AM stretch, and that this may represent a fundamental contributing factor in the development of VILI; however, to date, this has not been conclusively demonstrated.

Alveolar Epithelial Cells

Another potential source of cytokines are alveolar epithelial cells, which have been recently demonstrated to be a source of many cytokines and chemokines. Using *in situ* hybridization and immunohistochemistry it has been suggested that the expression of TNF-α is significantly increased in the airway and alveolar epithelium after injurious ventilation (157). It has been further demonstrated that primary cultured rat lung alveolar epithelial cells could produce TNF-α (158). Production of IL-8 has been found from a human lung carcinoma cell line (A549 cells) (159). Vlahakis et al recently cultured these human alveolar epithelial cells on a deformable silicoelastic membrane. When stretched to 130% of baseline for 48 hrs IL-8 secretion increased by up to 34% (160). Stud et al demonstrated that the physical stress exerted on the alveolar epithelial cells by a deposited asbestos fibers is greatly enhanced by cyclical stretch. Coating of fibers with fibronectin, a glycoprotein abundant in the alveolar lining fluid, significantly amplified the fiber-induced cell response. Furthermore, this response was inhibited by addition of an integrin-blocking agent to the culture medium, suggesting that adhesive interactions between protein-coated fibers and cell-surface molecules are involved in determining IL-8 secretion (161).

In an effort to further characterize the nature of the stretch response in epithelial cells, Morgan et al applied mechanical stretch to primary cultured fetal rat lung cells in an organotypic culture where the three dimensional culture system allowed for fetal lung cells to form alveolar-like structures. Using this system this group was able to demonstrate an additive affect between submaximal concentrations of LPS and mechanical stretch on production of MIP-2, a rodent homologue of human IL-8 (22) produced by rat lung alveolar epithelial cells (162). LPS alone was associated with increases in MIP-2 mRNA levels whilst stretch alone significantly increased MIP-2 production in the absence of *de novo* protein synthesis - implicating increased MIP-2 secretion as the possible effector mechanism of MIP-2 increase in epithelial cells activated by stretch (22). Since mechanical stretch is primarily applied to the alveolar septa, cytokines and chemokines produced from these cells could play an important role in mediating VILI.

Blood-Gas Barrier

The role of the blood-gas barrier in regulating biotrauma, albeit fundamental, is very poorly understood. In a recent review on pulmonary capillary stress failure, West pointed to the profound cytopathological changes associated with pulmonary microvascular hypertension (163). West suggested that the membrane's strength comes from type IV collagen in the basement membrane and that its structure is continually regulated in response to capillary wall stress. Despite extensive literature on vascular remodeling in pulmonary arteries and vein, little is known about the possible remodeling of pulmonary capillaries. Experiments by West and colleagues have shown a generalized organ-specific response after localized (unilateral) application of mechanical force. Increased lung inflation over a 4 h period resulted in increased gene expression for $\alpha 1$ (III) and $\alpha 2$ (IV) procollagens, fibronectin, basic fibroblast growth factor, and TGF- β_1 in both over-inflated and normally inflated control lungs. Similar patterns of gene expression were obtained in experiments in which the capillary transmural pressure was increased (163).

Evidence to suggest that mechanical stretch alters the ability of biological structures to heal from mechanical injury comes from experiments by Waters and colleagues. In a series of experiments, this group demonstrated that cyclic mechanical strain inhibits wound closure in airway epithelium (164). Repair and injury of the airway epithelium seems to occur continuously; and the repair mechanisms depend on arachidonic acid metabolism. Exposing alveolar epithelial cells to cyclical stretch led to down regulation of the synthesis of prostaglandins (PG) E2, PGI2, and thromboxane A2 (71). Cyclic stretch induced inactivation of cyclo-oxygenase, this was presumably mediated by oxidants. Keratinocyte growth factor (KGF) conferred protection from hyperoxia-induced lung injury (which appears to occur via stretch-regulated channels) (165). This may be dependent on the activation of PKC and involve processes that rely on cytoskeletal stabilization (57). Although these studies were performed on alveolar epithelial cells, the implications pertain to our understanding of the maintenance of the blood-gas barrier integrity, especially in the context of mechanical ventilation induced injury.

In more recent experiments to determine the correlation between ultrastructural and physiological changes in blood-gas barrier function in lungs transiently exposed to very high vascular pressures, Maron et al demonstrated that although endothelial and epithelial breaks were occasionally observed in some experiments, their incidence was not increased in the high-pressure group (166). These data indicate that the increased transvascular water and protein flux observed in the lungs treated with high pressure, occurred through pathways of a size not resolvable by electron microscopy.

Generalization of the Inflammatory Response

It is thought that as compartmentalization of the local pulmonary response is lost, systemic release of inflammatory mediators promotes the massive inflammatory response that underlies MSOF (2, 6, 8, and Figure 4). This is rapidly followed by the generation of an equally dramatic compensatory anti-inflammatory reaction that is designed to down-regulate and attenuate the pro-inflammatory response (2, 6, and Chapter 17). Loss of appropriate immune modulation, or persistence of inflammatory injury, appears to be involved in the inability of organisms to bring about resolution of the pro-inflammatory response and ultimately death (6, 91).

In animal models, Chiumello et al (85) and more recently Haitsma et al (86) demonstrated that ventilatory strategies, which are known to induce VILI, disturb the compartmentalization of early response cytokines. Furthermore, lung protective strategies may reduce leakage of cytokines across the alveolar capillary membrane. The data presented by Haitsma et al are also the first to suggest that cytokines may leak from the systemic circulation into the alveolar space (possibly explaining how patients with sepsis may develop ARDS). In animals, this may occur as quickly as in 20 minutes, thus indicating that even a short period of ventilation can lead to translocation of cytokines.

Clinical evidence in support of this model for the development of MSOF in ventilated ARDS patients became available when Ranieri et al demonstrated that the concentrations of pro-inflammatory cytokines in both

BAL fluid and plasma could indeed be decreased in patients ventilated with a lung protective strategy (87). Patients in the lung protective strategy group had reductions in both their BAL and serum concentrations of polymorphonuclear cells, pro-inflammatory, and anti-inflammatory cytokines (87). In a follow-up of this study the authors demonstrated that changes in plasma concentrations of some of these mediators correlated with changes in the development of organ dysfunction (101).

Mechanical ventilation may also contribute to systemic inflammation by increasing translocation of intra-alveolar pathogens from the airspace into the circulation (Figure 4). Ventilatory models which allow end expiratory collapse have been shown to induce bacterial translocation from the lung to the systemic circulation when very high tidal volumes are used (94, 167); even strategies that use relatively normal tidal volumes can induce endotoxin translocation from the lung to the systemic circulation (168). Evidence that important inflammatory mediators can escape the confines of the lung provides important clues to mechanisms that may potentially lead to the development of MSOF in VILI. Furthermore, in a recently presented experimental study, the use of lung protective strategies with low tidal volume delayed bacteremia and consequent distal injury (169).

Moreover, injurious ventilatory regimens have recently been shown to be more deleterious when applied to injured or infected lungs (169). The prevailing theory is that this phenomenon (the "two hit hypothesis") is due to the fact that infectious/inflammatory insults act in a cumulative fashion. The initial stimulus generates an acute inflammatory reaction, akin to "priming," that is apparently contained or restrained within the confines of the lung. The second insult, however, seems to mediate the loss of this compartmentalization or control and promotes the initiation of an inflammatory response that escapes regulatory mechanisms, and may ultimately lead to MSOF and/or death.

A number of investigators have shown that a high tidal volume alone (5) or the use of a ventilatory strategy without PEEP at low lung volumes (170) can increase capillary permeability and that this is both required and sufficient for translocation of, at least cytokines, from the lung into the circulation. (171, 172). However, recent data support the evidence that two hits appear to be needed (at least for tidal volumes that are possible *in vivo* without markedly affecting hemodynamics) since mechanical ventilation with high tidal volumes alone did not result in higher cytokine levels in the plasma of anaesthetized patients with healthy lungs when compared to a lung-protective strategy (173). Consequently, prior lung injury seems to be required, in conjunction with mechanical ventilation, for marked elevation in cytokine levels to be detected.

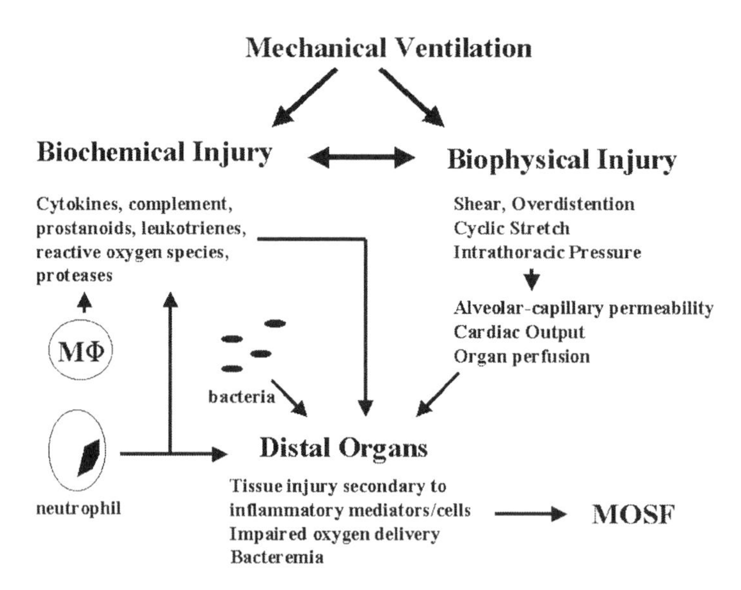

Figure 4. Postulated mechanisms whereby mechanical ventilation may contribute to MSOF.

Conclusion

Mechanical ventilation is clearly an indispensable tool for patients who require ventilatory support; but, if we are to succeed in eliminating its iatrogenic consequences, we must understand the molecular and cellular effects of exposing the lung to the forces generated by mechanical ventilation. It is also true that, if correct, this new conceptualization of VILI could lead to a paradigm shift in which therapies to prevent VILI are not solely based on changes in ventilatory strategies to limit mechanical injury, but are also aimed at constraining and/or modulating the inflammatory response. Moreover, further studies have now demonstrated that a patient's genetic make-up may predict response to sepsis. Mira et al have shown that specific polymorphisms of the TNFα gene predisposes patients to an increase risk of death (174); and, Arbour and colleagues have recently demonstrated that differences in responsiveness to inhaled LPS in humans may be due to differences in common mutations of the TLR4 gene (175). If similar polymorphisms are important in the context of biotrauma, then in the future, intensivists may choose to use different anti-mediator therapy based on a patient's genotypic profile.

Furthermore, in the case of VILI, we are in a unique position of knowing exactly when the inflammatory stimulus will begin (on initiation of mechanical ventilation) and hence could start anti-inflammatory therapy soon after, or even before, the initiating stimulus begins (176). Two papers have specifically addressed the issue of immunotherapy for VILI. Imai et al pre-treated rabbits with polyclonal anti-TNFα antibody. Animals subsequently underwent conventional mechanical ventilation. Pretreatment with intra-tracheally delivered polyclonal anti-TNFα antibody improved oxygenation, respiratory compliance, decreased leukocyte infiltration, and ameliorated pathological findings (177). In another study Narimanbekov, et al used recombinant IL-1 receptor antagonist prior to exposure to conventional mechanical ventilation. BAL fluid of rabbits pre-treated if IL receptor antagonist showed lower albumin and elastase concentrations, as well as lower neutrophil counts (95).

In this chapter we have attempted to review the literature which suggests that cells respond to mechanical stimuli, and provide a perspective as to how this may relate to VILI. Continued efforts in basic research aimed at elucidating the processes by which cells respond to stimuli and in clinical studies addressing these concepts are still required. Increasing our knowledge and understanding of basic cell physiology and molecular biology is critical if new approaches to diseases are to become the reality of intensive care in the future.

REFERENCES

1. Fothergill J. (1745) Observations on a case published in the last volume of the medical essays, & c. of recovering a man dead in appearance, by distending the lungs with air. *Philos Trans R Soc Lond* 43, 275-281
2. Tremblay, L.N., and Slutsky, A.S. (1998) Ventilator-induced injury: from barotrauma to biotrauma. *Proc Assoc Am Physicians* 110, 482-488
3. Montgomery, A.B., Stager, M.A., Carrico, C.J., and Hudson, L.D. (1985) Causes of mortality in patients with the adult respiratory distress syndrome. *Am Rev Respir Dis* 132, 485-489
4. Dreyfuss, D., Soler, P., and Saumon, G. (1992) Spontaneous resolution of pulmonary edema caused by short periods of cyclic overinflation. *J Appl Physiol* 72, 2081-2089
5. Dreyfuss, D., Soler, P., and Saumon, G. (1995) Mechanical ventilation-induced pulmonary edema. Interaction with previous lung alterations. *Am J Respir Crit Care Med* 151, 1568-1575
6. Slutsky, A.S., and Tremblay, L.N. (1998) Multiple system organ failure. Is mechanical ventilation a contributing factor? *Am J Respir Crit Care Med* 157, 1721-1725
7. ARDS Network. (2000) Ventilation with lower tidal volumes as compared with traditional tidal volumes for acute lung injury and the acute respiratory distress syndrome. The Acute Respiratory Distress Syndrome Network *N Engl J Med* 342, 1301-1308
8. Slutsky, AS. (1999) Lung injury caused by mechanical ventilation. *Chest* 116, 9S-15S
9. John, E., McDevitt, M., Wilborn, W., and Cassady, G. (1982) Ultrastructure of the lung after ventilation. *Br J Exp Pathol* 63, 401-407
10. Parker, J.C., Hernandez, L.A., and Peevy, K.J. (1993) Mechanisms of ventilator-induced lung injury. *Crit Care Med* 21, 131-143

11. Fu, Z., Costello, M.L., Tsukimoto, K., Prediletto, R., Elliott, A.R., Mathieu-Costello, O., and West, J.B. (1992) High lung volume increases stress failure in pulmonary capillaries. *J Appl Physiol* 73, 123-133
12. Dreyfuss, D., and Saumon, G. (1998) Ventilator-induced lung injury: lessons from experimental studies. *Am J Respir Crit Care Med* 157, 294-323
13. Liu, M., and Post, M. (2000) Invited review: mechanochemical signal transduction in the fetal lung. *J Appl Physiol* 89, 2078-2084
14. Liu, M., Tanswell, A.K., and Post, M. (1999) Mechanical force-induced signal transduction in lung cells. *Am J Physiol* 277, L667-L683
15. Wirtz, H.R., and Dobbs, L.G. (2000) The effects of mechanical forces on lung functions. *Respir Physiol* 119, 1-17
16. dos Santos, C.C., and Slutsky, A.S. (2000) Invited review: mechanisms of ventilator-induced lung injury: a perspective. *J Appl Physiol* 89, 1645-1655
17. Sachs, F. (1992) Stretch-sensitive ion channels: an update. *Soc Gen Physiol Ser* 47, 241-260
18. Ghazi, A., Berrier, C., Ajouz, B., and Besnard, M. (1998) Mechanosensitive ion channels and their mode of activation. *Biochimie* 80, 357-362
19. Parker, J.C., Ivey, C.L., and Tucker, J.A. (1998) Gadolinium prevents high airway pressure-induced permeability increases in isolated rat lungs. *J Appl Physiol* 84, 1113-1118
20. Liu, M., Xu, J., Tanswell, A.K., and Post, M. (1994) Inhibition of mechanical strain-induced fetal rat lung cell proliferation by gadolinium, a stretch-activated channel blocker. *J Cell Physiol* 161, 501-507
21. Xu, J., Liu, M., Liu, J., Caniggia, I., and Post, M. (1996) Mechanical strain induces constitutive and regulated secretion of glycosaminoglycans and proteoglycans in fetal lung cells. *J Cell Sci* 109, 1605-1613
22. Mourgeon, E., Isowa, N., Keshavjee, S., Zhang X., Slutsky, A.S., and Liu, M. (2000) Mechanical stretch stimulates macrophage inflammatory protein-2 secretion from fetal rat lung cells. *Am J Physiol Lung Cell Mol Physiol* 279, L699-L706
23. Martin, D.K., Bootcov, M.R., Campbell, T.J., French, P.W., and Breit, S.N. (1995) Human macrophages contain a stretch-sensitive potassium channel that is activated by adherence and cytokines. *J Membr Biol* 147, 305-315
24. Waters, C.M., Ridge, K.M., Sunio, G., Venetsanou, K., and Sznajder, J.I. (1999) Mechanical stretching of alveolar epithelial cells increases Na(+)-K(+)- ATPase activity. *J Appl Physiol* 87, 715-721
25. Hong, K., and Driscoll, M. (1994) A transmembrane domain of the putative channel subunit MEC-4 influences mechanotransduction and neurodegeneration in C. elegans. *Nature* 367, 470-473
26. Huang, M., and Chalfie, M. (1994) Gene interactions affecting mechanosensory transduction in Caenorhabditis elegans. *Nature* 367, 467-470
27. Lai, C.C., Hong, K., Kinnell, M., Chalfie, M., and Driscoll, M (1996) Sequence and transmembrane topology of MEC-4, an ion channel subunit required for mechanotransduction in Caenorhabditis elegans. *J Cell Biol* 133, 1071-1081
28. Vlahakis, N., and Hubmayr, R. (2000) Cellular responses to mechanical stress: plasma membrane stress failure in alveolar epithelial cells. *J Appl Physiol* 89, 2490-2496
29. Sayeed, M.M. (1996) Alterations in calcium signaling and cellular responses in septic injury. *New Horiz* 4, 72-86
30. Hinman, L.E., Beilman, G.J., Groehler, K.E., and Sammak, P.J. (1997) Wound-induced calcium waves in alveolar type II cells. *Am J Physiol* 273, L1242-L1248
31. Grembowicz, K.P., Sprague, D., and McNeil, P.L. (1999) Temporary disruption of the plasma membrane is required for c-fos expression in response to mechanical stress. *Mol Biol Cell* 10, 1247-1257

32. Bajpai, A., Andrews, G.K., and Ebner, K.E. (1989) Induction of c-fos mRNA in rat lymphoma Nb-2 cells. *Biochem Biophys Res Commun* 165, 1359-1363
33. Ghosh, A., and Greenberg, M.E. (1995) Calcium signaling in neurons: molecular mechanisms and cellular consequences. *Science* 268, 239-247
34. Tremblay, L., Valenza, F., Ribeiro, S.P., Li., J., and Slutsky, A.S. (1997) Injurious ventilatory strategies increase cytokines and c-fos m-RNA expression in an isolated rat lung model. *J Clin Invest* 99, 944-952
35. Dawes, N.J., Cox, V.M., Park, K.S., Nga, H., and Goldspink, D.F. (1996) The induction of c-fos and c-jun in the stretched latissimus dorsi muscle of the rabbit: responses to duration, degree and re-application of the stretch stimulus. *Exp Physiol* 81, 329-339
36. Sadoshima, J., and Izumo, S. (1993) Mechanotransduction in stretch-induced hypertrophy of cardiac myocytes. *J Recept Res* 13, 777-794
37. Kawata, A., and Mikuni-Takagaki, Y. (1998) Mechanotransduction in stretched osteocytes--temporal expression of immediate early and other genes. *Biochem Biophys Res Commun* 246, 404-408
38. Roelofsen, J., Klein-Nulend, J., and Burger, E.H. (1995) Mechanical stimulation by intermittent hydrostatic compression promotes bone-specific gene expression in vitro. *J Biomech* 28, 1493-1503
39. Ballermann, B.J., Dardik, A., Eng, E., and Liu, A. (1998) Shear stress and the endothelium. *Kidney Int Suppl* 67, S100-S108
40. McNeil, P.L., and Steinhardt, R.A. (1997) Loss, restoration, and maintenance of plasma membrane integrity. *J Cell Biol* 137, 1-4
41. Muthukrishnan, L., Warder, E., and McNeil, P.L. (1991) Basic fibroblast growth factor is efficiently released from a cytosolic storage site through plasma membrane disruptions of endothelial cells. *J Cell Physiol* 148, 1-16
42. Ingber, D.E. (1997) Tensegrity: the architectural basis of cellular mechanotransduction. *Annu Rev Physiol* 59, 575-599
43. Tschumperlin, D.J., Fredberg, J.J., and Drazen, J.M. (2000) Mechanotransduction via specific cell-matrix interactions in airway epithelial cells. *Am J Respir Crit Care Med* 161, A259
44. Wang, N., Butler, J.P., and Ingber, D.E. (1993) Mechanotransduction across the cell surface and through the cytoskeleton. *Science* 260, 1124-1127
45. Peake, M.A., Cooling, L.M., Magnay, J.L., Thomas, P.B., and El Haj, A.J. (2000) Selected contribution: regulatory pathways involved in mechanical induction of c-fos gene expression in bone cells. *J Appl Physiol* 89, 2498-2507
46. Hubmayr, R.D., Shore, S.A., Fredberg, J.J., Planus, E., Panettieri, R.A., Moller, W., Heyder, J., and Wang, N. (1996) Pharmacological activation changes stiffness of cultured human airway smooth muscle cells. *Am J Physiol* 271, C1660-C1668
47. Bhullar, I.S., Li, Y.S., Miao, H., Zandi, E., Kim, M., Shyy, J.Y., and Chien, S. (1998) Fluid shear stress activation of IkappaB kinase is integrin-dependent. *J Biol Chem* 273, 30544-30549
48. Oberholzer, A., Oberholzer, C., and Moldawer, L.L. (2000) Cytokine signaling--regulation of the immune response in normal and critically ill states. *Crit Care Med* 28, N3-12
49. Keane, M.P., and Strieter, R.M. (2000) Chemokine signaling in inflammation. *Crit Care Med* 28, N13-N26
50. Ingber, D.E. (1997) Tensegrity: the architectural basis of cellular mechanotransduction. *Annu Rev Physiol* 59, 575-599
51. Russo, L.A., Rannels, S.R., Laslow, K.S., and Rannels, D.E. (1989) Stretch-related changes in lung cAMP after partial pneumonectomy. *Am J Physiol* 257, E261-E268

52. Liu, M., Xu, J., Souza, P., Tanswell, B., Tanswell, A.K., and Post. M. (1995) The effect of mechanical strain on fetal rat lung cell proliferation: comparison of two- and three-dimensional culture systems. *In Vitro Cell Dev Biol Anim* 31, 858-866
53. Liu, M., Xu, J., Liu, J., Kraw, M.E., Tanswell, A.K., and Post, M. (1995) Mechanical strain-enhanced fetal lung cell proliferation is mediated by phospholipase C and D and protein kinase C. *Am J Physiol* 268, L729-L738
54. Chess, P.R., Toia, L., and Finkelstein, J.N. (2000) Mechanical strain-induced proliferation and signaling in pulmonary epithelial H441 cells. *Am J Physiol Lung Cell Mol Physiol* 279, L43-L51
55. Quinn, D., Tager, A., Joseph, P.M., Bonventre, J.V., Force, T., and Hales, C.A. (1999) Stretch-induced mitogen-activated protein kinase activation and interleukin-8 production in type II alveolar cells. *Chest* 116, 89S-90S
56. Kito, H., Chen, E.L., Wang, X., Ikeda, M., Azuma, N., Nakajima, N., Gahtan, V. and Sumpio, B.E. (2000) Role of mitogen-activated protein kinases in pulmonary endothelial cells exposed to cyclic strain. *J Appl Physiol* 89, 2391-2400
57. Waters, C.M., Savla, U., and Panos, R.J. (1997) KGF prevents hydrogen peroxide-induced increases in airway epithelial cell permeability. *Am J Physiol* 272, L681-L689
58. Wang, H.C., Zentner, M.D., Deng, H.T., Kim, K.J., Wu, R., Yang, P.C., and Ann, D.K. (2000) Oxidative stress disrupts glucocorticoid hormone-dependent transcription of the amiloride-sensitive epithelial sodium channel alpha-subunit in lung epithelial cells through ERK-dependent and thioredoxin-sensitive pathways. *J Biol Chem* 275, 8600-8609
59. Dlugosz, J.A., Munk, S., Kapor-Drezgic, J., Goldberg, H.J., Fantus, I.G., Scholey, J.W., and Whiteside, C.I. (2000) Stretch-induced mesangial cell ERK1/ERK2 activation is enhanced in high glucose by decreased dephosphorylation. *Am J Physiol Renal Physiol* 279, F688-F697
60. Irigoyen, J.P., Besser, D., and Nagamine, Y. (1997) Cytoskeleton reorganization induces the urokinase-type plasminogen activator gene via the Ras/extracellular signal-regulated kinase (ERK) signaling pathway. *J Biol Chem* 272, 1904-1909
61. Reusch, H.P., Chan, G., Ives, H.E., and Nemenoff, R.A. (1997) Activation of JNK/SAPK and ERK by mechanical strain in vascular smooth muscle cells depends on extracellular matrix composition. *Biochem Biophys Res Commun* 237, 239-244
62. Nguyen, H.T., Adam, R.M., Bride, S.H., Park, J.M., Peters, C.A., and Freeman, M.R. (2000) Cyclic stretch activates p38 SAPK2-, ErbB2-, and AT1-dependent signaling in bladder smooth muscle cells. *Am J Physiol Cell Physiol* 279, C1155-C1167
63. Pan, J., Fukuda, K., Saito, M., Matsuzaki, J., Kodama, H., Sano, M., Tkahashi, T., Kato, T., and Ogawa, S. (1999) Mechanical stretch activates the JAK/STAT pathway in rat cardiomyocytes. *Circ Res* 84, 1127-1136
64. Bhattacharya, S., Ying, X., Fu, C., Patel, R., Kuebler, W., Greenberg, S. and Bhattacharya, J. (2000) alpha(v)beta(3) integrin induces tyrosine phosphorylation-dependent Ca(2+) influx in pulmonary endothelial cells. *Circ Res* 86, 456-462
65. Liu, M., Qin, Y., Liu, J., Tanswell, A.K., and Post, M. (1996) Mechanical strain induces pp60src activation and translocation to cytoskeleton in fetal rat lung cells. *J Biol Chem* 271, 7066-7071
66. Parker, J.C., Ivey, C.L., and Tucker, A. (1998) Phosphotyrosine phosphatase and tyrosine kinase inhibition modulate airway pressure-induced lung injury. *J Appl Physiol* 85, 1753-1761
67. MacGillivray, M.K., Cruz, T.F., and McCulloch, C.A. (2000) The recruitment of the interleukin-1 (IL-1) receptor-associated kinase (IRAK) into focal adhesion complexes is required for IL-1beta -induced ERK activation. *J Biol Chem* 275, 23509-23515
68. Ranjan, V., Xiao, Z., and Diamond, S.L. (1995) Constitutive NOS expression in cultured endothelial cells is elevated by fluid shear stress. *Am J Physiol* 269, H550-H555

69. Uematsu, M., Ohara, Y., Navas, J.P., Nishida, K. Murphy, T.J., Alexander, R.W. Nerem, R.M., and Harrison, D.G. (1995) Regulation of endothelial cell nitric oxide synthase mRNA expression by shear stress. *Am J Physiol* 269, C1371-C1378
70. Awolesi, M.A., Sessa, W.C., and Sumpio, B.E. (1995) Cyclic strain upregulates nitric oxide synthase in cultured bovine aortic endothelial cells. *J Clin Invest* 96, 1449-1454
71. Savla, U., Sporn, P.H., and Waters, C.M. (1997) Cyclic stretch of airway epithelium inhibits prostanoid synthesis. *Am J Physiol* 273, L1013-L1019
72. Blackwell, T.S., and Christman, J.W. (1997) The role of nuclear factor-kappa B in cytokine gene regulation. *Am J Respir Cell Mol Biol* 17, 3-9
73. Lentsch, A.B., Czermak, B.J., Bless, N.M., Van Rooijen, N., and Ward, P.A. (1999) Essential role of alveolar macrophages in intrapulmonary activation of NF-kappaB. *Am J Respir Cell Mol Biol* 20, 692-698
74. Du, W., Mills, I., and Sumpio, B.E. (1995) Cyclic strain causes heterogeneous induction of transcription factors, AP-1, CRE binding protein and NF-κB, in endothelial cells: species and vascular bed diversity. *J Biomech* 28, 1485-1491
75. Schwartz, M.D., Moore, E.E., Moore, F.A., Shenkar, R., Moine, P., Haenel, J.B. and Abraham, E. (1996) Nuclear factor-kappa B is activated in alveolar macrophages from patients with acute respiratory distress syndrome. *Crit Care Med* 24, 1285-1292
76. Pugin, J., Dunn, I., Jolliet, P., Tassaux, D., Magnenat, J.L., Nicod, L.P. and Chevrolet, J.C. (1998) Activation of human macrophages by mechanical ventilation in vitro. *Am J Physiol* 275, L1040-L1050
77. Held, H.D., Boettcher, S., Hamann, and L., Uhlig, S. (2001) Ventilation-Induced chemokine and cytokine release is associated with activation of nuclear factor-kappaB and is blocked by steroids. *Am J Respir Crit Care Med* 163, 711-716
78. Beutler, B. (2000) Endotoxin, toll-like receptor 4, and the afferent limb of innate immunity. *Curr Opin Microbiol* 3, 23-28
79. van Deventer, S.J. (2000) Cytokine and cytokine receptor polymorphisms in infectious disease. *Intensive Care Med* 26, S98-102
80. Poltorak, A., He, X., Smirnova, I., Liu, M.Y., Huffel, C.V., Du, X, Birdwell, D., Alejos, E., Silva, M., Galanos, C., Freudenberg, M., Ricciardi-Castagnoli, P., Layton, B., and Beutler, B. (1998) Defective LPS signaling in C3H/HeJ and C57BL/10ScCr mice: mutations in Tlr4 gene. *Science* 282, 2085-2088
81. Takata, M., Abe, J., Tanaka, H., Kitano, Y., Doi, S., Kohsaka, T., and Miyasaka, K. (1997) Intraalveolar expression of tumor necrosis factor-alpha gene during conventional and high-frequency ventilation. *Am J Respir Crit Care Med* 156, 272-279
82. Imai,Y., Kawano, T., Miyasaka, K., Takata, M., Imai, T., and Okuyama, K. (1994) Inflammatory chemical mediators during conventional ventilation and during high frequency oscillatory ventilation. *Am J Respir Crit Care Med* 150, 1550-1554
83. von Bethmann,A.N., Brasch, F., Nusing, R., Vogt, K., Volk, H.D., Muller, K.M., Wendel, A., and Uhlig, S. (1998) Hyperventilation induces release of cytokines from perfused mouse lung. *Am J Respir Crit Care Med* 157, 263-272
84. Kawano, T., Mori, S., Cybulsky, M., Burger, R., Ballin, A., Cutz, E., and Bryan, A.C. (1987) Effect of granulocyte depletion in a ventilated surfactant-depleted lung. *J Appl Physiol* 62, 27-33
85. Chiumello, D., Pristine, G., and Slutsky, A.S. (1999) Mechanical ventilation affects local and systemic cytokines in an animal model of acute respiratory distress syndrome. *Am J Respir Crit Care Med* 160, 109-116
86. Haitsma, J.J., Uhlig, S., Goggel, R., Verbrugge, S.J., Lachmann, U., and Lachmann, B. (2000) Ventilator-induced lung injury leads to loss of alveolar and systemic compartmentalization of tumor necrosis factor-alpha. *Intensive Care Med* 26, 1515-1522

87. Ranieri, V.M., Suter, P.M., Tortorella, C, De Tullio, R., Daye, J.M., Brienza, A, Bruno, F. and Slutsky, A.S. (1999) Effect of mechanical ventilation on inflammatory mediators in patients with acute respiratory distress syndrome: a randomized controlled trial. *JAMA* 282, 54-61
88. Dreyfuss, D., and Saumon, G. (1998) From ventilator-induced lung injury to multiple organ dysfunction? *Intensive Care Med* 24, 102-104
89. Millar, A.B., Foley, N.M., Singer, M., Johnson, N.M., Meager, A., and Rook, G.A. (1989) Tumour necrosis factor in bronchopulmonary secretions of patients with adult respiratory distress syndrome. *Lancet* 2, 712-714
90. Moldawer, L.L., and Minter, R.M. (2000) Tumor necrosis factor-alpha and the development of multiple organ failure. *Crit Care Med* 28, 2158-2159
91. Bone, R.C. (1996) Toward a theory regarding the pathogenesis of the systemic inflammatory response syndrome: what we do and do not know about cytokine regulation. *Crit Care Med* 24, 163-172
92. Sadikot, R.T., Christman, J.W., and Blackwell, T.S. (2000) Chemokines and chemokine receptors in pulmonary diseases. *Curr Opin Investig Drugs* 1, 314-320
93. Pugin, J., Ricou, B., Steinberg, K.P., Suter, P.M., and Martin, T.R. (1996) Proinflammatory activity in bronchoalveolar lavage fluids from patients with ARDS, a prominent role for interleukin-1. *Am J Respir Crit Care Med* 153, 1850-1856
94. Verbrugge, S.J., Sorm, V., van't Veen, A., Mouton, J.W, Gommers, D., and Lachmann, B. (1998) Lung overinflation without positive end-expiratory pressure promotes bacteremia after experimental Klebsiella pneumoniae inoculation. *Intensive Care Med* 24, 172-177
95. Narimanbekov, I.O., and Rozycki, H.J. (1995) Effect of IL-1 blockade on inflammatory manifestations of acute ventilator-induced lung injury in a rabbit model. Exp Lung Res 21, 239-254
96. Shibata, Y., Nakamura, H., Kato, S., and Tomoike, H (1996) Cellular detachment and deformation induce IL-8 gene expression in human bronchial epithelial cells. *J Immunol* 156, 772-777
97. Ding, A.H., Porteu, F., Sanchez, E., and Nathan, C.F. (1990) Shared actions of endotoxin and taxol on TNF receptors and TNF release. *Science* 248, 370-372
98. Manie, S., Schmid-Alliana, A., Kubar, J., Ferrua, B., and Rossi, B. (1993) Disruption of microtubule network in human monocytes induces expression of interleukin-1 but not that of interleukin-6 nor tumor necrosis factor-alpha. Involvement of protein kinase A stimulation. *J Biol Chem* 268, 13675-13681
99. Bedard, M., McClure, C.D., Schiller, N.L., Francoeur, C., Cantin, A., and Denis, M. (1993) Release of interleukin-8, interleukin-6, and colony-stimulating factors by upper airway epithelial cells: implications for cystic fibrosis. *Am J Respir Cell Mol Biol* 9, 455-462
100. Okada, M., Matsumori, A., Ono, K, Furukawa, Y., Shioi ,T., Iwasaki, A., Matsushima, K. and Sasayama, S. (1998) Cyclic stretch upregulates production of interleukin-8 and monocyte chemotactic and activating factor/monocyte chemoattractant protein-1 in human endothelial cells. *Arterioscler Thromb Vasc Biol* 18, 894-901
101. Ranieri, V.M., Giunta, F., Suter, P.M., and Slutsky, A.S. (2000) Mechanical ventilation as a mediator of multisystem organ failure in acute respiratory distress syndrome. *JAMA* 284, 43-44
102. Kuninaka, S., Yano, T., Yokoyama, H., Fukuyama, Y., Terazaki, Y., Uehara, T., Kanematsu, T., Asoh, H., and Ichinose, Y. (2000) Direct influences of pro-inflammatory cytokines (IL-1beta, TNF-alpha, IL-6) on the proliferation and cell-surface antigen expression of cancer cells. *Cytokine* 12, 8-11
103. Akira, S., Hirano, T., Taga, T., and Kishimoto, T. (1990) Biology of multifunctional cytokines: IL 6 and related molecules (IL 1 and TNF). *FASEB J* 4, 2860-2867

104. Villavicencio, R.T., Liu, S., Kibbe, M.R., Williams, D.L., Ganster, R.W., Dyer, K.F., Tweardy, D.J., Billiar, T.R., and Pitt, B.R. (2000) Induced nitric oxide inhibits IL-6-induced stat3 activation and type II acute phase mRNA expression. *Shock* 13, 441-445
105. Xing, Z., Gauldie, J., Cox, G., Baumann, H., Jordana, M., Lei, X.F., and Achong, M.K. (1998) IL-6 is an antiinflammatory cytokine required for controlling local or systemic acute inflammatory responses. *J Clin Invest* 101, 311-320
106. Bonfield, T.L., Konstan, M.W., Burfeind, P., Panuska, J.R., Hilliard, J.B., and Berger, M. (1995) Normal bronchial epithelial cells constitutively produce the anti-inflammatory cytokine interleukin-10, which is downregulated in cystic fibrosis. *Am J Respir Cell Mol Biol* 13, 257-261
107. Wanidworanun, C., and Strober. W. (1993) Predominant role of tumor necrosis factor-alpha in human monocyte IL-10 synthesis. *J Immunol* 151, 6853-6861
108. Friedman, G., Jankowski, S., Marchant, A., Goldman, M., Kahn, R.J., and Vincent, J.L. (1997) Blood interleukin 10 levels parallel the severity of septic shock. *J Crit Care* 12, 183-187
109. Donnelly, S.C., Strieter, R.M., Reid, P.T., Kunkel, S.L., Burdick, M.D., Armstrong, I., Mackenzie, A., and Haslett, C. (1996) The association between mortality rates and decreased concentrations of interleukin-10 and interleukin-1 receptor antagonist in the lung fluids of patients with the adult respiratory distress syndrome. *Ann Intern Med* 125, 191-196
110. Gazzinelli, R.T., Oswald, I.P., James, S.L., and Sher, A. (1992) IL-10 inhibits parasite killing and nitrogen oxide production by IFN- gamma-activated macrophages. *J Immunol* 148, 1792-1796
111. Lo, C.J., Fu, M., and Cryer, H.G. (1998) Interleukin 10 inhibits alveolar macrophage production of inflammatory mediators involved in adult respiratory distress syndrome. *J Surg Res* 79, 179-184
112. Xing, Z., Ohkawara, Y., Jordana, M., Graham, F.L, and Gauldie, J. (1997) Adenoviral vector-mediated interleukin-10 expression in vivo: intramuscular gene transfer inhibits cytokine responses in endotoxemia. *Gene Ther* 4, 140-149
113. Esmon, C.T., Taylor, F.B. Jr., and Snow, T. (1991) Inflammation and Coagulation: linked processes potentially regulated through a common pathway mediated by protein C. *Thromb Haemostasis* 66, 160-165
114. Chan, A.K., Baranowski, B., Berry, L., Liu, M., Rafii, B., Post, M., O'Brodovich, H., Monagle, P., and Andrew, M. (1998) Influence of mechanical stretch on thrombin regulation by fetal mixed lung cells. *Am J Respir Cell Mol Biol* 19, 419-425
115. Ruwhof, C., and van der, L.A. (2000) Mechanical stress-induced cardiac hypertrophy: mechanisms and signal transduction pathways. *Cardiovasc Res* 47, 23-37
116. Cattaruzza, M., Dimigen, C., Ehrenreich, H., and Hecker, M. (2000) Stretch-induced endothelin B receptor-mediated apoptosis in vascular smooth muscle cells. *FASEB J* 14, 991-998
117. Behnia, R., Molteni, A., Waters, C.M., Panos, R.J., Ward, W.F., Schnaper, H.W., and TS'Ao, C.H. (1996) Early markers of ventilator-induced lung injury in rats. *Ann Clin Lab Sci* 26, 437-450
118. Simma, B., Gulberg, V., Schobel, P., Trawoger, R., Ulmer, H., Gerbes, A.L., and Putz, G. (2000) High-frequency oscillatory ventilation does not decrease endothelin release in lung-lavaged rabbits. *Scand J Clin Lab Invest* 60, 213-220
119. Bernard, G., Vincent, J., Laterre, P., LaRosa, S., Dhainaut, J., Lopes-Rodriguez, A., Steingrub, J.S., Garber, G.E., Helterbrand, J.D., Ely, E.W., and Fisher, C.J. (2001) Efficacy and safety of recombinant human activated protein c for severe sepsis. *N Engl J Med* 344, 699-709

120. White, B., Schmidt, M., Murphy, C., Livingstone, W., O'Toole, D., Lawler, M., O'Neill, L., Kelleher, D., Schwarz, H.P., and Smith, O.P. (2000) Activated protein C inhibits lipopolysaccharide-induced nuclear translocation of nuclear factor kappaB (NF-kappaB) and tumour necrosis factor alpha (TNF-alpha) production in the THP-1 monocytic cell line. *Br J Haematol* 110, 130-134
121. Esmon, C.T. (1987) The regulation of natural anticoagulant pathways. *Science* 235, 1348-1352
122. Uchiba, M., Okajima, K., Murakami, K., Johno, M., Okabe, H., and Takatsuki, K. (1996) Recombinant thrombomodulin prevents endotoxin-induced lung injury in rats by inhibiting leukocyte activation. *Am J Physiol* 271, L470-L475
123. Grinnell, B.W., Hermann, R.B., and Yan, S.B. (1994) Human protein C inhibits selectin-mediated cell adhesion: role of unique fucosylated oligosaccharide. *Glycobiology* 4, 221-225
124. Bevilacqua, M.P., Stengelin, S., Gimbrone, M.A, and Seed, B. (1989) Endothelial leukocyte adhesion molecule 1: an inducible receptor for neutrophils related to complement regulatory proteins and lectins. *Science* 243, 1160-1165
125. Kuroki, Y., Tsutahara, S., Shijubo, N., Takahashi, H., Shiratori, M., Hattori, A., Honda, Y., Abe, S., and Akino, T. (1993) Elevated levels of lung surfactant protein A in sera from patients with idiopathic pulmonary fibrosis and pulmonary alveolar proteinosis. *Am Rev Respir Dis* 147, 723-729.
126. Chida, S., Phelps, D.S., Soll, R.F., and Taeusch, H.W. (1991) Surfactant proteins and anti-surfactant antibodies in sera from infants with respiratory distress syndrome with and without surfactant treatment. *Pediatrics* 88, 84-89
127. Honda, Y., Kuroki, Y., Matsuura, E. Nagae, H. , Takahashi, H., Akino, T., and Abe, S. (1995) Pulmonary surfactant protein D in sera and bronchoalveolar lavage fluids. *Am J Respir Crit Care Med* 152, 1860-1866
128. Veldhuizen, R.A., Tremblay, L.N., Govindarajan, A., van Rozendaal, B.A., Haagsman, H.P., and Slutsky, A.S. (2000) Pulmonary surfactant is altered during mechanical ventilation of isolated rat lung. *Crit Care Med* 28, 1545-2551
129. Verbrugge, S.J., Vazquez, D.A., Gommers, D., Neggers, S.J,, Sorm, V., Bohm, S.H., and Lachmann, B. (1998) Exogenous surfactant preserves lung function and reduces alveolar Evans blue dye influx in a rat model of ventilation-induced lung injury. *Anesthesiology* 89, 467-474
130. Verbrugge, S.J., Bohm, S.H., Gommers, D., Zimmerman, L.J., and Lachmann, B. (1998) Surfactant impairment after mechanical ventilation with large alveolar surface area changes and effects of positive end-expiratory pressure. *Br J Anaesth* 80, 360-364
131. Taskar, V., John, J., Evander, E., Robertson, B., and Jonson, B. (1997) Surfactant dysfunction makes lungs vulnerable to repetitive collapse and reexpansion. *Am J Respir Crit Care Med* 155, 313-320
132. Rose, F., Zwick, K., Ghofrani, H.A., Sibelius, U., Seeger, W., Walmrath, D., and Grimminger, F. (1999) Prostacyclin enhances stretch-induced surfactant secretion in alveolar epithelial type II cells. *Am J Respir Crit Care Med* 160, 846-851
133. Rose, F., Kurth-Landwehr, C., Sibelius, U., Reuner, K.H., Aktories, K., Seeger, W., and Grimminger, F. (1999) Role of actin depolymerization in the surfactant secretory response of alveolar epithelial type II cells. *Am J Respir Crit Care Med* 159, 206-212
134. Floros, J., and Karinch, A.M. (1995) Human SP-A: then and now. *Am J Physiol* 268, L162-L165
135. Wright, J.R. (1997) Immunomodulatory functions of surfactant. *Physiol Rev* 77, 931-962
136. LeVine, A.M., Kurak, K.E., Bruno, M.D., Stark, J.M., Whitsett, J.A., and Korfhagen. T.R. (1998) Surfactant protein-A-deficient mice are susceptible to Pseudomonas aeruginosa infection. *Am J Respir Cell Mol Biol* 19, 700-708

137. Cochrane, C.G., and Revak, S.D. (1991) Pulmonary surfactant protein B (SP-B): structure-function relationships. *Science* 254, 566-568
138. Whitsett, J.A., Nogee, L.M., Weaver, T.E., and Horowitz, A.D. (1995) Human surfactant protein B: structure, function, regulation, and genetic disease. *Physiol Rev* 75, 749-757
139. Arias-Diaz, J., Garcia-Verdugo, I., Casals, C., Sanchez-Rico, N., Vara, E., and Balibrea, J.L. (2000) Effect of surfactant protein A (SP-A) on the production of cytokines by human pulmonary macrophages. *Shock* 14, 300-306
140. Chandel, N., and Sznajder, J. (2000) Stretching the lung and programmed cell death. *Am J of Physiol Lung Cell Mol Physiol* 279, L1003-L1004
141. Edwards, Y., Sutherland, L., and Murray, A. (2000) NO protects alveolar type II cells from stretch-induced apoptosis. A novel role for macrophages in the lung. *Am J Physiol Lung Cell Mol Physiol* 279, L1236-L1242
142. Davis, D.W., Weidner, D.A., Holian, A., and McConkey, D.J. (2000) Nitric oxide-dependent activation of p53 suppresses bleomycin-induced apoptosis in the lung. *J Exp Med* 192, 857-869
143. Artlich, A., Adding, C., Agvald, P., Persson, M.G., Lonnqvist, P.A., and Gustafsson, L.E. (1999) Exhaled nitric oxide increases during high frequency oscillatory ventilation in rabbits. *Exp Physiol* 84, 959-969
144. Bannenberg, G.L., and Gustafsson, L.E. (1997) Stretch-induced stimulation of lower airway nitric oxide formation in the guinea-pig: inhibition by gadolinium chloride. *Pharmacol Toxicol* 81, 13-18
145. Ashino, Y., Ying, X., Dobbs, L.G., and Bhattacharya, J. (2000) [Ca(2+)](i) oscillations regulate type II cell exocytosis in the pulmonary alveolus. *Am J Physiol Lung Cell Mol Physiol* 279, L5-13
146. Kim, Y.M., Bombeck, C.A., and Billiar, T.R. (1999) Nitric oxide as a bifunctional regulator of apoptosis. *Circ Res* 84, 253-256
147. Tanaka, Y., Kanai ,Y., Okada, Y., Nonaka, S., Takeda, S., Harada, A., and Hirokawa, N. (1998) Targeted disruption of mouse conventional kinesin heavy chain, kif5B, results in abnormal perinuclear clustering of mitochondria. *Cell* 93, 1147-1158
148. Downey, G.P., Dong, Q., Kruger, J., Dedhar, S., and Cherapanov, V. (1999) Regulation of neutrophil activation in acute lung injury. *Chest* 116, 46S-54S
149. Hamilton, P.P., Onayemi, A., Smyth, J.A., Gillan, J.E., Cutz, E., Froese, A.B., Bryan, A.C. (1983) Comparison of conventional and high-frequency ventilation: oxygenation and lung pathology. *J Appl Physiol* 55, 131-138
150. Zhang, F.X., Kirschning, C.J., Mancinelli, R., Xu, X.P., Jin, Y., Faure, E., Mantovani, A., Rothe, M., Muzio, M., and Arditi, M. (1999) Bacterial lipopolysaccharide activates nuclear factor-kappaB through interleukin-1 signaling mediators in cultured human dermal endothelial cells and mononuclear phagocytes. *J Biol Chem* 274, 7611-7614
151. Stenmark, K.R., and Mecham, R.P. (1997) Cellular and molecular mechanisms of pulmonary vascular remodeling. *Annu Rev Physiol* 59, 89-144
152. Davies, P.F. (1995) Flow-mediated endothelial mechanotransduction. *Physiol Rev* 75. 519-560
153. Zimmerman, G.A., Albertine, K.H., Carveth, H.J., Gill, E.A., Grissom, C.K., Hoidal, J.R., Imaizumi, T., Maloney, C.G., McIntyre, T.M., Michael, J.R., Orme, J.F., Prescott, S.M., and Topham, M.S. (1999) Endothelial activation in ARDS. *Chest* 116, 18S-24S
154. Beck, G.C., Yard, B.A., Breedijk, A.J., van Ackern, K., and Van Der Woude, F.J. (1999) Release of CXC-chemokines by human lung microvascular endothelial cells (LMVEC) compared with macrovascular umbilical vein endothelial cells. *Clin Exp Immunol* 118, 298-303
155. Dunn, I., and Pugin, J. (1999) Mechanical ventilation of various human lung cells in vitro: identification of the macrophage as the main producer of inflammatory mediators. *Chest* 116, 95S-97S

156. Pugin, J., Verghese, G., Widmer, M.C., and Matthay, M.A. (1999) The alveolar space is the site of intense inflammatory and profibrotic reactions in the early phase of acute respiratory distress syndrome. *Crit Care Med* 27, 304-312
157. Tremblay, L., Miatto, D., Hamid, Q., and Slutsky, A.S. (1997) Changes in cytokine expression secondary to injurious mechanical ventilation strategies in an ex vivo lung model. *Intensive Care Med* 23, S3
158. McRitchie, D.I., Isowa, N., Edelson, J.D., Xavier, A.M., Cai, L., Man, H.Y., Wang, Y.T., Keshavjee, S.H., Slutsky, A.S., and Liu, M. (2000) Production of tumour necrosis factor alpha by primary cultured rat alveolar epithelial cells. *Cytokine* 12, 644-654
159. Standiford, T.J., Kunkel, S.L., Basha, M.A., Chensue, S.W., Lynch, J.P., III, Toews, G.B., Westwick, J., and Strieter, R.M. (1990) Interleukin-8 gene expression by a pulmonary epithelial cell line. A model for cytokine networks in the lung. *J Clin Invest* 86, 1945-1953
160. Vlahakis, N.E., Schroeder, M.A., Limper, A.H., and Hubmayr, R.D. (1999) Stretch induces cytokine release by alveolar epithelial cells in vitro. *Am J Physiol* 277, L167-L173
161. Tsuda, A., Stringer, B.K., Mijailovich, S.M., Rogers, R.A., Hamada, K., and Gray, M.L. (1999) Alveolar cell stretching in the presence of fibrous particles induces interleukin-8 responses. *Am J Respir Cell Mol Biol* 21, 455-462
162. Xavier, A.M., Isowa, N., Cai, L., Dziak, E., Opas, M., McRitchie, D.I., Slutsky, A.S., Keshavjee, S.H., and Liu, M. (1999) Tumor necrosis factor-alpha mediates lipopolysaccharide-induced macrophage inflammatory protein-2 release from alveolar epithelial cells. Autoregulation in host defense. *Am J Respir Cell Mol Biol* 21, 510-520
163. West, J.B. (2000) Cellular Response to Mechanical Stress: Pulmonary Capillary Stress Failure. *J Appl Physiol* 89, 2483-2489
164. Savla, U., Appel, H.J. Sporn, P.H., and Waters, C.M. (2001) Prostaglandin E(2) regulates wound closure in airway epithelium. *Am J Physiol Lung Cell Mol Physiol* 280, L421-L431
165. Waters, C.M., and Savla, U. (1999) Keratinocyte growth factor accelerates wound closure in airway epithelium during cyclic mechanical strain. *J Cell Physiol* 181, 424-432
166. Maron, M.B., Fu, Z., Mathieu-Costello, O., and West, J.B. (2001) Effect of high transcapillary pressures on capillary ultrastructure and permeability coefficients in dog lung. *J Appl Physiol* 90, 638-648
167. Nahum, A., Hoyt, J., Schmitz, L., Moody, J., Shapiro, R., and Marini, J.J. (1997) Effect of mechanical ventilation strategy on dissemination of intratracheally instilled Escherichia coli in dogs. *Crit Care Med* 25, 1733-1743
168. Murphy, D.B., Cregg, N., Tremblay, L., Engelberts, D., Laffey, J.G., Slutsky, A.S., Romaschin, A, and Kavanagh, B.P. (2000) Adverse ventilatory strategy causes pulmonary-to-systemic translocation of endotoxin. *Am J Respir Crit Care Med* 162, 27-33
169. Savel, R.H., Yao, E.C., and Gropper, M.A. (2001) Protective effects of low tidal volume ventilation in a rabbit model of Pseudomonas aeruginosa-induced acute lung injury. *Crit Care Med* 29, 392-398
170. Muscedere, J.G., Mullen, J.B., Gan, K., and Slutsky, A.S. (1994) Tidal ventilation at low airway pressures can augment lung injury. *Am J Respir Crit Care Med* 149, 1327-1334
171. Tutor, J.D., Mason, C.M., Dobard, E., Beckerman, R.C., Summer, W.R., and Nelson, S. (1994) Loss of compartmentalization of alveolar tumor necrosis factor after lung injury. *Am J Respir Crit Care Med* 149, 1107-1111

172. Debs, R.J., Fuchs, H.J., Philip, R., Montgomery, A.B., Brunette, E.N., Liggitt, D., Patton, J.S., and Shellito, J.E. (1988) Lung-specific delivery of cytokines induces sustained pulmonary and systemic immunomodulation in rats. *J Immunol* 140, 3482-3488
173. Wrigge, H., Zinserling, J., Stuber, F., von Spiegel, T., Hering, R., Wetegrove, S., Hoeft, A., and Putensen, C. (2000) Effects of mechanical ventilation on release of cytokines into systemic circulation in patients with normal pulmonary function. *Anesthesiology* 93, 1413-1417
174. Mira, J.P., Cariou, A., Grall, F., Delclaux, C., Losser, M.R., Heshmati, F., Cheval, C., Monchi, M., Teboul, J.L., Riche, F., Leleu, G., Arbibe, L., Mignon, A., Delpech, M., and Dhainaut, J.F. (1999) Association of TNF2, a TNF-alpha promoter polymorphism, with septic shock susceptibility and mortality: a multicenter study. *JAMA* 282, 561-568
175. Arbour, N.C., Lorenz, E., Schutte, B.C., Zabner, J., Kline, J.N., Jones, M., Frees, K., Watt, J.L., and Schwartz, D.A. (2000) TLR4 mutations are associated with endotoxin hyporesponsiveness in humans. *Nat Genet* 25, 187-191
176. Slutsky, A.S. (2001) Basic Science in Ventilator-induced Lung Injury. Implications for the bedside. *Am J Respir Crit Care Med* 163, 599-600
177. Imai, Y., Kawano, T., Iwamoto, S., Nakagawa, S., Takata, M., Miyasaka, K. (1999) Intratracheal anti-tumor necrosis factor-alpha antibody attenuates ventilator-induced lung injury in rabbits. *J Appl Physiol* 87, 510-515

INDEX